Means
ADA Compliance
Pricing Guide
2nd Edition
Updated to 2004 ADA

Cost Estimates for More Than 70 Common Modifications

A Collaboration between
Adaptive Environments Center and
RSMeans Engineering Staff

Means
ADA Compliance
Pricing Guide *2nd Edition*

Updated to 2004 ADAAG

Cost Estimates for More Than 70 Common Modifications

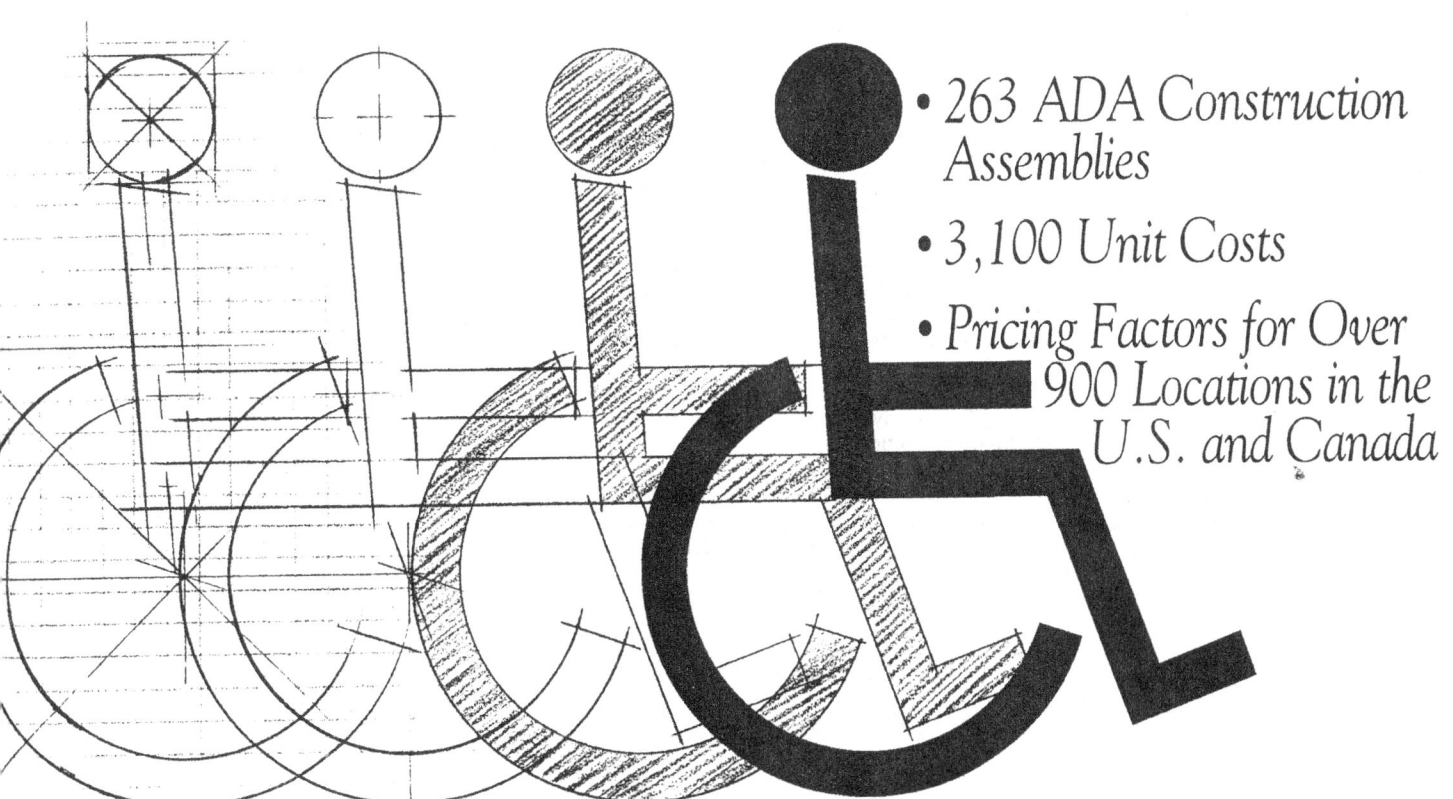

- 263 ADA Construction Assemblies
- 3,100 Unit Costs
- Pricing Factors for Over 900 Locations in the U.S. and Canada

A Collaboration between Adaptive Environments Center and RSMeans Engineering Staff

RSMeans

RSMeans

Copyright © 2004
A division of Reed Construction Data
Construction Publishers & Consultants
63 Smiths Lane
Kingston, MA 02364-3008
781-422-5000
www.rsmeans.com

Managing Editor: Mary Greene. Editor: Andrea Sillah.
Production Manager: Michael Kokernak. Production Coordinator: Marion Schofield.
Composition: Sheryl Rose. Proofreader: Robin Richardson.
Book and cover design: Sheryl Rose.

Printed in the United States of America

10 9 8 7 6 5 4 3 2

Library of Congress Catalog Number Pending

ISBN 978-0-87629-739-1

Table of Contents

About the Authors

Adaptive Environments Center is a private, nonprofit education and consulting organization whose architects and access experts have designed and planned accessibility for commercial, municipal, transportation, residential, and institutional facilities. The organization has surveyed hundreds of sites for ADA compliance and developed guidelines for facility managers and design professionals on planning accessible facilities. Steve Demos, RA, was responsible for editing the text for the second edition, first written by Elaine Ostroff and Joshua Barnett in 1994. He also updated the original illustrations by Rex Pace.

RSMeans, leading provider of construction cost information in North America, supplied the cost data for this publication. RSMeans produces 27 annually updated cost data books, electronic products, and over 60 construction reference books. In addition, RSMeans presents seminars on construction cost-related topics throughout the country, and offers consulting services to design, construction, and facilities professionals in the public and private sectors. Means' cost data is the industry standard, relied on by over 150,000 construction professionals.

Introduction

The New 2004 ADA

In July 2004, a new version of the *Americans with Disabilities Act Accessibility Guidelines for Buildings and Facilities* (ADAAG) was published. This document, not formally revised since the ADA was first enacted in 1990, has new requirements and guidelines, as well as a completely new format that is more consistent with the way information is organized in the building trades.

Means ADA Compliance Pricing Guide has also been completely revised and updated in this edition—not only to include the 2004 ADAAG changes, but to provide current cost estimates for remodeling projects required to meet the ADA. This revised edition also includes new building technologies and materials.

Changes in the 2004 ADA guidelines include:
- A decimal-based, numeric-only numbering system, consistent with building codes. Wherever possible, the ADAAG format also parallels ICC/ANSI A117.1 (International Code Council's *Standard on Accessible & Usable Buildings & Facilities*)in its chapter and section numbering.
- Changes in substance, scoping, and technical provisions—to better meet the needs of people with disabilities, to recognize technological developments, and to reconcile differences from national consensus standards.

- New sections covering accessible residential housing, additional recreational facilities and parks, judicial and legislative facilities, and detention and correctional facilities. New standards have been introduced that set minimum requirements for children's environments.
- Non-mandatory information set off as "advisory," accompanied by information and commentary.
- Fully explained requirements that were formerly covered in illustrations only.
- Scoping requirements that specify what must be made accessible—clearly separated from "technical requirements" (how access is to be achieved).
- Technical sections organized into chapters according to the type of element or space, such as plumbing components, communication elements, and recreation facilities.
- Special occupancy chapters (except for transportation facilities), integrated into the main body of the document.

Note: As of the publication of this book, the new format of the ADAAG was still under review by the U.S. Department of Justice and was thus not considered the law. Full adoption is expected in 2005.

The Purpose of This Publication

Two of the most frequently asked questions about the Americans with Disabilities Act are: "What do I have to do?" and "How much will it cost?" Not knowing the actual cost of building modifications for accessibility compliance greatly impedes the implementation of ADA, since people often assume they can't afford it. This publication provides that information, to help facility managers, owners, contractors, designers, and building users budget the renovations needed to provide accessibility.

How the Book Is Organized

This book is a guide to typical ADA compliance modifications—what's involved and how much they will cost. Like all renovations, many access modifications require on-site construction estimating or design services in addition to actual construction/installation. The book will help identify these kinds of situations.

The first step in budgeting for ADA compliance projects is to survey the facility to identify barriers to access. Once a survey has been conducted, this book can be used to estimate the cost of barrier removal and to help prioritize the order in which projects will be completed. The model projects in this book not only help identify what needs to be done, but also the construction tasks, materials, labor, and total costs.

The model projects are organized according to the typical route of travel from the parking lot up to and through a facility. To help correlate your projects with the ADA Accessibility Guidelines (ADAAG), a cross-reference chart is included at the back of the book, showing how each project correlates to specific ADAAG sections.

The main body of the book consists of the model projects with estimates. Each project is a single access modification that might be constructed in a facility. The projects fall into three categories:

- General Access Modifications: Typical ADAAG compliance access modifications.
- Specialized Access Modifications: Access elements included in specific types of facilities.
- Case Studies: Examples of the barrier removal and access construction process. These illustrate how a range of solutions to barrier removal might be devised.

Individual Projects

Each individual project is divided into three parts: illustration, text, and cost estimates—to convey the use of the design element, regulations, construction, and cost information. The illustration shows a typical access element with relevant information, such as installation, dimensions, or use. The text gives an overview of the requirements, design considerations, and level of difficulty involved for that project. The estimate lists the materials and labor involved for construction of the project in a typical situation, often with several variations in design or materials, based on varying site conditions or other options.

All projects contain the following information:

- ADAAG references: Where a particular element might be required by ADAAG, the specific

ADAAG technical standards for compliance are listed.
- Design requirements, suggestions, and discussion of the particular project's application or limits.
- Key items: A summary of materials and/or products necessary to execute the project.
- Level of difficulty for performing the work required to construct the modification: A rough approximation of the expense and labor involved in executing a particular access modification in a typical situation.

Low = minimal cost, usually doesn't require a building permit, and may be performed by building maintenance staff.

Moderate = more difficult, possibly involving skilled tradespeople and a building permit.

High = expensive, involving structural and/or utility work, specialized tradespeople, and building permits.

- Cost estimates for each accessibility project.

Additional References

Following the model projects with estimates are the **case studies**. These are two complete projects, a bathroom renovation and an entry modification for a historic facility. Each case study shows the evaluation process assessing the ADA requirements and the building site conditions and layout, then devising the best solution.

Part Four, **Unit Costs**, which follows, provides over 75 pages of RSMeans cost data—hundreds of individual construction line items that you can use to make adjustments, if necessary, to the model project estimates in the main part of the book. For example, in project No. 21, "Install New Door," it is assumed that some demolition is required of an exterior wall, which may be constructed of

horizontal wood siding with metal studs and gypsum wallboard on the interior. If, in fact, your particular wall is made of different materials, such as brick or concrete, you can substitute the appropriate demolition line item from the Unit Cost section for the demolition line item in the project. At the beginning of the Unit Cost section is a full explanation of how to find and apply the information you need.

To account for regional variations in construction costs, we have included **Location Factors** for 900 cities throughout the U.S. and Canada. This section provides percentage multipliers that you can apply to the total cost (including contractor's overhead and profit) for each of the model estimates. For example, if the total cost estimate for a project is $1,000, and you are located in Atlanta, Georgia, your location factor is .89; therefore, your project cost is $890.

Following the Unit Cost section are **Resources**, government agency addresses and phone numbers for obtaining technical assistance on ADA issues. This section also lists non-government checklists and publications that may be useful in performing facility accessibility surveys and obtaining additional technical assistance.

At the end of the book, a **glossary** of terms defines ADA and access terminology, and an abbreviations list clarifies design and construction terms used in the estimates.

Reminders

Each situation is different. While the most likely projects for ADA compliance are featured, detailed, and estimated based on normal working conditions, it is not possible to cover every circumstance requiring access modifications. The projects included represent the most common building element modifications needed to bring a facility into compliance with a particular requirement of ADAAG. Additional modifications not included in this book might be necessary to bring a facility into compliance with ADAAG, or to make it more accessible.

State access regulations were already in force in most states when the ADA was passed, and many municipalities have additional accessibility requirements. Some local regulations may be stricter than ADAAG (for instance, requiring ramps to be 48" wide instead of 36"). Some states require a variance for a platform lift; most prohibit them on designated fire egress stairs. Plumbing codes in some localities prohibit a unisex accessible restroom. In case of conflict between regulations, the stricter requirements govern the design. State and local codes must always be investigated prior to executing any construction.

The estimates are intended as a guide to budgeting access modifications. They are not intended to be absolutely comprehensive or definitive. While many alternatives in materials and site conditions have been presented, there are still others that can be used if they comply with ADAAG. More expensive materials, such as those often used in historic situations, will, of course, affect the total installation price.

Prices vary between cities and regions. Use the Location Factors at the back of this book, with its 900 zip code locations, to adjust costs to your area conditions. Material costs vary according to supply and demand, usually up (but occasionally down). Labor costs also vary widely, depending on the difficulty of a particular renovation, prevailing local wage laws, and the availability of labor.

The size of the job affects the cost. Construction unit costs are always less when renovation is more extensive and more modifications are done at once, due to factors such as bulk purchasing of materials and the contractor's travel and setup time. This should be considered when deciding whether or not to proceed with more than one modification for access. In the long run, hiring a contractor to build a ramp and widen three doorways at the same time is usually less expensive than doing them at separate times.

In the project estimates, where a total unit cost such as *per square foot* or *per linear foot* is given, this price assumes a quantity of work sufficient to cover contractor's costs and a fair profit. For projects of minimal size, the contractor's markup might be significantly higher. As with all renovations, a range of bids from different contractors can help determine the best price. Again, the pricing in the book cannot be absolute; a contractor's estimate can vary considerably from a price obtained from this (or any other) book due to a number of nonquantifiable factors.

The main goal is to eliminate discrimination based on accessibility and to create a facility usable by a larger group of people with a wider range of abilities. Even though ADAAG is not a building code, the same consideration exists when applying the standards as with a building code: these are only minimum requirements. It is often possible, and desirable, to exceed the minimum standards given.

Exceeding the minimum requirement can also make facilities maintenance less critical in ensuring compliance. (Even an inch or less of wear and tear over the years can result in non-compliance.)

The ADA guidelines are not to be interpreted as a design guide, but as the *minimum level of accessibility* compliance. Actual design will be determined by the individual situation.

Above all, input from people with disabilities is key to the successful design of access modifications. It is possible (and all too common) to have an element that is compliant, but does not work for an individual. Whenever possible, a design should be reviewed by potential users who have disabilities to determine if it is, in fact, the optimal solution for removal of a particular barrier.

Summary

The new 2004 ADA guidelines address an ongoing need for enhanced compliance. The model projects in this book show that barrier removal is not complicated, and that removing a barrier for a particular group of people removes it for everyone. Modifications such as curb cuts, ramps, level door handles, paddle sink faucets, audible elevator signals, and visual fire alarms are just some of the accessibility projects that make a facility easier and safer for all users. The access renovation projects included here will help make your facility more compliant with ADAAG, accessible by people with disabilities, and more open to everyone.

Part One
Getting Started

This section introduces two important concepts that will apply to all of the projects with estimates that follow in Part Two. First, there is an overview of the Americans with Disabilities Act, which outlines the accessibility requirements for public accommodations and the roles of the various federal agencies who issue and enforce the standards. The basic requirements for complying with the new 2004 *Americans with Disabilities Act Accessibility Guidelines for Buildings and Facilities* are spelled out for new construction, existing facilities, and renovations. This section also describes the types of facilities covered by the ADA and how the law applies to each.

The second component of Part One is guidance on how to use the 74 projects in Part Two to arrive at an estimated cost for your facility's compliance modifications. This section defines the elements of each project and offers guidance to help apply the information given to arrive at accurate cost estimates for your own particular material requirements and site conditions.

Overview of the ADA

What is the ADA?

The Americans with Disabilities Act (ADA) is a federal civil rights act enacted in 1990, prohibiting discrimination against people with disabilities. The main premise of the ADA is that accessibility barriers must be removed in all places of public accommodation when readily achievable, or in other words, easily accomplished without excessive difficulty or expense.

The Americans with Disabilities Act Accessibility Guidelines for Buildings and Facilities (ADAAG), updated in 2004, requires that all public facilities comply with its accessibility requirements—for both new construction and alterations. The five types of facilities, or "titles," are:

- Title I: Employment
- Title II: State and Local Government
- Title III: Public Accommodations and Commercial Facilities
- Title IV: Telecommunications
- Title V: Miscellaneous provisions of the law.

(See the "What Facilities are Covered by ADA?" section for more on compliance requirements of specific facilities.)

While the Department of Justice (DOJ) oversees the enforcement of the ADA, an independent federal agency—the Architectural and Transportation Barriers Compliance Board (known as the U.S. Access Board)—is charged with compiling the *minimum* design guidelines that DOJ can adopt, the *Americans with Disabilities Act Accessibility Guidelines for Buildings and Facilities.* Once these are adopted by the DOJ, they become law, along with additional regulations determined by the DOJ. For this reason, it is important to also refer to the DOJ Final Rule for the latest specific compliance requirements.

Since the ADA is an anti-discrimination civil rights act and not a building code, there is more involved with compliance than just meeting the minimum technical requirements. (For full background on the ADA, contact the ADA Technical Assistance Centers found in the Resources section.)

What is ADAAG?

In the early 1990s, the U.S. Access Board issued the first *Americans with Disabilities Act Accessibility Guidelines,* setting forth the minimum design requirements necessary to comply with Titles I, II, and III. While there have been minor modifications from time to time over the last decade, it remained relatively unchanged until July, 2004. At that time, the format for ADAAG was replaced with one that is consistent with ICC A117.1, the accessibility guidelines used in the International Building Code.

Hereinafter, in the *Means ADA Compliance Pricing Guide,* any reference to "ADAAG" refers to the 2004 version. To distinguish between the old and new format, the 1990 version will be referred to as the "original ADAAG."

Additional technical sections are being developed that will gradually become incorporated into ADAAG. Currently under development are additional scoping and technical requirements for children's environments, additional recreational facilities, parks and outdoor developed spaces, and work undertaken in public rights of way.

The original ADAAG was based on the American National Standards Institute's (ANSI) *Accessible and Usable Buildings and Facilities,* which, in its current form, is incorporated into all the national building codes. There are differences between ANSI's requirements and ADAAG, however, so it shouldn't be assumed that compliance with ICC A117.1 or a model building code is the same as compliance with ADAAG.

Some facilities do not fall under the ADA, but are covered by the *Architectural Barriers Act* (ABA). However, ADAAG now uses the same technical criteria for both the ADA and the ABA; only the scoping requirements differ. The new ADAAG has handled this by having separate scoping sections. Chapters F1 and F2 are for facilities under the ABA, Chapters 1 and 2 are for facilities under the ADA, and the remaining chapters are for both.

While ADAAG provides specific design requirements and will be your primary means for meeting the ADA, ADAAG Section 103, "Equivalent Facilitation," states, "Nothing in these guidelines is intended to prevent the use of designs or technologies as alternatives to those prescribed in this document, provided they provide equivalent or superior accessibility and usability." Hence, a designer may deviate from the requirements as long as the requirements are met. As stated before, ADAAG sets the minimum; you are encouraged to provide greater access when you can. Often it costs little or nothing more.

What Facilities Are Covered by ADA?

All types of facilities are covered by the ADA, with the exception of privately-owned permanent housing, religious facilities, and private clubs, which are exempt from Title III. Although ADAAG is written like a building code, it has no sign-off. This means that facility owners are responsible for determining just how their facilities are covered under ADA, and which modifications might be required.

Most states already have accessibility regulations, often based on ICC A117.1, that are enforced as part of the local building codes. When local access codes give greater access, these always govern.

New Construction and Additions

For new construction, both publicly used and employee-only spaces must comply with ADAAG. Only "non-occupiable" spaces, such as elevator pits, machinery spaces, and the like, are exempt. (See ADAAG 203, General Exceptions.) Employee work areas in particular are tricky (ADAAG 203.9): one must be able to approach, enter, and exit the facility, as well as circulate between work stations and have an accessible means of egress from within the work space. A work station must also have audible and visual alarms.

Alterations and Renovations

Any alterations to existing facilities, both public and employee-only spaces, must comply with ADAAG, unless it is "technically infeasible" to do so. Alterations to an "area containing a primary function" trigger accessibility requirements for modifying the path of travel to the area.

Existing Facilities

- *Employee-only facilities or spaces:* Under Title I, employers with 15 or more employees must make "reasonable accommodations" for employees. Reasonable accommodations are made according to individual need and may include architectural modifications, but no modifications are required if not requested.
- *Government facilities:* Under Title II, state and local governments must have all their programs accessible. Modifications to existing buildings are required when administrative changes to programs are not sufficient to create access.
- *Public accommodations (privately owned businesses serving the public):* Under Title III, businesses must facilitate use of their services by removing existing barriers when it is readily achievable to do so— "easily accomplished with little expense." Readily achievable modifications are required even when no other renovations are planned.

With such broad requirements, complications arise in deciding exactly what is "the maximum extent feasible," what are "reasonable accommodations," what is "an area of primary function" and whether removal of a barrier is "readily achievable," since each varies greatly depending on particular circumstances. (Refer to the Glossary at the back of the book for the definitions given in ADAAG.)

As written, the regulations allow facility owners or managers some leeway in deciding how to comply with the different provisions and how these might apply to their facilities. At the same time, they also place responsibility for compliance on the facility owner. (For complete information, refer to the technical assistance manuals identified in the resource section at the back of this book. This section also gives information on access resources and accessibility checklists to help identify existing barriers.)

Title I, Employment

- *New Construction, Additions, and Alterations:* Design of employee areas for new construction, additions, and alterations is covered in Title II and III. ADAAG provides the design requirements.
- *Existing Facilities:* Employees with disabilities in companies with more than 15 employees may request that the employer provide "reasonable accommodation" to allow them to carry out their essential job functions. A "reasonable accommodation" may or may not be architectural in nature, depending on the individual's needs. While what determines a "reasonable accommodation" is worked out between employer and employee, criteria have been set by the federal Equal Employment Opportunity Commission (EEOC). Reasonable accommodations need not be expensive; examples are raising a desk to allow knee space for a wheelchair user, providing lever door handles, increasing lighting levels, or installing an accessible parking space.

Title II, State and Local Government Services

- *New Construction and Additions:* Design of new facilities for state or local government agencies is covered by ADAAG. All new construction and additions must be fully in compliance.
- *Alterations:* All alterations to spaces used by the public and employee-only areas must comply with ADAAG unless it is "technically infeasible" to do so. "Technically infeasible" is defined as having little likelihood of being done, because a major structural member would have to be moved or because an existing physical or site constraint prohibits compliance. If technically infeasible, the alteration must comply "to the maximum extent feasible."
- *Existing Facilities:* Title II requires state and local governments to make their programs accessible, which often might not require alterations to an existing facility. For instance, if a town hall is inaccessible, it might be possible to hold meetings in an accessible high school. If, however, a facility is unique in its program (such as one library that serves a given area), the building has to be made accessible in order to achieve program access.

Title III, Public Accommodations and Commercial Facilities Operated by Private Entities

Title III distinguishes between privately-owned businesses that invite the public in to purchase goods and services (public accommodations) and those that don't (commercial facilities). Title III regulations cover twelve categories of public accommodations:

- *Places of lodging* (inns, hotels, motels, or other places of lodging)
- *Establishments serving food or drink* (restaurants, bars, or other establishments serving food or drink)
- *Places of exhibition or entertainment* (movie theaters, theaters, concert halls, stadiums, or other places of exhibition or entertainment)
- *Places of public gathering* (auditoriums, convention centers, lecture halls, or other places of public gathering)
- *Sales or rental establishments* (bakeries, grocery stores, clothing stores, hardware stores, shopping centers, or other sales or rental establishments)
- *Service establishments* (laundromats, dry cleaners, banks, barber shops, beauty salons, travel services, shoe repair services, funeral parlors, gas stations, offices of accountants or lawyers, pharmacies, insurance offices, professional offices of health care providers, hospitals, or other service establishments)
- *Stations used for public transportation* (terminals, depots, or other stations used for specified public transportation)
- *Places for public display or collection* (museums, libraries, galleries, or other places of public display or collection)
- *Places of recreation* (parks, zoos, amusement parks, or other places of recreation)
- *Places of education* (preschools, elementary, secondary, undergraduate, or postgraduate private schools, or other places of education)
- *Social service establishments* (daycare centers, senior citizen centers, homeless shelters, food banks, adoption agencies, or other social service centers)
- *Places of exercise and recreation* (gymnasiums, health spas, bowling alleys, golf courses, or other places of exercise or recreation)

All such establishments must comply with the requirements for public accommodations; only private clubs and religious establishments are exempt (but any public accommodations leasing spaces from them must comply).

Business establishments that do not fall into any of the above categories are classified by Title III as "commercial facilities." If part of the facility serves the public (such as a tour of a factory), that portion of the facility falls under the requirements for a public accommodation. Even if a new building or addition will not be open to the public, it must comply with the requirements for commercial facilities.

- *New Construction and Additions:* New public accommodations and commercial facilities must fully comply with ADAAG. In a general office area, an employee must be able to approach, enter, and exit the area, and circulate between individual work places within the area if they are built-in. Individual employee-only work spaces (such as lab stations) do not have to comply, but they do have to be on an accessible route of travel. Individual work spaces are modified on an as-needed basis under Title I, Employment. Only non-occupiable spaces, such as catwalks and elevator pits, machinery rooms, and the like, are exempted.
- *Alterations:* Alterations to public and employee-only areas must comply with ADAAG unless it is "technically infeasible" to do so. Even then, the alteration must comply "to the maximum extent feasible."

Alterations to an area containing a primary function trigger additional accessibility requirements in both public accommodations and commercial facilities. A primary function is defined as "a major activity for which the facility is intended," such as a bank's customer service area, or the dining area of a cafeteria. Spaces such as ancillary rooms, entrances, and restrooms are not areas of primary function, meaning that alterations to these kinds of spaces do not trigger additional accessibility requirements. If an area of primary function is being altered, the path of travel to the area of primary function must be brought into compliance with ADAAG at a cost up to 20% of the total cost of the area's renovation.

For leased places of public accommodation that are being renovated, ADA does not state who is directly responsible for removing barriers, but places the responsibility for ADA compliance on the contractual agreement between landlord and tenant.

- *Existing Facilities:* For public accommodations where no renovations are planned, all barriers must be removed if it is readily achievable to do so. What is readily achievable for a national chain store might not be readily achievable for a small grocery store. Determining what barrier removal measures would be considered readily achievable must be on a case-by-case basis. Factors to consider include the nature and cost of the remedial action, the organization's financial resources, the size of the organization (number of facilities and employees), and type of operation.

The Department of Justice regulations help determine what modifications might be undertaken, and recommend the following order of priorities in removing barriers:

1. Access into the facility.
2. Access to where goods and services are made available to the public.
3. Access to and within restrooms if they are for public use.
4. Other amenities.

These recommendations, given in section 36.304(c) of the Title III regulations, can vary in different facilities depending on what service is being provided. Examples of readily achievable modifications are installing grab bars in a restroom, adding a lever to a door knob, or putting up Braille signage. Modifications that might be readily achievable depending on the individual circumstances include installing a ramp up a small flight of stairs, installing a platform lift, or replacing a sink. Installing a new elevator probably would not be considered a readily achievable modification except for a large facility. Barrier removal that is not readily achievable can be addressed at a later date during building modernization or as part of a facility's ongoing obligation to remove barriers.

What is the Process for Complying with ADA?

There is a basic process for complying with ADA:

- Learning about the requirements of ADA and how they apply to a facility or program;
- Conducting a survey to identify barriers;
- Establishing a list of potential modifications for barrier removal, including changes to policies, facilities, and cost estimates; and finally,
- Removing existing barriers.

Surveying the Facility

Surveying the facility is the first step in costing accessibility modifications. ADA regulations do not specify one particular method of identifying barriers, but surveying with an accessibility checklist can be very useful. (Several are listed in the Resources section at the back of the book). When surveying a facility for accessibility, it is vital to remember that wheelchair access is only one factor to consider. To comply with ADA, people with other mobility impairments, such as balance or stamina problems, and people with visual, hearing, or cognitive impairments, must also be accommodated. For instance, ADAAG includes requirements that improve access for people who are blind (raised character and Braille signage), people with hearing impairments (visual fire alarms), people with low fine motor control (lever handles or paddle faucets), people with limited reach (control heights), and many others.

Surveying a facility for accessibility involves identifying spaces, routes of travel, and individual items which are not usable by persons with disabilities and may not be in compliance with ADAAG. To determine which barriers can be removed, it is important to know the entire list of access issues within a facility. Knowing the areas of the facility that need to be surveyed is key. Public areas in public accommodations need to be surveyed, obviously, but employee areas also need to be surveyed even though changes may not be required until an employee must be accommodated. The existing barriers then need to be prioritized as to severity in impeding access and ease of removal. The DOJ recommendations for barrier removal are just that. If an entrance is not fully in compliance but is still usable as an accessible entrance, and there is no accessible restroom in the facility, the restroom would be the higher priority in terms of barrier removal, even though it is listed as a lower priority. The ultimate decision on priorities rests with management.

Involving people with disabilities is critical in setting priorities for barrier identification and removal. To quote directly from the Department of Justice preamble to the ADA Title III Final Rule, discussion of Section 36.304 Removal of Barriers: "The Department recommends that this process include appropriate consultation with individuals with disabilities or organizations representing them. A serious effort at self-assessment and consultation can diminish the threat of litigation and save resources. Such a plan, if appropriately designed and diligently executed, could serve as evidence of a good faith effort to comply."

As stated, this publication is not a survey tool. In some cases, such as an inaccessible entrance that can be ramped, a survey might not be necessary, but even in these cases, other issues could exist (and, in fact, probably do) that should be identified. Again, the most valuable information comes from building users.

Using the Projects for Cost Estimating

Project Budgeting

The model projects in this book cover the most needed modifications for ADA compliance—both large and small. Once an accessibility barrier has been identified, various design solutions should be tested. For instance, if a flight of stairs presents a barrier, several critical questions need to be asked: Can a ramp be built, and if so, is there enough room? Would the ramp block the stairs? Could an electric lift be installed on the stairs, and if so, is there enough room? Would a lift impede required fire egress? Are there other solutions, such as re-grading a sidewalk? There is always more than one way of doing what is needed. The next step is comparing the costs of the various design solutions.

As with any renovation, there are two main influences on the cost of ADA modifications: technical requirements and the design. The design and materials decisions should be made by the facility owner or manager. (The model projects and supplemental cost data in this book cover a range of material alternatives.) For example, while a ramp may be required to have a specific slope, it could be constructed of wood, concrete, or granite—a design decision based on existing conditions and the project budget. As with all renovations, budget is a major factor in determining design.

Applying the Projects

Each model project specifies the standards to which that particular modification needs to be designed. Each includes:

- An **illustration** showing how a modification would be used in a typical application.
- **ADAAG references**—the regulation sections in the 2004 ADAAG that apply to the project.
- **"Where Applicable"** —under what circumstances the modification is required by the ADAAG. For instance, the "Construct New Pathways" project specifies that all new public and common-use pathways must comply with the listed sections of ADAAG.
- **Design requirements**—a summary of the minimum technical standards to which the project must conform to be in compliance with the ADAAG.
- **Design suggestions**—useful recommendations. In some cases, the suggestions may exceed the ADA's minimum technical requirements, with the aim of ensuring long-term compliance. "Best use" criteria should be considered from the outset, prior to design. For instance, one design suggestion for a ramp would be that it have a slope shallower than 1:12 (the maximum slope allowed by

ADAAG), thereby making it easier to use by a wider range of people. While this modification would increase the ramp's length, and probably its cost, it may better conform with the existing grades, and hence won't look as "tacked on."

The list of design suggestions provided for any project is by no means comprehensive, and should be supplemented by suggestions from potential facility users, who are often the best source. The constraints imposed by both existing conditions and the budget will help determine if it is possible to apply any of the recommendations.

- **Key Items**—the materials and construction tasks needed to execute the design. This list may vary depending on the particulars of the situation. For example, a ramp always needs a slip-resistant surface, structure, and rails, but it may also include a set of steps (this is recommended, since some people find stairs easier to use than a ramp), an overhang either over the entrance or the entire ramp, or a heated surface to melt snow. These features should be evaluated, and a full list of chosen elements completed prior to costing.
- **Level of Difficulty**—how difficult the project is to undertake. This will help determine whether it will

be necessary to hire an outside contractor or consult a design professional. This section also offers guidance on which, if any, specialized trades may be needed. In most cases, a building permit will be necessary prior to making any modifications, and in all cases, the project will need to comply with all local building and fire safety codes, regardless of the level of difficulty or the involvement of a designer on the project. The possible levels of difficulty—low, moderate, and high—are defined below.

Low = minimal cost, usually doesn't require a building permit, and may be performed by building maintenance staff.

Moderate = more difficult, possibly involving skilled tradespeople and a building permit.

High = expensive, involving structural and/or utility work, specialized tradespeople, and building permits.

- **Estimate(s)**—at least one complete cost breakdown with a total cost for the project. There are often several estimates to cover a number of design alternatives. Factored into the cost is an estimate of the overhead and profit a contractor is

likely to apply. (Overhead includes items such as the contractor's time spent visiting the site, planning and estimating the job, and a percentage of the miscellaneous costs of doing business.) The costs can be tailored to your specific location in the country using the "Location Adjustment Factors" section at the end of the book.

In many cases (such as in the estimates for various ramp configurations) a cost per unit is also given, which can be applied to the same project in a different size. For each ramp estimate, for example, a total project installation cost is given. Underneath that cost is a "per linear foot" cost that can be multiplied by the number of feet your ramp requires. For instance, if a ramp will be 24' long and constructed of wood, the total cost per linear foot for this type of ramp would be multiplied by 24.

Other units might also be used, depending on the project. For instance, widening an existing concrete pathway is estimated per square foot of added area. If an existing 60' long pathway is to be widened by 1', multiply the cost per square foot for that modification by 60.

Example

The following example shows how project estimates may be combined to produce a total modification cost.

One item identified in an accessibility survey of an existing public facility is a 60' long concrete pathway from an accessible doorway to a street crossing. The pathway has a running slope of less than 1:20 and a cross-slope of less than 1:50. It is in good condition, but is surprisingly narrow (30") and needs to be widened. Also, there is no curb ramp.

It was decided that the walk should be widened to 48" for its entire length (even though 36" would have met the ADAAG requirement), and a curb ramp installed. To obtain a cost estimate for this task, the relevant model projects from this publication are:

Project 5: Construct Graded Entrance Pathway, Estimate 1: Widen Existing Concrete Pathway, and

Project 3: Install or Modify Curb Ramps, Estimate 1: Install New Curb Cut (Concrete Sidewalk)

Adjustments to modification totals can be expected, based on a specific site and the total amount of construction work being performed at one time.

Part Two
Project Estimates

This section contains 74 ADA modification project descriptions and their corresponding cost estimates. The model projects cover everything from installing ramps and walkways and widening doorways, to installing elevators, relocating light switches, adding signage, remodeling bathrooms and kitchens, and many more. The order of the projects is based on the way one would normally approach and move through a building, from the parking lot to pathways and ramps, into the building interior, and then to the surrounding grounds, such as playgrounds, swimming pools, and hiking trails.

Many of the model projects contain multiple estimates, so you can select the one that is most appropriate for your particular site, design, and budget. The additional cost data and location factors at the back of the book will further help you refine the project costs to your circumstances and location. If your project calls for special materials or extra work not described in the model project in Part Two, refer to the Unit Costs section in Part Four, where you will find over 3,000 construction line items that can be used to adjust your estimate.

1
Install Accessible Parking Spaces

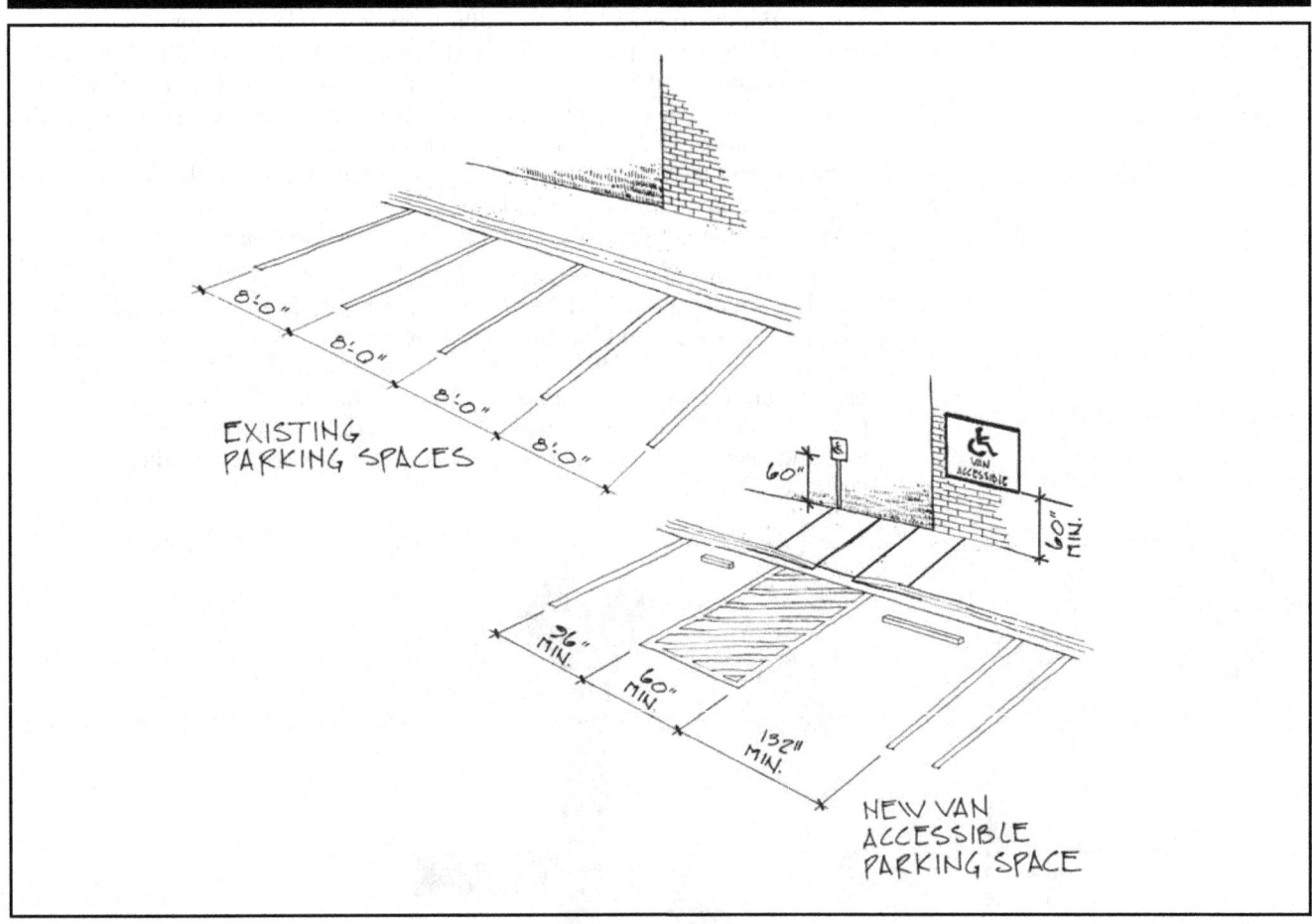

EXISTING PARKING SPACES

8'-0"
8'-0"
8'-0"
8'-0"

60"

60" MIN.

36" MIN.
60" MIN.
132" MIN.

NEW VAN ACCESSIBLE PARKING SPACE

Parking spaces are often the first part of an accessible route of travel. Proper design and location can ensure the safety of *all* people using them, and often can mean the difference between an accessible facility that cannot be used and an accessible and inaccessible facility.

ADAAG References
208, Parking Spaces
216.5, Parking Signs
502, Parking Spaces
(See also Projects 3 through 7 for sidewalks and curb ramps.)

Where Applicable
All parking lots and parking structures for residents, visitors, or employees.

Design Requirements

- Number of accessible spaces as designated by ADAAG Table 208.2. Hospital outpatient facilities must have 10% accessible; rehabilitation facilities specializing in treating mobility conditions and outpatient physical therapy facilities must have 20% accessible.
- The number of parking spaces calculated separately for separate lots. Accessible spaces can be dispersed among multiple lots with accessible entrances.
- One designated van-accessible space for every six accessible spaces (never less than one).
- Spaces on an accessible route.
- Spaces located closest to accessible entrance to the building, or to the lot entrance if no particular building is served.
- Access aisle: 60" wide, minimum, two spaces can share access aisles.
- Car spaces: 96" wide, minimum.
- Van-accessible spaces: 132" wide, minimum. If the access aisle is 96" wide, the van space can be 96" wide, minimum. 98" minimum height clearance in space and route to space.
- Surfaces of parking and aisle: Stable, slip resistant, maximum slope of 2% (1:50) in all directions.
- Signage with accessibility symbol; 60" to bottom of sign, minimum. Van-accessible spots must have an additional sign, "Van Accessible."

Design Suggestions

ADAAG's illustrations show perpendicular spaces, but it is possible to use angled or parallel spaces if access aisle requirements can be met. For angled van parking, the access aisle must be on the right. Check the route of travel before locating the space. Spaces should be located so that people with disabilities are not forced into vehicular traffic.

If a curb adjacent to a sidewalk is being used as a wheel stop, cars will hang over the sidewalk. Accordingly, the sidewalk may have to be modified to make it at least 54" wide. Alternatively, if the sidewalk is wider than 36" but less than 54", wheel stops can be installed 18" from the curb.

Sidewalk construction with a curb ramp may be necessary, and usually the sidewalk for parking areas is narrow. Because of the narrow sidewalk, a parallel curb ramp may be required.

Vans equipped with wheelchair lifts typically exit to the passenger side, though some exit at the rear. If only one van space is required, the wide access aisle should be on the passenger side of the space. When vans share an aisle, it is expected that one will back into the parking slot. Aisles for vans should be striped to discourage others from parking in them.

Locate accessibility signs so that they are always visible to the car's driver and are not blocked by SUVs, poles, and so forth. The recommended height is 80" to the bottom of the sign.

Key Items

Paving material (asphalt or concrete), striping paint, signs, and sidewalk and curb ramp material (if necessary).

Level of Difficulty

Low to moderate. Trades required: re-striping, ground markings, or installing signage can be done by facility staff or landscape contractors. Installing a curb ramp is usually done by a concrete contractor.

Estimates

Add new accessible spaces

Description	Quantity	Unit	Labor-Hours	Material
Saw cutting, concrete, per inch of depth (20 linear feet, 4" deep)	80.000	L.F.	1.306	28.00
Site demolition, remove granite curb	14.000	L.F.	0.933	0.00
Site demolition, remove concrete sidewalk	5.340	S.Y.	0.838	0.00
Site demolition, remove asphalt paving	1.670	S.Y.	0.123	0.00
Painted line removal	100.000	L.F.	2.667	0.00
Install granite curbing, 6" x 18"	14.000	L.F.	1.493	171.50
Gravel base, 4" deep	48.000	S.F.	0.461	18.72
Install 4" concrete sidewalk, broom finish	48.000	S.F.	1.920	58.08
Miscellaneous asphalt patching	1.670	S.Y.	0.891	5.53
Line painting, latex, yellow, 4" wide	20.000	L.F.	0.048	3.00
Line painting, gore lines	100.000	S.F.	1.600	113.00
Install signs	2.000	Ea.	0.914	63.00
Install sign post	1.000	Ea.	0.160	14.35
Totals			**13.354**	**475.18**

Total for two accessible spaces including
general contractor's overhead and profit **$1,817**

Re-stripe existing parking lot

Description	Quantity	Unit	Labor-Hours	Material
Painted line removal	263.000	L.F.	7.014	0.00
Line painting, latex, yellow, 4" wide	20.000	L.F.	0.048	3.00
Line painting, gore lines	500.000	S.F.	8.000	565.00
Totals			**15.062**	**568.00**

Total per space including general
contractor's overhead and profit **$21**

Total per one hundred spaces including
general contractor's overhead and profit **$2,118**

Notes

2
Create Accessible Passenger Drop-off

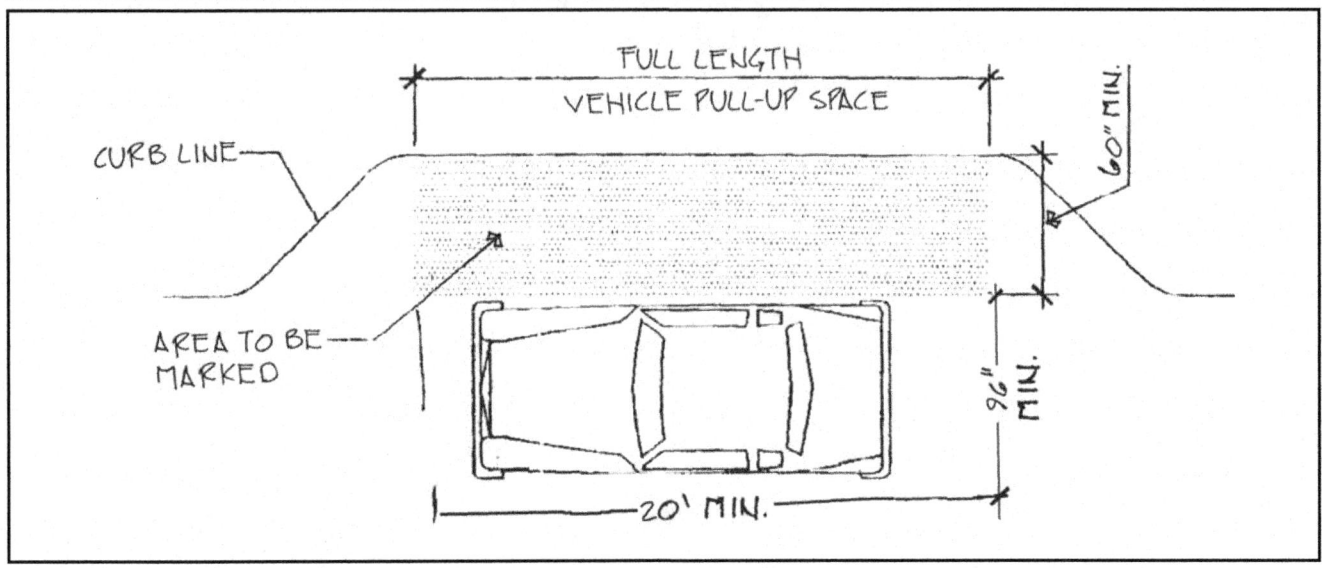

A facility that requires vehicular access but does not have an accessible passenger drop-off can force a user with a disability to either get out into traffic or not use the building at all. A safe, accessible drop-off creates an area where all users can get in or out of their cars or vans, and be directly on an accessible route.

ADAAG References
209, Passenger Loading Zones and Bus Stops
302, Floor or Ground Surfaces
403, Walking Surfaces
503, Passenger Loading Zones
(See also Projects 3 through 7 for sidewalks and curb ramps.)

Where Applicable
At medical and long-term care facilities, valet parking, and bus stops only used by designated or specified public transportation vehicles. At other facilities where passenger loading zones are provided.

Design Requirements
- Vehicle pull-up space: 96" minimum width by 20' minimum length; 114" height clearance minimum at space and along vehicular route to and from the space.
- Vehicle space and access aisle surface: Stable, slip resistant, 2% [1:50] slope maximum in all directions.

- Access aisle: Adjacent to vehicle space, 60" wide by 20' long and marked to discourage parking in them.
- Access to an accessible route. (A curb ramp or modifications to a sidewalk may be necessary.)

Design Suggestions
Visibility is important at a drop-off. Both the vehicle space and the access aisle should be clearly striped. Locate drop-offs as near as possible to accessible entrances, especially where accessible parking is not available.

If there is a curb between the access aisle and the adjacent sidewalk or accessible route, a curb ramp is required. Frequently, the accessible route is less than 11' wide, which means that a perpendicular curb ramp cannot be used, and a parallel or combination curb ramp is required.

Key Items

Paving material(s), striping, signage.

Level of Difficulty

Moderate to high, depending on existing site conditions. Requires planning, paving, and possibly extensive sidewalk modifications.

May require design drawings and approval by local building department or town Department of Public Works.

Estimates

Stripe vehicle space and access aisle, install pole-mounted signage, install concrete curb cut

Description	Quantity	Unit	Labor-Hours	Material
Saw cutting, concrete, per inch of depth (20 linear feet, 4″ deep)	80.000	L.F.	1.306	28.00
Site demolition, remove granite curb	14.000	L.F.	0.933	0.00
Site demolition, remove concrete sidewalk	5.340	S.Y.	0.838	0.00
Site demolition, remove asphalt paving	1.670	S.Y.	0.123	0.00
Install granite curbing, 6″ x 18″	14.000	L.F.	1.493	171.50
Gravel base, 4″ thick	48.000	S.F.	0.461	18.72
Install 4″ concrete sidewalk, broom finish	48.000	S.F.	1.920	58.08
Miscellaneous asphalt patching	1.670	S.Y.	0.891	5.53
Line painting, gore lines	100.000	S.F.	1.600	113.00
Install sign	1.000	Ea.	0.457	31.50
Install sign post	1.000	Ea.	0.160	14.35
Totals			**10.182**	**440.68**

Total per each including general contractor's overhead and profit **$1,561**

Stripe vehicle space and access aisle, install pole-mounted signage, install asphalt curb cut

Description	Quantity	Unit	Labor-Hours	Material
Saw cutting, asphalt, to 3″ deep	20.000	L.F.	0.305	5.20
Site demolition, remove granite curb	14.000	L.F.	0.933	0.00
Site demolition, remove asphalt paving	7.000	S.Y.	0.438	0.00
Install granite curbing, 6″ x 18″	14.000	L.F.	1.493	171.50
Gravel base, 4″ thick	48.000	S.F.	0.461	18.72
Asphalt sidewalk, 2-1/2″ thick	5.340	S.Y.	0.388	22.64
Miscellaneous asphalt patching	1.670	S.Y.	0.891	5.53
Line painting, gore lines	100.000	S.F.	1.600	113.00
Install sign	1.000	Ea.	0.457	31.50
Install sign post	1.000	Ea.	0.160	14.35
Totals			**7.126**	**382.44**

Total per each including general contractor's overhead and profit **$1,227**

3
Install or Modify Curb Ramps

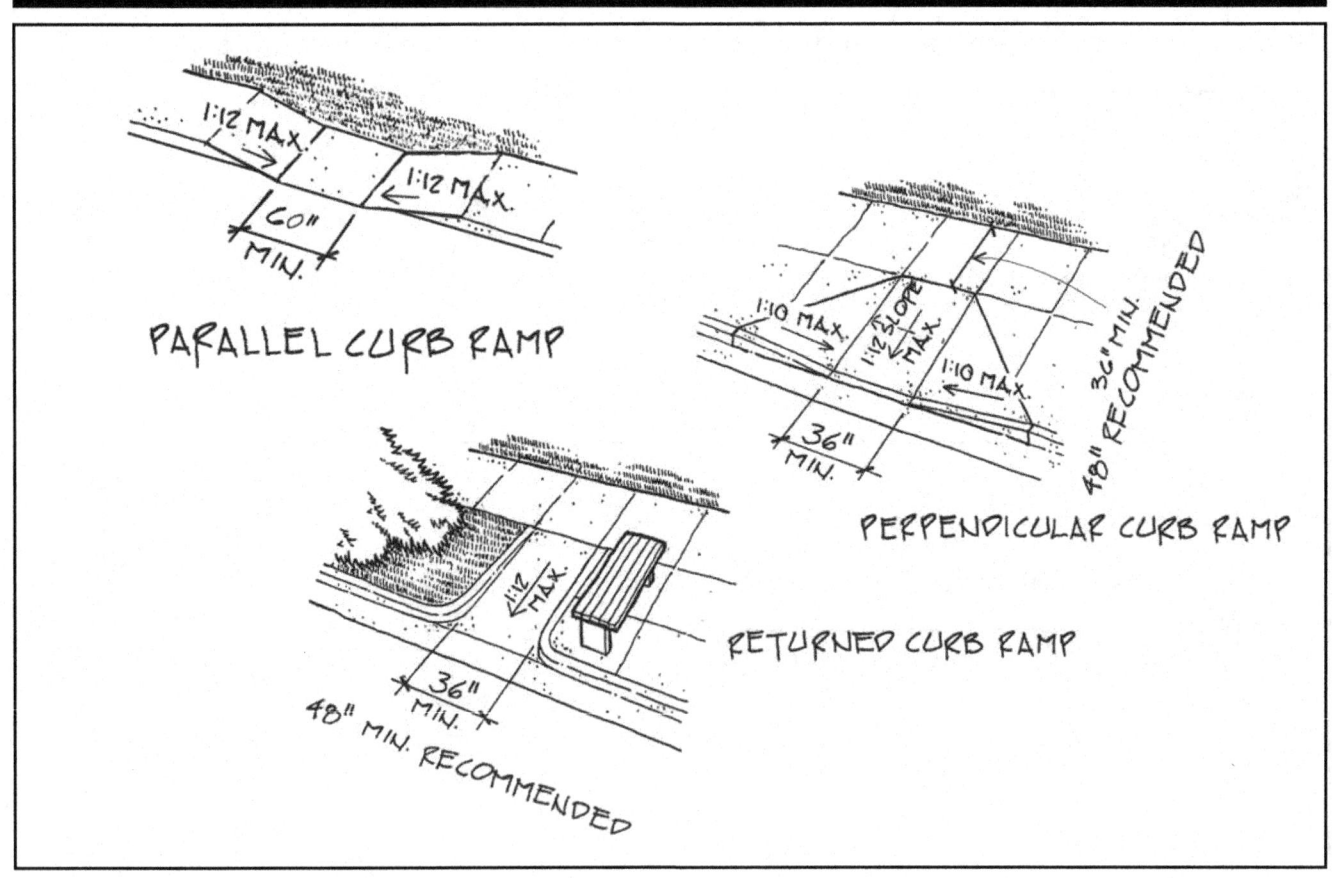

PARALLEL CURB RAMP

1:12 MAX.
1:12 MAX.
60" MIN.

PERPENDICULAR CURB RAMP

1:10 MAX.
1:12 SLOPE MAX.
1:10 MAX.
36" MIN.
36" MIN. RECOMMENDED
48" MIN. RECOMMENDED

RETURNED CURB RAMP

1:12 MAX.
36" MIN.
48" MIN. RECOMMENDED

Installation of curb ramps is the access modification most vital to all people, not just those with mobility impairments. To someone using a wheelchair, a curb without a curb ramp is as difficult to use as a curb four feet high would be to a walking person. Curb ramps also remove a barrier for people with baby strollers or shopping carts, or for those on bicycles. Proper design and location of curb ramps is essential in creating an exterior accessible route of travel.

(Note: ADAAG uses the term "curb ramp" to describe any sloped surface at a curb used to create access for people. It uses "curb cut" to refer to all other cuts in the curb, such as for driveways.)

ADAAG References
302, Floor or Ground Surfaces
303, Changes in Level
405.2–.5, Ramps
405.10, Ramps
406, Curb Ramps

Where Applicable

Wherever a pedestrian route crosses a curb with a vertical drop—at corners, at driveways, and so forth. If a curb ramp is installed on one side of a street, a reciprocal ramp must be installed on the other side.

Design Requirements

- Surface: Stable, firm, slip-resistant with no grates or utilities in ramp portion or in landings.
- Ramp width: 36" minimum; 60" or wider if it is possible.
- Landing at top, 36" long (48" recommended) and as wide as the ramp portion.
- Lip flush with street paving (flush = 1/4" max.).
- Sides of curb ramp: Returned vertical sides allowed only where pedestrians do not walk across the curb ramp. Otherwise, flared sides are required. Slope of flared sides: 1:10 maximum if there is a 36" (48" recommended) long landing at top of curb ramp; 1:12 if no landing.
- Ramp portion located within crosswalk (if crosswalk provided), not to be obstructed by vehicles.
- Diagonal curb ramps at corners: 36" minimum wide by 48" minimum

long clear space at base (toe) that is behind projected curb lines, 24" minimum length of vertical curb on each side within crosswalk.
- No obstructions, such as utility poles, blocking accessible approach to top landing.

Design Suggestions

When installing curb ramps on a street corner, two per corner are recommended. This allows direct crossing of the street without entering traffic, which can be a danger with a single curb ramp located diagonally on the corner. It is strongly recommended not to use a diagonal curb ramp for corner radii less than 15' (impossible to fit the toe landing in without protruding into the path of oncoming vehicles). When possible, install a wide curb ramp, since most people prefer to use the curb ramp rather than stepping up or down. It is also important to ensure adequate drainage (e.g., a curb inlet just uphill from the curb ramp), since the base of a curb ramp is frequently a natural low point.

A landing (48" recommended) at the top allows people using the sidewalk to continue along their path without encountering the severe cross slope of

the curb ramp. When the sidewalk is less than 7', use a parallel ramp.

Only slip-resistant materials should be used, such as thermal finished stone, broom-finished concrete, or unglazed pavers. Concrete or asphalt are the recommended paving materials. It helps if the curb ramp paving material is different from the sidewalk; the concrete can be stained (rather than painted) to provide visual contrast. If pavers are used for the curb ramp, they should be installed over a concrete base to minimize future heaving. The paving material can be scored parallel to the slope to create additional slip-resistance. This is difficult with asphalt or stone, but easier with concrete. (Concrete must, however, be maintained to prevent cracking.)

Key Items

Paving materials.

Level of Difficulty

Moderate to high. May require heavy demolition. Must ensure no disruption of under-street utilities. May require permission from local public works department. Trades involved: paving contractor.

3. Install or Modify Curb Ramps

Estimates

Install new curb cut (concrete sidewalk)

Description	Quantity	Unit	Labor-Hours	Material
Saw cutting, concrete per inch of depth (20 linear feet, 4" deep)	80.000	L.F.	1.306	28.00
Site demolition, remove stone curb	14.000	L.F.	0.933	0.00
Site demolition, remove concrete paving	5.340	S.Y.	0.838	0.00
Site demolition, remove asphalt paving	1.670	S.Y.	0.123	0.00
Install granite curbing, 6" x 18"	14.000	L.F.	1.493	171.50
Install gravel base, 4" thick	48.000	S.F.	0.461	18.72
Install concrete walk, 4" thick, broom finish	48.000	S.F.	1.920	58.08
Miscellaneous asphalt patching	1.670	S.Y.	0.891	5.53
Totals			**7.965**	**281.83**

Total per each including general contractor's overhead and profit **$1,090**

Install new curb cut (asphalt sidewalk)

Description	Quantity	Unit	Labor-Hours	Material
Saw cutting, asphalt	80.000	L.F.	1.219	20.80
Site demolition, remove stone curb	14.000	L.F.	0.933	0.00
Site demolition, remove asphalt paving	7.000	S.Y.	0.517	0.00
Install granite curbing, 6" x 18"	14.000	L.F.	1.493	171.50
Install gravel base, 4" thick	48.000	S.F.	0.461	18.72
Install 2-1/2" asphalt sidewalk	5.340	S.Y.	0.388	22.64
Miscellaneous asphalt patching	1.670	S.Y.	0.891	5.53
Totals			**5.902**	**239.19**

Total per each including general contractor's overhead and profit **$860**

Install new curb cut (brick paver sidewalk on concrete base)

Description	Quantity	Unit	Labor-Hours	Material
Remove brick pavers	9.000	S.Y.	0.800	0.00
Site demolition, remove concrete base	6.000	S.Y.	0.960	0.00
Site demolition, remove stone curb	14.000	L.F.	0.933	0.00
Site demolition, remove asphalt paving	1.670	S.Y.	0.123	0.00
Install granite curbing, 6" x 18"	14.000	L.F.	1.493	171.50
Install 4" concrete mud slab	81.000	S.F.	3.240	98.01
Install 4" x 8" x 2-1/4" brick pavers	81.000	S.F.	11.781	228.42
Miscellaneous asphalt patching	1.670	S.Y.	0.891	5.53
Totals			**20.221**	**503.46**

Total per each including general contractor's overhead and profit **$2,233**

Install new curb cut (brick paver sidewalk on stone dust base)

Description	Quantity	Unit	Labor-Hours	Material
Remove brick pavers	9.000	S.Y.	0.800	0.00
Hand excavation of paver base	0.450	C.Y.	0.900	0.00
Site demolition, remove stone curb	14.000	L.F.	0.933	0.00
Site demolition, remove asphalt	1.670	S.Y.	0.123	0.00
Install granite curbing, 6" x 18"	14.000	L.F.	1.493	171.50
Install stone dust base	6.000	S.Y.	0.046	15.60
Install 4" x 8" x 2-1/4" brick pavers	81.000	S.F.	11.781	228.42
Miscellaneous asphalt patching	1.670	S.Y.	0.891	5.53
Totals			**16.967**	**421.05**

Total per each including general
contractor's overhead and profit **$1,868**

Patch lip at base of curb cut (asphalt paving)

Description	Quantity	Unit	Labor-Hours	Material
Site demolition, remove asphalt	1.670	S.Y.	0.123	0.00
Miscellaneous asphalt patching	1.670	S.Y.	0.891	5.53
Totals			**1.014**	**5.53**

Total per each including general
contractor's overhead and profit **$75**

Install flared sides, existing curb cut (concrete sidewalk)

Description	Quantity	Unit	Labor-Hours	Material
Saw cutting, concrete per inch of depth (16 linear feet, 4" deep)	64.000	L.F.	1.045	22.40
Site demolition, remove stone curb	25.000	L.F.	1.667	0.00
Site demolition, remove concrete paving	3.340	S.Y.	0.524	0.00
Site demolition, remove asphalt paving	1.670	S.Y.	0.123	0.00
Miscellaneous fill at curb removal	12.500	S.F.	0.100	6.63
Install granite curbing, 6" x 18"	14.000	L.F.	1.493	171.50
Install gravel base, 4" thick	30.000	S.F.	0.288	11.70
Install concrete walk, 4" thick, broom finish	30.000	S.F.	1.200	36.30
Miscellaneous asphalt patching	1.670	S.Y.	0.891	5.53
Totals			**7.331**	**254.06**

Total per each including general
contractor's overhead and profit **$994**

3. Install or Modify Curb Ramps *(continued)*

Install flared sides, existing curb cut (asphalt sidewalk)

Description	Quantity	Unit	Labor-Hours	Material
Saw cutting, asphalt	64.000	L.F.	0.975	16.64
Site demolition, remove stone curb	25.000	L.F.	1.667	0.00
Site demolition, remove asphalt paving	5.000	S.Y.	0.369	0.00
Miscellaneous fill at curb removal	12.500	S.F.	0.100	6.63
Install granite curbing, 6" x 18"	14.000	L.F.	1.493	171.50
Install gravel base, 4" thick	30.000	S.F.	0.288	11.70
Install 2-1/2" asphalt sidewalk	3.340	S.Y.	0.243	14.16
Miscellaneous asphalt patching	1.670	S.Y.	0.891	5.53
Totals			**6.026**	**226.16**

Total per each including general contractor's overhead and profit **$847**

Install new curb ramp

Description	Quantity	Unit	Labor-Hours	Material
Install concrete curb ramp, up to 6" thick, broom finish	54.000	S.F.	2.541	101.52
Totals			**2.541**	**101.52**

Total per each including general contractor's overhead and profit **$340**

Notes

4
Construct New Pathway

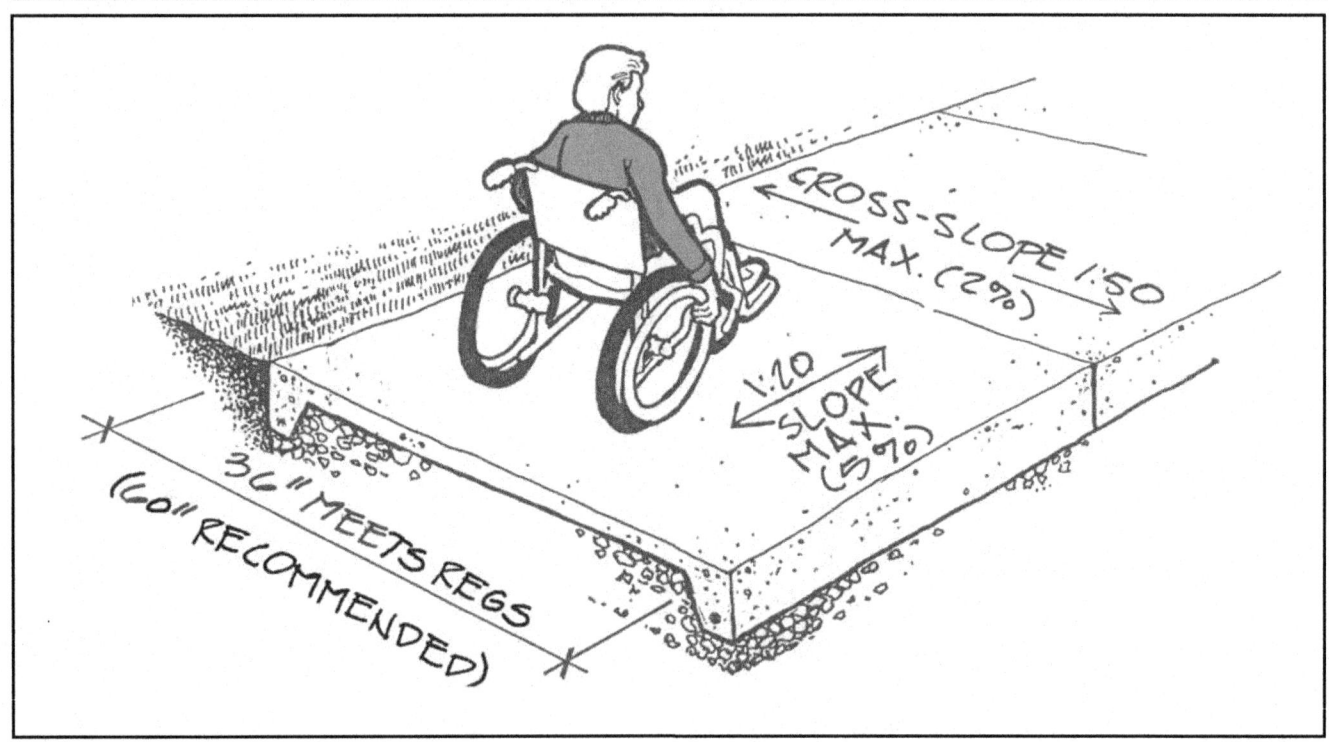

New pathways are needed to create an accessible route where there is an inaccessible route of travel, such as a stepped or steep walkway. New accessible pathways are also useful where there is no designated route of travel, such as in parks or recreation areas with soft surfaces, or from exit doors. Accessible pathways provide access not only for people who use wheelchairs, but also for those with rolling suitcases, baby carriages, and hand trucks.

ADAAG References
302, Floor or Ground Surfaces
303, Changes in Level
307, Protruding Objects
402, Accessible Routes
403, Walking Surfaces

Where Applicable
New sidewalks; wherever a new exterior accessible door or route of travel is installed at entrance or exit doors.

Design Requirements
- Surface: Stable, firm, slip-resistant with openings no wider than 1/2" in the direction of travel in items such as grates, sidewalk cracks, ornamental features, etc. Elongated openings should be placed so that the long dimension is perpendicular to the dominant travel direction.
- Width: 36" minimum, with 60" by 60" passing spots every 200', maximum.

- Level change: 0" to 1/4", vertical; 1/4" to 1/2", beveled at maximum slope of 1:2; 1/2" to 6", slope of 1:12, maximum; greater than 6", meet requirements for a ramp. (See ADAAG 405.)
- Slopes: Maximum 5% running slope (1:20). Maximum 2% cross-slope (1:50).
- Protruding objects: 80" minimum height clearance. No objects 27" above finished floor (a.f.f.) or higher protruding more than 4" into pathway without warnings below.

Design Suggestions

As much as possible, the accessible route should coincide with the route used by others. It should at least begin and end on the other route. The best solution is to make the accessible route the most convenient for everyone so that everyone uses the same route.

Avoid unit pavers (e.g., bricks) if possible. If used, install unit pavers on a firm base. A firm substrate concrete is preferred, although it is more expensive than tamped sand or stone dust.

Avoid placing utilities (e.g., drainage gratings) in the path of travel, even if the clearance dimensions comply. Similarly, locate amenities such as phones adjacent to, rather than in, the path of travel, but within reach from the accessible path.

Thirty-six inches meets the minimum width requirement, but does not allow for a companion alongside or a person using an assistive animal, and the narrow width will usually result in mud paths along each side of the walk. Thirty-six inches should not be used unless absolutely necessary. Forty-eight inches should be

considered the functional minimum width, but 60" is recommended, because it allows passing, signs, and general convenience.

By the same token, 5% is the maximum running slope for a pathway (unless a ramp with handrails is provided), but shallower slopes are easier to use.

Key Items

Paving materials, substrate.

Level of Difficulty

Moderate to high, depending on material used. Pavers or asphalt are easier to install than concrete. Usually involves some landscaping of areas adjacent to the path. Trades involved: paving contractors, masons, landscapers.

Estimates

Install three-foot-wide concrete pathway

Description	Quantity	Unit	Labor-Hours	Material
Excavating	0.110	C.Y.	0.020	0.00
Gravel base, 4" thick	3.000	S.F.	0.029	1.17
Install 4" concrete sidewalk, broom finish	3.000	S.F.	0.120	3.63
Install topsoil, 4" deep	0.230	S.Y.	0.002	0.58
Install sod	0.002	M.S.F.	0.004	0.43
Totals			**0.175**	**5.81**

Total per linear foot including general contractor's overhead and profit _____ **$23**

4. Construct New Pathway *(continued)*

Install three-foot-wide asphalt pathway

Description	Quantity	Unit	Labor-Hours	Material
Excavating	0.100	C.Y.	0.018	0.00
Gravel base, 4" thick	3.000	S.F.	0.029	1.17
Install 2-1/2" thick asphalt sidewalk	0.340	S.Y.	0.025	1.44
Install topsoil, 4" deep	0.230	S.Y.	0.002	0.58
Install sod	0.002	M.S.F.	0.004	0.43
Totals			**0.078**	**3.62**

Total per linear foot including general contractor's overhead and profit **$11**

Install three-foot-wide brick paver pathway over tamped earth base

Description	Quantity	Unit	Labor-Hours	Material
Excavating	0.080	C.Y.	0.014	0.00
Gravel base, 4" thick	3.000	S.F.	0.029	1.17
Install 4" x 8" x 2-1/4" brick pavers	3.000	S.F.	0.436	8.46
Install topsoil, 4" deep	0.230	S.Y.	0.002	0.58
Install sod	0.002	M.S.F.	0.004	0.43
Totals			**0.485**	**10.64**

Total per linear foot including general contractor's overhead and profit **$50**

Install three-foot-wide brick paver pathway over stone dust base

Description	Quantity	Unit	Labor-Hours	Material
Excavating	0.080	C.Y.	0.014	0.00
Install stone dust base	0.340	S.Y.	0.003	0.88
Install 4" x 8" x 2-1/4" brick pavers	3.000	S.F.	0.436	8.46
Install topsoil, 4" deep	0.230	S.Y.	0.002	0.58
Install sod	0.002	M.S.F.	0.004	0.43
Totals			**0.459**	**10.35**

Total per linear foot including general contractor's overhead and profit **$48**

Install three-foot-wide brick paver pathway over concrete base

Description	Quantity	Unit	Labor-Hours	Material
Excavating	0.140	C.Y.	0.025	0.00
Install 4" concrete mud slab	3.000	S.F.	0.120	3.63
Gravel base, 4" thick	3.000	S.F.	0.029	1.17
Install 4" x 8" x 2-1/4" brick pavers	3.000	S.F.	0.436	8.46
Install topsoil, 4" deep	0.230	S.Y.	0.002	0.58
Install sod	0.002	M.S.F.	0.004	0.43
Totals			**0.616**	**14.27**

Total per linear foot including general contractor's overhead and profit **$65**

Copyright Reed Construction Data, 2004

Notes

5
Construct Graded Entrance Pathway

Where an entrance is not on the same grade as an accessible route of travel, constructing a graded pathway may be an alternative to a ramp. Since graded pathways have a slope no greater than 1:20, they have the added advantage of being usable by people with a wider range of abilities. Graded pathways can be created through careful landscaping, which can be useful in situations where a ramp might be visually disruptive (as in some historic facilities).

ADAAG References
302, Surfaces
303, Changes in Level
307, Protruding Objects
402, Accessible Routes
403, Walking Surfaces

Where Applicable
Existing routes of travel or a new exterior route of travel that can be constructed at a slope of 1:20 or less.

Design Requirements
- Surface: Stable, firm, slip-resistant with openings no wider than 1/2" in the direction of travel. No gratings with openings greater than 1/2" across in the lesser dimension.
- Width: 36" minimum, with 60" by 60" passing spots every 200', maximum.
- Level change: 0" to 1/4", vertical; 1/4" to 1/2", beveled at 1:2 maximum slope; 1/2" to 6", slope of 1:12, maximum; greater than 6", meet requirements for a ramp (ADAAG 405 or 406).
- Slopes: Maximum 5% running slope (1:20). Maximum 2% cross-slope (1:50).
- Protruding objects: 80" minimum height clearance. No objects 27" a.f.f. or higher protruding more than 4" into pathway without warnings below.

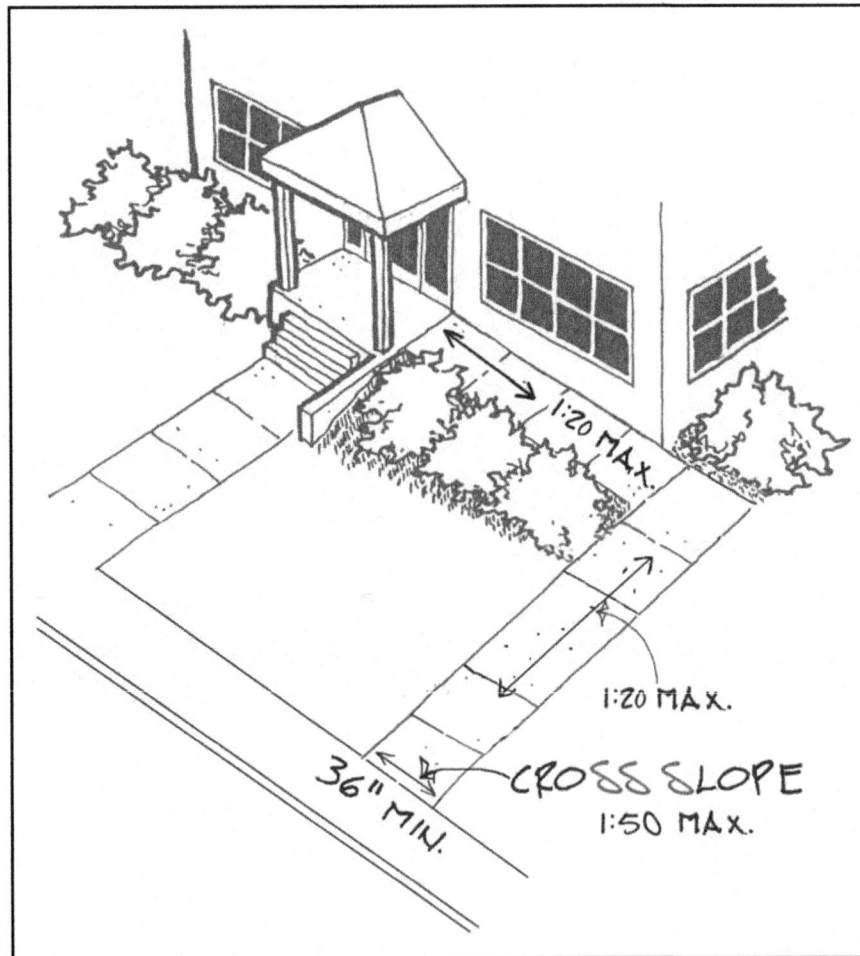

Design Suggestions

Pathways graded at a slope less than 1:20 can avoid the use of ramps. In new construction or additions, the first floor elevation can often be set to accommodate a sloped pathway. In renovations, a graded pathway can be advantageous where a ramp would be difficult to incorporate into the existing building style, as in historic structures, or where a ramp would be prohibitively expensive. For historic buildings, the pathway may have to be located away from the building to avoid damage to existing features. Generally, pathways are more usable by people with a wider range of mobility impairments than ramps and are therefore a recommended design option for creating access. However, some people with stamina difficulties find ramps easier to use; other people prefer to use stairs.

It is important to consider the distance between one end of the route of travel and the entrance. Excessively long pathways can create circuitous accessible routes of travel, and in such situations, ramps might be preferable. When there is not enough space for a 5% path, a combination of a 5% path and short ramp should be considered.

Use the 36" width cited in ADAAG only if there is insufficient room for a wider path; 48" is sufficient for short distances (up to 50'). Sixty inches is convenient for passing. By the same token, the 5% slope is a maximum; shallower slopes are easier to use. Avoid unit pavers (e.g., bricks) if possible. Concrete is preferred. If pavers are used, install unit pavers on a firm base; avoid placing utilities (e.g., drainage gratings) in the path of travel, even if the dimensions comply. Similarly, locate amenities, such as phones, drinking fountains, and ATMs, adjacent to, rather than in, the path of travel but within reach of the accessible path. Ensure that drainage is sufficient to prevent puddling on the pathway.

Key Items

Paving materials, substrate.

Level of Difficulty

Moderate to high, depending on the material. Pavers or bituminous materials are easier to install than concrete. Usually involves landscape planning. Trades involved: paving contractors, masons, landscapers.

Estimates

Install three-foot-wide concrete pathway on graded ramp (total rise, 1 foot)

Description	Quantity	Unit	Labor-Hours	Material
Gravel fill	0.300	Ton	0.000	6.45
Gravel delivery charge	0.200	C.Y.	0.019	0.00
Hand spread gravel	0.300	Ton	0.164	0.00
Compaction (4 passes)	0.200	C.Y.	0.011	0.00
Gravel base, 4" thick	3.000	S.F.	0.029	1.17
Install 4" concrete sidewalk, broom finish	3.000	S.F.	0.120	3.63
Install topsoil, 4" deep	0.500	S.Y.	0.005	1.27
Install sod	0.005	M.S.F.	0.009	1.09
Totals			0.357	13.61

Total per linear foot including general contractor's overhead and profit _____ **$47**

5. Construct Graded Entrance Pathway *(continued)*

Install three-foot-wide asphalt pathway on graded ramp (total rise, 1 foot)

Description	Quantity	Unit	Labor-Hours	Material
Gravel fill	0.300	Ton	0.000	6.45
Gravel delivery charge	0.200	C.Y.	0.019	0.00
Hand spread gravel	0.300	Ton	0.164	0.00
Compaction (4 passes)	0.200	C.Y.	0.011	0.00
Gravel base, 4" thick	3.000	S.F.	0.029	1.17
Install 2-1/2" thick asphalt sidewalk	0.340	S.Y.	0.025	1.44
Install topsoil, 4" deep	0.500	S.Y.	0.005	1.27
Install sod	0.005	M.S.F.	0.009	1.09
Totals			**0.262**	**11.42**

Total per linear foot including general contractor's overhead and profit **$37**

Install three-foot-wide asphalt paver pathway over tamped earth base on graded ramp (total rise, 1 foot)

Description	Quantity	Unit	Labor-Hours	Material
Gravel fill	0.300	Ton	0.000	6.45
Gravel delivery charge	0.200	C.Y.	0.019	0.00
Hand spread gravel	0.300	Ton	0.164	0.00
Compaction (4 passes)	0.200	C.Y.	0.011	0.00
Gravel base, 4" thick	3.000	S.F.	0.029	1.17
Install 8" x 8" x 2" asphalt block pavers	3.000	S.F.	0.369	16.95
Install topsoil, 4" deep	0.500	S.Y.	0.005	1.27
Install sod	0.005	M.S.F.	0.009	1.09
Totals			**0.606**	**26.93**

Total per linear foot including general contractor's overhead and profit **$85**

Install three-foot-wide brick paver pathway over tamped earth base on graded ramp (total rise, 1 foot)

Description	Quantity	Unit	Labor-Hours	Material
Gravel fill	0.300	Ton	0.000	6.45
Gravel delivery charge	0.200	C.Y.	0.019	0.00
Hand spread gravel	0.300	Ton	0.164	0.00
Compaction (4 passes)	0.200	C.Y.	0.011	0.00
Gravel base, 4" thick	3.000	S.F.	0.029	1.17
Install 4" x 8" x 2-1/4" brick pavers	3.000	S.F.	0.436	8.46
Install topsoil, 4" deep	0.500	S.Y.	0.005	1.27
Install sod	0.005	M.S.F.	0.009	1.09
Totals			**0.673**	**18.44**

Total per linear foot including general contractor's overhead and profit **$75**

Install three-foot-wide brick paver pathway over stone dust base on graded ramp (total rise, 1 foot)

Description	Quantity	Unit	Labor-Hours	Material
Gravel fill	0.300	Ton	0.000	6.45
Gravel delivery charge	0.200	C.Y.	0.019	0.00
Hand spread gravel	0.300	Ton	0.164	0.00
Compaction (4 passes)	0.200	C.Y.	0.011	0.00
Install stone dust base	0.340	S.Y.	0.003	0.88
Install 4" x 8" x 2-1/4" brick pavers	3.000	S.F.	0.436	8.46
Install topsoil, 4" deep	0.050	S.Y.	0.000	0.13
Install sod	0.005	M.S.F.	0.009	1.09
Totals			**0.642**	**17.01**

Total per linear foot including general contractor's overhead and profit **$71**

Install three-foot-wide brick paver pathway over concrete base on graded ramp (total rise, 1 foot)

Description	Quantity	Unit	Labor-Hours	Material
Gravel fill	0.300	Ton	0.000	6.45
Gravel delivery charge	0.200	C.Y.	0.019	0.00
Hand spread gravel	0.300	Ton	0.164	0.00
Compaction (4 passes)	0.200	C.Y.	0.011	0.00
Install 4" concrete mud slab	3.000	S.F.	0.120	3.63
Gravel base, 4" thick	3.000	S.F.	0.029	1.17
Install 4" x 8" x 2-1/4" brick pavers	3.000	S.F.	0.436	8.46
Install topsoil, 4" deep	0.050	S.Y.	0.000	0.13
Install sod	0.005	M.S.F.	0.009	1.09
Totals			**0.788**	**20.93**

Total per linear foot including general contractor's overhead and profit **$88**

Copyright Reed Construction Data, 2004

6
Modify Existing Pathway

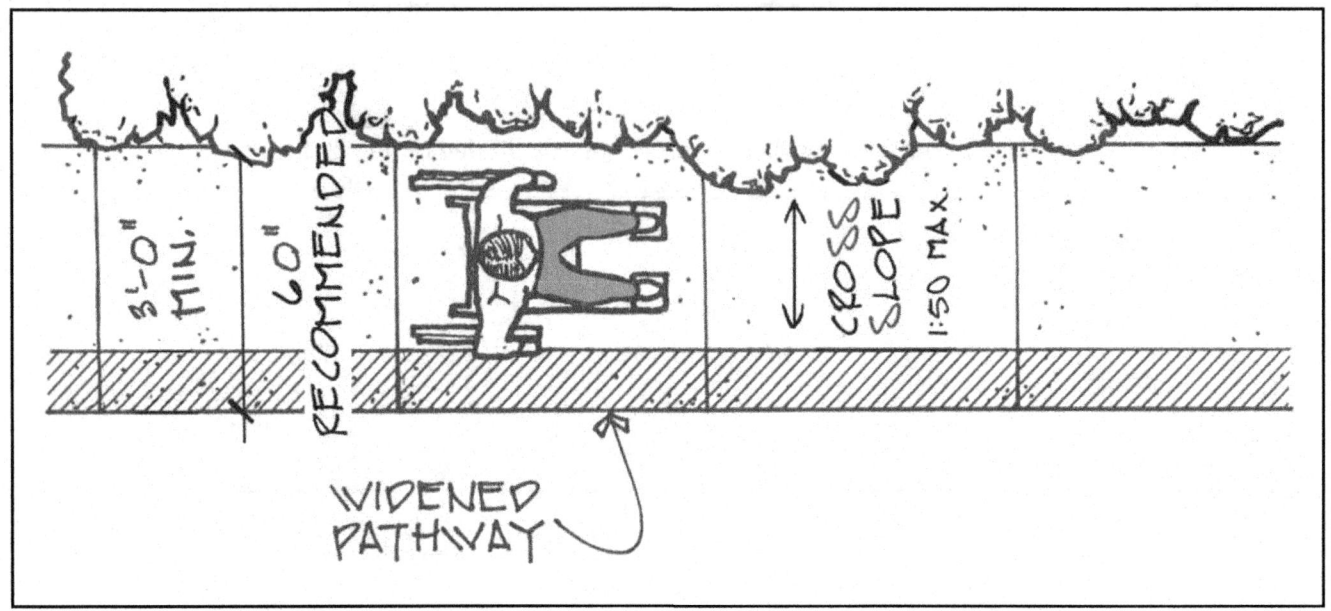

WIDENED PATHWAY

Narrow, uneven, or steeply pitched pathways can be difficult for anyone to use and impassable for people with mobility or visual impairments—folks who cannot maneuver around or see a large crack in the walkway. Fixing an existing path of travel to make it level and smooth is often one of the easiest and most useful modifications that can be done to create an accessible facility.

ADAAG References

302, Floor or Ground Surfaces
303, Changes in Level
307, Protruding Objects
402, Accessible Routes
403, Walking Surfaces

Where Applicable

Wherever an exterior pathway needs to be widened, smoothed, re-graded, or resurfaced.

Design Requirements

Surface: Stable, firm, slip-resistant, with openings no wider than 1/2" in the direction of travel. No gratings with openings greater than 1/2" across in the lesser dimension.
Width: 36" minimum, with 60" by 60" passing spots every 200', maximum.
Level change: 0" to 1/4", vertical; 1/4" to 1/2", beveled at maximum slope of 1:2; 1/2" to 6", slope of

1:12, maximum; greater than 6", meet requirements for a ramp (ADAAG 405 or 406).
Slopes: Maximum 5% running slope (1:20). Maximum 2% cross-slope (1:50).
Protruding objects: 80" minimum height clearance. No objects higher than 27" a.f.f. that protrude more than 4" into pathway without warnings below 27".

Design Suggestions

Thirty-six inches meets the minimum width requirement, but does not allow for a companion alongside or person using an assistive animal, and the narrow width will usually result in mud paths along each side of the walk. Thirty-six inches should not be used unless absolutely necessary. Forty-eight inches should be considered as the functional minimum width, and 60" is recommended because it allows passing, signs, and general convenience. Avoid unit pavers (e.g., bricks) if possible. If used, install unit pavers on a firm base, such as concrete, even though it is more expensive. Avoid placing utilities (e.g., drainage gratings) in the path of travel, even if the dimensions comply. Similarly, locate amenities such as phones, drinking fountains, and fire pull alarms, adjacent to, rather than in, the path of travel, but within reach of the accessible path.

Key Items

Paving materials, substrate.

Level of Difficulty

Moderate, depending on the materials used. Trades involved: paving contractors, masons, landscapers.

Estimates

Widen existing concrete pathway

Description	Quantity	Unit	Labor-Hours	Material
Excavating	0.037	C.Y.	0.007	0.00
Gravel base, 4" thick	1.000	S.F.	0.010	0.39
Install 4" concrete sidewalk, broom finish	1.000	S.F.	0.040	1.21
Install topsoil, 4" deep	0.110	S.Y.	0.001	0.28
Install sod	0.001	M.S.F.	0.002	0.22
Totals			0.060	2.10

Total per square foot including general contractor's overhead and profit **$7**

Widen existing asphalt pathway

Description	Quantity	Unit	Labor-Hours	Material
Excavating	0.034	C.Y.	0.006	0.00
Gravel base, 4" thick	1.000	S.F.	0.010	0.39
Install 2-1/2" thick asphalt sidewalk	0.110	S.Y.	0.008	0.47
Install topsoil, 4" deep	0.110	S.Y.	0.001	0.28
Install sod	0.001	M.S.F.	0.002	0.22
Totals			0.027	1.36

Total per square foot including general contractor's overhead and profit **$4**

Widen existing asphalt block paver pathway

Description	Quantity	Unit	Labor-Hours	Material
Excavating	0.028	C.Y.	0.005	0.00
Gravel base, 4" thick	1.000	S.F.	0.010	0.39
Install 8" x 8" x 2" asphalt block pavers	1.000	S.F.	0.123	5.65
Install topsoil, 4" deep	0.110	S.Y.	0.001	0.28
Install sod	0.001	M.S.F.	0.002	0.22
Totals			0.141	6.54

Total per square foot including general contractor's overhead and profit **$20**

6. Modify Existing Pathway *(continued)*

Widen existing brick paver pathway

Description	Quantity	Unit	Labor-Hours	Material
Excavating	0.028	C.Y.	0.005	0.00
Gravel base, 4″ thick	1.000	S.F.	0.010	0.39
Install 4″ x 8″ x 2-1/4″ brick pavers	1.000	S.F.	0.145	2.82
Install topsoil, 4″ deep	0.110	S.Y.	0.001	0.28
Install sod	0.001	M.S.F.	0.002	0.22
Totals			**0.163**	**3.71**

Total per square foot including general contractor's overhead and profit **$17**

Add concrete to correct pathway cross slope

Description	Quantity	Unit	Labor-Hours	Material
Install concrete walk, up to 4″ thick, broom finish	1.000	S.F.	0.040	1.21
Totals			**0.040**	**1.21**

Total per square foot including general contractor's overhead and profit **$4**

Remove concrete pathway, add gravel to correct pathway cross slope, install new concrete pathway

Description	Quantity	Unit	Labor-Hours	Material
Concrete sidewalk demolition	0.110	S.Y.	0.018	0.00
Gravel base, 4″ thick	1.000	S.F.	0.010	0.39
Install concrete walk, up to 4″ thick, broom finish	1.000	S.F.	0.040	1.21
Totals			**0.068**	**1.60**

Total per square foot including general contractor's overhead and profit **$7**

Patch existing concrete pathway (4″ thick)

Description	Quantity	Unit	Labor-Hours	Material
Saw cutting, per inch of depth (6 linear feet, 4″ deep)	24.000	L.F.	0.392	8.40
Concrete demolition	1.000	S.Y.	0.160	0.00
Install concrete pathway, up to 4″ thick, broom finish	9.000	S.F.	0.360	10.89
Totals			**0.912**	**19.29**

Total per square foot including general contractor's overhead and profit **$11**

Total per 3′ x 3′ section including general contractor's overhead and profit **$102**

Copyright Reed Construction Data, 2004

Patch existing asphalt pathway (2-1/2″ thick)

Description	Quantity	Unit	Labor-Hours	Material
Saw cutting (3 L.F. per S.F., 1″ deep)	7.500	L.F.	0.114	1.95
Asphalt demolition	0.110	S.Y.	0.008	0.00
Miscellaneous asphalt patching	0.110	S.Y.	0.059	0.36
Totals			0.181	2.31

Total per square yard including general contractor's overhead and profit **$18**

Patch existing asphalt paver pathway

Description	Quantity	Unit	Labor-Hours	Material
Remove damaged asphalt block pavers	0.110	S.Y.	0.010	0.00
Install new asphalt block pavers	1.000	S.F.	0.123	5.65
Totals			0.133	5.65

Total per square foot including general contractor's overhead and profit **$18**

Patch existing brick paver pathway

Description	Quantity	Unit	Labor-Hours	Material
Remove brick pavers	0.110	S.Y.	0.010	0.00
Install 4″ x 8″ x 2-1/4″ brick pavers	1.000	S.F.	0.145	2.82
Totals			0.155	2.82

Total per square foot including general contractor's overhead and profit **$16**

Relocate objects in path (e.g., bench, bolted to surface)

Description	Quantity	Unit	Labor-Hours	Material
Remove bench	1.000	Ea.	0.133	0.00
Re-install bench	1.000	Job	2.000	0.00
Totals			2.133	0.00

Total per each including general contractor's overhead and profit **$124**

Install detectable warning in concrete pathway at traffic crossing or drop off

Description	Quantity	Unit	Labor-Hours	Material
Saw cutting (3 L.F. per S.F., 1″ deep)	12.000	L.F.	0.196	4.20
Concrete demolition	0.110	S.Y.	0.018	0.00
Install 4″ concrete sidewalk	1.000	S.F.	0.040	1.21
Install patterned surface finish	1.000	S.F.	0.020	0.00
Totals			0.274	5.41

Total per square foot including general contractor's overhead and profit **$31**

Copyright Reed Construction Data, 2004

7
Install Gratings that Meet ADAAG

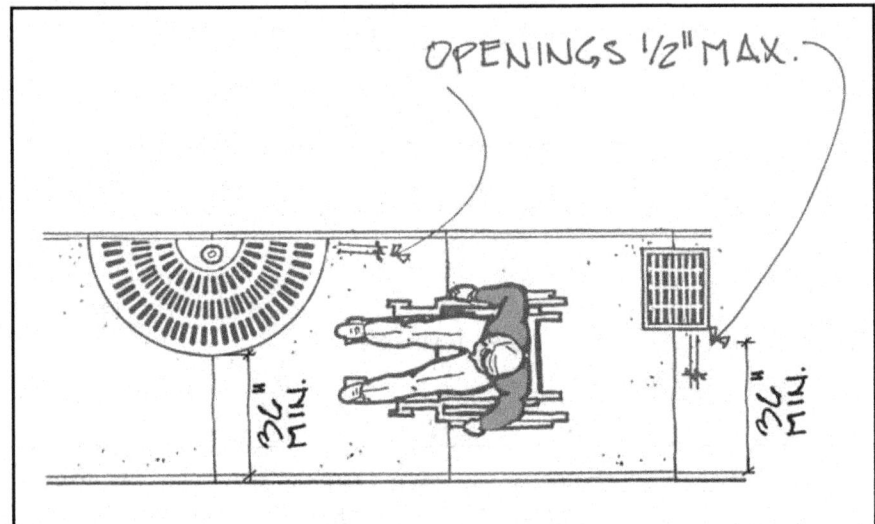

OPENINGS 1/2" MAX.

36" MIN.

36" MIN.

A feature as seemingly innocuous as a grate can pose a severe inconvenience, if not an outright safety hazard, to someone with a mobility or visual impairment. Grates with large openings in a path of travel act as a barrier to people who use wheelchairs, canes, walkers, or crutches, or who have balance impairments. They can also be dangerous for people wearing high-heeled shoes. Attention to details like grating location and design helps to create a truly accessible facility.

ADAAG References
302.3, Openings
303.2, Vertical Changes in Level

Where Applicable
Any grating located within a common pedestrian route of travel.

Design Requirements
- No openings greater than 1/2" across in the lesser dimension.
- Grates with elongated holes placed so that the long dimension is perpendicular to the direction of travel.

Design Suggestions
Wherever possible, locate grates (and, in fact, all surface utilities) away from the most commonly used routes of travel, with top of grate and edges flush with surrounding paving. Grates and utilities usually rest on a substructure, which will settle/heave differently from the path, resulting over time in a vertical displacement greater than 1/4". Accordingly, even grates in compliance should not impinge into the required 36" clear width. If holes in tree grates are larger than allowed by ADAAG, an inexpensive, though not ideal, solution would be to fill holes flush with the surrounding surface, either with gravel or a wood chip and sand mixture, to provide a relatively smooth surface. (This is useful no

matter how big the grate openings are.) Storm grates and ventilation grilles often have specific requirements for the amount of open area, and may need to be re-sized prior to replacement with an accessible grating.

Key Items
Metal gratings.

Level of Difficulty
Low to moderate. Existing gratings can sometimes be replaced by facility maintenance crews. A manufacturer's representative or civil or structural engineer may be needed to verify opening dimensions, structural characteristics, and total square inches of opening. New gratings are installed by a contractor.

Estimates

Replace trench drain

Description	Quantity	Unit	Labor-Hours	Material
Remove trench drains	1.000	Ea.	0.137	0.00
Install new trench drains, modular, 12″ x 12″	1.000	Ea.	2.000	345.00
Totals			2.137	345.00

Total per linear foot of 1 foot wide trench drain including general contractor's overhead and profit **$749**

Replace 5′ diameter cast iron tree grate

Description	Quantity	Unit	Labor-Hours	Material
Remove tree grate	1.000	Ea.	0.333	0.00
Install cast iron tree grate, 5′ diameter	1.000	Ea.	0.960	835.00
Totals			1.293	835.00

Total per each including general contractor's overhead and profit **$1,517**

Replace catch basin frame and cover

Description	Quantity	Unit	Labor-Hours	Material
Remove catch basin frame and cover	1.000	Ea.	1.846	0.00
Install new catch basin frame and cover	1.000	Ea.	3.077	231.00
Totals			4.923	231.00

Total per each including general contractor's overhead and profit **$763**

Install new curb cut (brick paver sidewalk on concrete base)

Description	Quantity	Unit	Labor-Hours	Material
Remove brick pavers	9.000	S.Y.	0.800	0.00
Site demolition, remove concrete base	6.000	S.Y.	0.960	0.00
Site demolition, remove stone curb	14.000	L.F.	0.933	0.00
Site demolition, remove asphalt paving	1.670	S.Y.	0.123	0.00
Install granite curbing, 6″ x 18″	14.000	L.F.	1.493	171.50
Install 4″ concrete mud slab	81.000	S.F.	3.240	98.01
Install 4″ x 8″ x 2-1/4″ brick pavers	81.000	S.F.	11.781	228.42
Miscellaneous asphalt patching	1.670	S.Y.	0.891	5.53
Totals			**20.221**	**503.46**

Total per each including general contractor's overhead and profit **$2,233**

Install new curb cut (brick paver sidewalk on stone dust base)

Description	Quantity	Unit	Labor-Hours	Material
Remove brick pavers	9.000	S.Y.	0.800	0.00
Hand excavation of paver base	0.450	C.Y.	0.900	0.00
Site demolition, remove stone curb	14.000	L.F.	0.933	0.00
Site demolition, remove asphalt	1.670	S.Y.	0.123	0.00
Install granite curbing, 6″ x 18″	14.000	L.F.	1.493	171.50
Install stone dust base	6.000	S.Y.	0.046	15.60
Install 4″ x 8″ x 2-1/4″ brick pavers	81.000	S.F.	11.781	228.42
Miscellaneous asphalt patching	1.670	S.Y.	0.891	5.53
Totals			**16.967**	**421.05**

Total per each including general contractor's overhead and profit **$1,868**

Patch lip at base of curb cut (asphalt paving)

Description	Quantity	Unit	Labor-Hours	Material
Site demolition, remove asphalt	1.670	S.Y.	0.123	0.00
Miscellaneous asphalt patching	1.670	S.Y.	0.891	5.53
Totals			**1.014**	**5.53**

Total per each including general contractor's overhead and profit **$75**

Install flared sides, existing curb cut (concrete sidewalk)

Description	Quantity	Unit	Labor-Hours	Material
Saw cutting, concrete per inch of depth (16 linear feet, 4" deep)	64.000	L.F.	1.045	22.40
Site demolition, remove stone curb	25.000	L.F.	1.667	0.00
Site demolition, remove concrete paving	3.340	S.Y.	0.524	0.00
Site demolition, remove asphalt paving	1.670	S.Y.	0.123	0.00
Miscellaneous fill at curb removal	12.500	S.F.	0.100	6.63
Install granite curbing, 6" x 18"	14.000	L.F.	1.493	171.50
Install gravel base, 4" thick	30.000	S.F.	0.288	11.70
Install concrete walk, 4" thick, broom finish	30.000	S.F.	1.200	36.30
Miscellaneous asphalt patching	1.670	S.Y.	0.891	5.53
Totals			**7.331**	**254.06**

Total per each including general contractor's overhead and profit **$994**

Install flared sides, existing curb cut (asphalt sidewalk)

Description	Quantity	Unit	Labor-Hours	Material
Saw cutting, asphalt	64.000	L.F.	0.975	16.64
Site demolition, remove stone curb	25.000	L.F.	1.667	0.00
Site demolition, remove asphalt paving	5.000	S.Y.	0.369	0.00
Miscellaneous fill at curb removal	12.500	S.F.	0.100	6.63
Install granite curbing, 6" x 18"	14.000	L.F.	1.493	171.50
Install gravel base, 4" thick	30.000	S.F.	0.288	11.70
Install 2-1/2" asphalt sidewalk	3.340	S.Y.	0.243	14.16
Miscellaneous asphalt patching	1.670	S.Y.	0.891	5.53
Totals			**6.026**	**226.16**

Total per each including general contractor's overhead and profit **$847**

Install new curb ramp

Description	Quantity	Unit	Labor-Hours	Material
Install concrete curb ramp, up to 6" thick, broom finish	54.000	S.F.	2.541	101.52
Totals			**2.541**	**101.52**

Total per each including general contractor's overhead and profit **$340**

8
Construct New Ramp: Straight

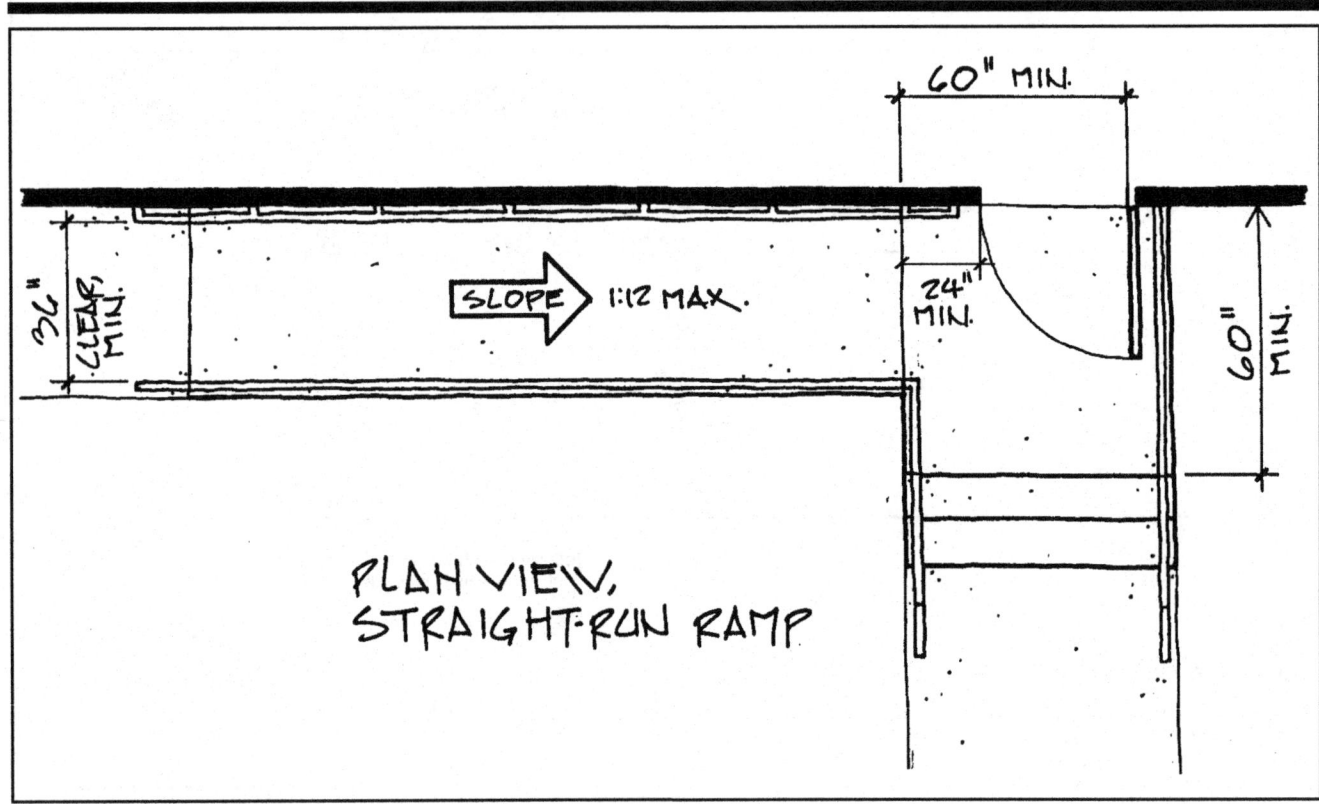

60" MIN.

SLOPE 1:12 MAX.

36" CLEAR MIN.

24" MIN.

60" MIN.

PLAN VIEW, STRAIGHT-RUN RAMP

Ramps are a common method of creating an accessible route, not only for people who use wheelchairs, but also for people with baby strollers, shopping carts, and almost any other device with wheels. A straight ramp is the easiest to use, since it requires no turns by the user. A long straight-run ramp can be a strong visual element, however. If aesthetic requirements are a consideration, an alternate configuration, such as a dog-leg ramp, might mitigate the visual impact.

ADAAG References
206.4, Entrances
207, Accessible Means of Egress
302, Floor or Ground Surfaces
404.2.4, Maneuvering Clearances
405, Ramps
505, Handrails

(Consult ADAAG 202.5 for special allowances for historic facilities and alterations.)
IBC 1007.5 Egress (2003 Edition)

Where Applicable
Entrances, exits, and routes of travel where steps or different levels exist. A straight ramp is usually employed when the vertical rise is 30" or less.

Design Requirements

Surface:
- Stable, firm, slip-resistant with openings no wider than 1/2" in the direction of travel and vertical displacements no greater than 1/4". Flush (1/4" or less) with adjacent paving. No pooling of water on ramp surface.

Width:
- 36" minimum between handrails or curbs.

Slope:
- 1:12 maximum (1" of rise for every foot of run). Any slope between 1:20 and 1:12 is considered a ramp. Maximum 2% cross-slope (1:50).
- Slope for existing condition: 1:8 maximum for a rise up to 3". 1:10 maximum for a rise between 3" and 6".

Landings:
- Located at base, at top, and after every 30", maximum, of rise; 60" deep, minimum; as wide as the ramp; level (2% max slope).
- Maximum vertical rise: 30" between landings.

Rails:
- On both sides, 34" to 38" above ramp surface, 1-1/2" from wall exactly, with no rotation in fittings.
- Between 1-1/4" and 2" diameter if round, or, if not round, the perimeter between 4" and 6-1/4" and the maximum width 2-1/4".
- Continuous rail without interruption by posts or other construction. (Brackets attached to the bottom are acceptable if they drop 1-1/2" or more before returning to the wall).
- 12" horizontal extensions at the top and bottom; return to wall, post, or floor/ground.

- No handrail is needed if the vertical rise is 6" or less and there is edge protection, or there is a shoulder that slopes away from the edge of the ramp at 1:10 maximum.

Edge protection:
- 4" high curbs or rails at any drop-offs on both sides of the ramp and landings unless adjacent surfaces continue for at least 12" beyond the handrails.

Design Suggestions

Always try to make the main or public entrance accessible. If that is not possible, modify another entrance, and install proper signage to direct people to the accessible entrance.

Before deciding on a ramp, first investigate whether a sloped walkway can be used. In new construction or additions, floor heights can be established so that neither steps nor ramps are necessary at entrances. Pathways at a slope of 1:20 or less might be possible, or if the space permits, several short ramps with a 1:20 path between.

It is recommended that the ramp be integrated into the existing building aesthetic, rather than constructed of completely different materials and details. Plantings or other landscaping can help integrate the ramp into the surrounding environment.

A ramped entrance or route should also have stairs adjacent to it, since stairs are easier for some people to use.

The 1:12 slope cited in ADAAG is a maximum, so even if a graded pathway is not possible, a ramp with a shallower slope might be.

All materials must be slip-resistant, a critical issue for wood ramps, which can be slippery when wet. Even though it may be expensive, consider the possibility of covering the ramp for weather protection. This helps keep the ramp free of water, snow, ice, and so forth, and makes maintenance easier. If the ramp cannot be protected, use a bottom rail 4" above the ramp surface rather than a curb, because it lets the rain water out and allows a person to push snow off the edge.

If children will be frequent users of the ramp, consider installing a second handrail a minimum of 9" below the top rail. Detail railings for continuity (Under-rail brackets work best.) Above all, remember that ramps are used by people with baby strollers, shopping carts, hand trucks, and so forth, and are a helpful addition to an existing facility. If the space permits, consider making interim landings wider, perhaps installing a bench, so that a person can rest before continuing up the ramp.

Key Items
- Ramp material: painted wood, treated lumber, concrete.
- Earthwork.
- Rails: pipe rails, metal rails, wood banisters or rails, uprights, attachments.
- Possibly, slip-resistant applied materials: sand paint, sandpaper strips.
- Possibly requires drainage. (See ADAAG 302.3, Openings, for grate requirements.)

Level of Difficulty

Moderate to high. A wood ramp requires skilled carpentry. Concrete requires foundation work. Design drawings and building permits may be required.

Estimates

Install new painted wood straight ramp

Description	Quantity	Unit	Labor-Hours	Material
Hand excavating for post footings	3.000	C.Y.	6.000	0.00
Concrete forms, 12" diameter tubes (16 forms, 4' deep)	64.000	L.F.	13.653	134.40
Hand backfilling around post footings	1.000	C.Y.	0.727	0.00
Concrete, material only	2.000	C.Y.	0.000	145.00
Place concrete for footings, direct chute	2.000	C.Y.	1.745	0.00
Install two-piece galvanized steel post foot	16.000	Ea.	0.985	86.40
4" x 4" post framing	0.170	M.B.F.	5.231	148.75
2" x 8" joist framing	0.380	M.B.F.	4.164	228.00
1" x 8" board decking	272.000	SF Flr.	4.352	217.60
2" x 6" railing enclosure	544.000	L.F.	6.963	282.88
Handrail stock	136.000	L.F.	13.600	165.92
Drilling bolt holes	408.000	Inch	7.254	0.00
Bolts, nuts & washers	102.000	Ea.	5.828	30.60
Handrail bracket	34.000	Ea.	5.667	153.34
Anchor layout and drilling	8.000	Ea.	1.280	0.56
1/2" anchors (attach wood framing to building)	4.000	Ea.	0.400	10.48
Painting (2 coats)	2160.000	L.F.	43.200	280.80
Totals			**121.049**	**1,884.73**

Total per linear foot including general contractor's overhead and profit **$191**

Total per each 60-foot ramp including general contractor's overhead and profit **$11,445**

Install new pressure-treated wood straight ramp

Description	Quantity	Unit	Labor-Hours	Material
Hand excavating for post footings	3.000	C.Y.	6.000	0.00
Concrete forms, 12" diameter tubes (16 forms, 4' deep)	64.000	L.F.	13.653	134.40
Hand backfilling around post footings	1.000	C.Y.	0.727	0.00
Place concrete for footings, direct chute	2.000	C.Y.	1.745	0.00
Concrete, material only	2.000	C.Y.	0.000	145.00
Install two-piece galvanized steel post foot	16.000	Ea.	0.985	86.40
4" x 4" post framing, w/pressure-treated lumber	0.170	M.B.F.	0.000	164.05
2" x 8" joist framing, w/pressure-treated lumber	0.380	M.B.F.	0.000	271.70
1" x 8" board decking	272.000	SF Flr.	4.975	217.60
2" x 6" railing enclosure	544.000	L.F.	0.000	416.16
Handrail stock	136.000	L.F.	13.600	165.92
Drilling bolt holes	408.000	Inch	7.254	0.00
Bolts, nuts & washers	102.000	Ea.	5.828	30.60
Handrail bracket	34.000	Ea.	5.667	153.34
Anchor layout and drilling	8.000	Ea.	1.280	0.56
1/2" anchors (attach wood framing to building)	4.000	Ea.	0.400	10.48
Totals			**62.114**	**1,796.21**

Total per linear foot including general contractor's overhead and profit **$124**

Total per each 60-foot ramp including general contractor's overhead and profit **$7,417**

Install new concrete straight ramp

Description	Quantity	Unit	Labor-Hours	Material
Excavation	90.000	C.Y.	13.334	0.00
Concrete forms, footings	272.000	SFCA	17.947	195.84
Concrete forms, walls	1824.000	SFCA	69.494	930.24
Reinforcing (@ 50 lbs./C.Y.)	0.830	Ton	8.853	448.20
Place concrete, direct chute	33.000	C.Y.	17.600	0.00
Concrete, material only	33.000	C.Y.	0.000	2,392.50
Backfilling	67.000	C.Y.	9.926	0.00
Gravel under slab	5.000	C.Y.	0.233	26.50
Slab for ramp	5.000	C.Y.	4.783	425.00
Aluminum pipe railing (2 rail)	136.000	L.F.	27.200	2,264.40
Totals			**169.370**	**6,682.68**

Total per linear foot including general contractor's overhead and profit $428

Total per each 60-foot ramp including general contractor's overhead and profit $25,674

Copyright Reed Construction Data, 2004

9
Construct New Ramp: Switch-Back

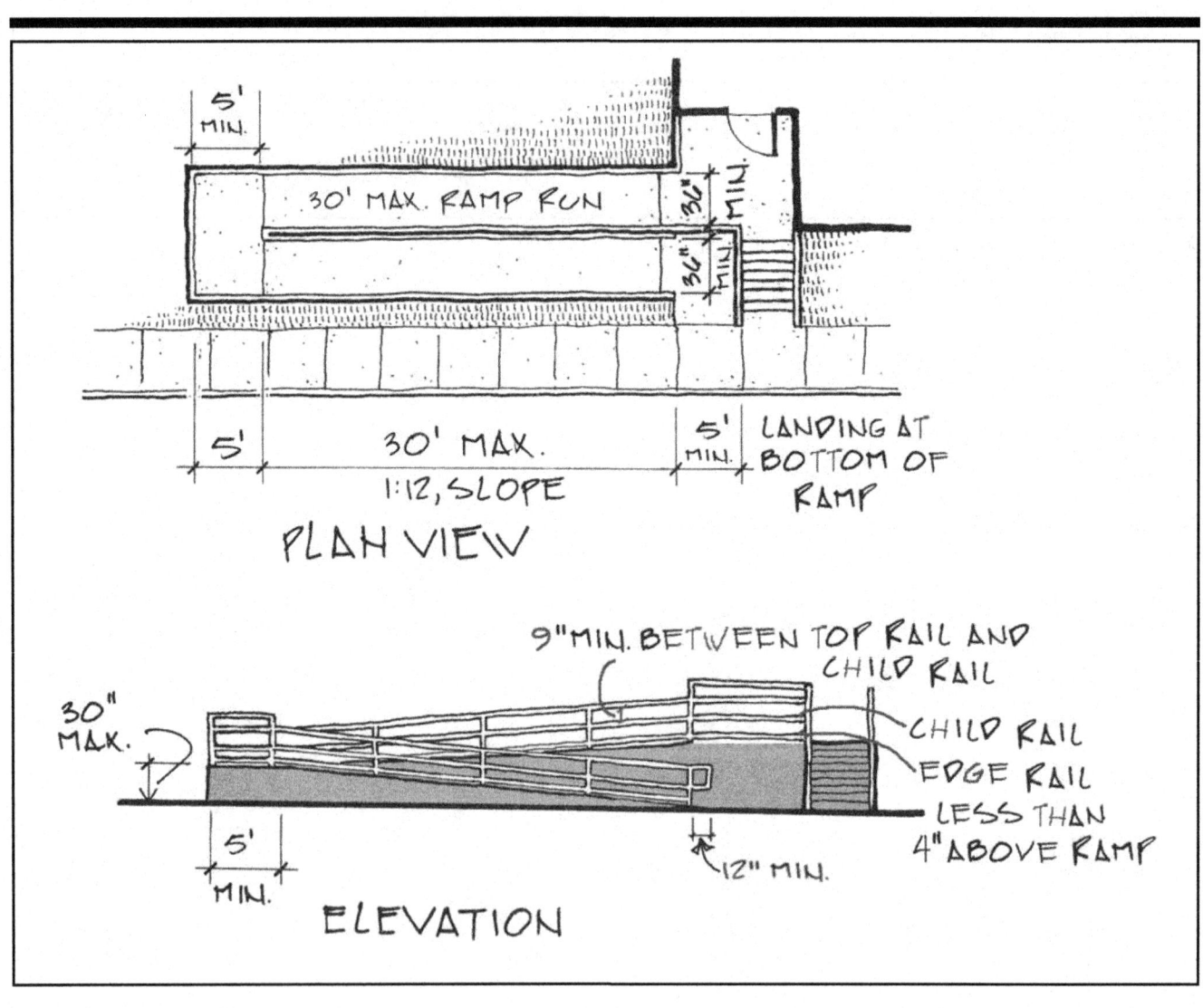

5' MIN.

30' MAX. RAMP RUN

36"

36" MIN.

30" MIN.

5'

30' MAX.
1:12, SLOPE

5' MIN.

LANDING AT BOTTOM OF RAMP

PLAN VIEW

30" MAX.

9" MIN. BETWEEN TOP RAIL AND CHILD RAIL

CHILD RAIL
EDGE RAIL
LESS THAN 4" ABOVE RAMP

5' MIN.

12" MIN.

ELEVATION

A switch-back ramp increases the potential height of a ramp by adding additional ramp length. A switch-back ramp can fit in a tighter area than a straight-run or dog-leg ramp, and may be useful in situations where there are obstructions or a lack of open land. Switch-back ramps also allow the ramp to leave a route at the bottom of a stair and rejoin the route at the top of the stair, so people using it can rejoin their companions.

ADAAG Reference
See Project 8 for references.

Where Applicable
Entrances and routes of travel where steps or different levels exist and usually the vertical rise is more than 30", or there is insufficient space for a straight or dog-leg ramp.

Design Requirements
See Project 8 for design requirements.

Design Suggestions
Covering a switch-back ramp can be difficult, so slip resistance and drainage are especially important. See Project 8 for other design suggestions.

Key Items
See Project 8 for Key items.

Level of Difficulty
Moderate to high. A wood ramp requires skilled carpentry. Concrete requires foundation work. May require design drawings and building permit.

Estimates

Install new painted wood switch-back ramp

Description	Quantity	Unit	Labor-Hours	Material
Hand excavating for post footings	3.000	C.Y.	6.000	0.00
Concrete forms, 12″ diameter tubes (16 forms, 4′ deep)	52.000	L.F.	11.093	109.20
Hand backfilling around post footings	1.000	C.Y.	0.727	0.00
Place concrete for footings, direct chute	2.000	C.Y.	1.745	0.00
Concrete, material only	2.000	C.Y.	0.000	145.00
Install two-piece galvanized steel post foot	16.000	Ea.	0.985	86.40
4″ x 4″ post framing	0.140	M.B.F.	4.308	122.50
2″ x 8″ joist framing	0.490	M.B.F.	5.370	294.00
1″ x 8″ board decking	312.000	SF Flr.	4.992	249.60
Handrail stock	144.000	L.F.	14.400	175.68
Drilling bolt holes	432.000	Inch	7.681	0.00
Bolts, nuts & washers	108.000	Ea.	6.171	32.40
Handrail bracket	36.000	Ea.	6.000	162.36
2″ x 6″ railing enclosure	564.000	L.F.	7.219	293.28
Anchor layout and drilling, per inch of depth (1/2″ diameter, 2″ deep)	16.000	Ea.	2.560	1.12
1/2″ anchors (for attachment of wood framing to building)	8.000	Ea.	0.800	20.96
Painting (2 coats)	2325.000	L.F.	46.500	302.25
Totals			**126.551**	**1,994.75**

Total per linear foot including general contractor's overhead and profit **$200**

Total per each 60-foot ramp including general contractor's overhead and profit **$12,020**

9. Construct New Ramp: Switch-Back *(continued)*

Install new pressure-treated wood switch-back ramp

Description	Quantity	Unit	Labor-Hours	Material
Hand excavating for post footings	3.000	C.Y.	6.000	0.00
Concrete forms, 12" diameter tubes (13 forms, 4' deep)	52.000	L.F.	11.093	109.20
Hand backfilling around post footings	1.000	C.Y.	0.727	0.00
Place concrete for footings, direct chute	2.000	C.Y.	1.745	0.00
Concrete, material only	2.000	C.Y.	0.000	145.00
Install two-piece galvanized steel post foot	13.000	Ea.	0.800	70.20
4" x 4" post framing, w/pressure-treated lumber	0.140	M.B.F.	0.000	135.10
2" x 8" joist framing, w/pressure-treated lumber	0.490	M.B.F.	0.000	350.35
1" x 8" board decking	312.000	SF Flr.	5.706	249.60
Handrail stock	144.000	L.F.	14.400	175.68
Drilling bolt holes	432.000	Inch	7.681	0.00
Bolts, nuts & washers	108.000	Ea.	6.171	32.40
Handrail bracket	36.000	Ea.	6.000	162.36
2" x 6" railing enclosure	564.000	L.F.	0.000	431.46
Anchor layout and drilling, per inch of depth (1/2" diameter, 2" deep)	16.000	Ea.	2.560	1.12
1/2" anchors (for attachment of wood framing to building)	8.000	Ea.	0.800	20.96
Totals			**63.683**	**1,883.43**

Total per linear foot including general contractor's overhead and profit	**$128**
Total per each 60-foot ramp including general contractor's overhead and profit	**$7,703**

Install new concrete switch-back ramp

Description	Quantity	Unit	Labor-Hours	Material
Excavation	60.000	C.Y.	8.889	0.00
Concrete forms, footings	248.000	SFCA	16.363	178.56
Concrete forms, walls	1752.000	SFCA	66.751	893.52
Reinforcing (@ 50 lbs./C.Y.)	0.800	Ton	8.533	432.00
Place concrete, direct chute	32.000	C.Y.	17.067	0.00
Concrete, material only	32.000	C.Y.	0.000	2,320.00
Backfilling	38.000	C.Y.	5.630	0.00
Gravel under slab	6.000	C.Y.	0.280	31.80
Slab for ramp, 4" thick	6.000	C.Y.	5.739	510.00
Aluminum pipe rail (2 rail)	150.000	L.F.	30.000	2,497.50
Totals			**159.252**	**6,863.38**

Total per linear foot including general contractor's overhead and profit	**$415**
Total per each 60-foot ramp including general contractor's overhead and profit	**$24,909**

Notes

10
Construct New Ramp: Dog-Leg

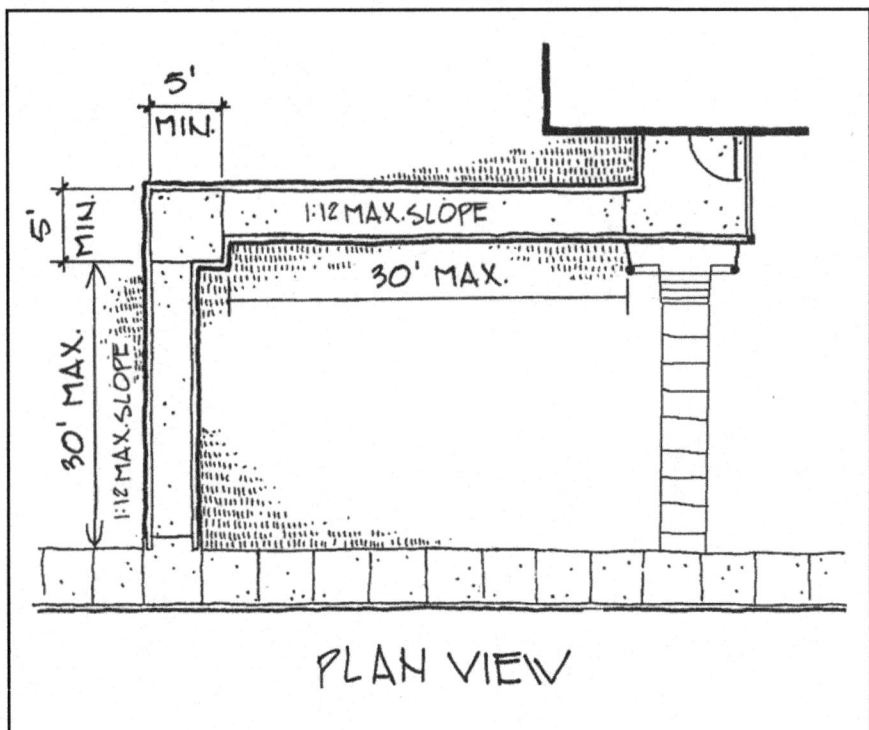

PLAN VIEW

In addition to all the general advantages of a ramp, a dog-leg ramp can double the height of a straight-run ramp, and can be a useful configuration for circumventing an obstacle, such as a tree. This type of ramp may also be installed at the corner of a building or along the edges of a courtyard.

ADAAG References
See Project 8 for references.

Where Applicable
Entrances and routes of travel where steps or different levels exist, or where there is sufficient room on the site or an object precludes the use of a straight or switch-back ramp. Usually when a dog-leg ramp is used, the vertical rise is more than 30". This type of ramp is often useful when the design can take advantage of a sloping site left to right.

Design Requirements
- Landings required at the intersection of 90° legs. 60" minimum in both directions.
- See Project 8 for other design requirements.

Design Suggestions
Dog-leg ramps must have at least one turn, which requires a landing larger than the minimum ramp width. This can be awkward visually and hard to build. Where possible, consider widening the ramp to be the same width as the landing.

Dog-leg ramps are often installed around the corner of a building, but this should not be done if it locates the base of the ramp away from the main approach. It is always desirable to locate the ramp entrance along the same path of travel as the typical entrance. Signage may be needed to direct people to the ramp entrance.

Dog-leg ramps can be used effectively when one leg can also serve as a landscape feature. In these situations, the bottom rail can be omitted if the surface adjacent to the ramp does not slope for at least one foot.

Covering a dog-legged ramp can be difficult, so slip resistance and drainage are especially important.

See Project 8 for other design suggestions.

Key Items

See Project 8 for key items.

Level of Difficulty

Moderate to high. A wood ramp requires skilled carpentry. Concrete requires foundation work. Design drawings and building permits may be required.

Estimates

Install new painted wood dog-leg ramp

Description	Quantity	Unit	Labor-Hours	Material
Hand excavating for post footings	3.000	C.Y.	6.000	0.00
Concrete forms, 12" diameter tubes (16 forms, 4' deep)	64.000	L.F.	13.653	134.40
Hand backfilling around post footings	1.000	C.Y.	0.727	0.00
Place concrete for footings, direct chute	2.000	C.Y.	1.745	0.00
Concrete, material only	2.000	C.Y.	0.000	145.00
Install two-piece galvanized steel post foot	16.000	Ea.	0.985	86.40
4" x 4" post framing	0.170	M.B.F.	5.231	148.75
2" x 8" joist framing	0.380	M.B.F.	4.164	228.00
1" x 8" board decking	272.000	SF Flr.	4.352	217.60
Handrail stock	136.000	L.F.	13.600	165.92
Drilling bolt holes	408.000	Inch	7.254	0.00
Bolts, nuts & washers	102.000	Ea.	5.828	30.60
Handrail bracket	34.000	Ea.	5.828	30.60
2" x 6" railing enclosure	544.000	L.F.	6.963	282.88
Anchor layout and drilling, per inch of depth (1/2" diameter, 2" deep)	8.000	Ea.	1.280	0.56
1/2" anchors (for attachment of wood framing to building)	4.000	Ea.	0.400	10.48
Painting (2 coats)	2160.000	L.F.	43.200	280.80
Totals			**121.049**	**1,884.73**

Total per linear foot including general contractor's overhead and profit	**$191**
Total per each 60-foot ramp including general contractor's overhead and profit	**$11,445**

Copyright Reed Construction Data, 2004

10. Construct New Ramp: Dog-Leg *(continued)*

Install new pressure-treated wood dog-leg ramp

Description	Quantity	Unit	Labor-Hours	Material
Hand excavating for post footings	3.000	C.Y.	6.000	0.00
Concrete forms, 12″ diameter tubes (16 forms, 4′ deep)	64.000	L.F.	13.653	134.40
Hand backfilling around post footings	1.000	C.Y.	0.727	0.00
Place concrete for footings, direct chute	2.000	C.Y.	1.745	0.00
Concrete, material only	2.000	C.Y.	0.000	145.00
Install two-piece galvanized steel post foot	16.000	Ea.	0.985	86.40
4″ x 4″ post framing, w/pressure-treated lumber	0.170	M.B.F.	0.000	164.05
2″ x 8″ joist framing, w/pressure-treated lumber	0.380	M.B.F.	0.000	271.70
1″ x 8″ board decking	272.000	SF Flr.	4.975	217.60
Handrail stock	136.000	L.F.	13.600	165.92
Drilling bolt holes	108.000	Inch	1.920	0.00
Bolts, nuts & washers	102.000	Ea.	5.828	30.60
Handrail bracket	34.000	Ea.	5.667	153.34
2″ x 6″ railing enclosure	544.000	L.F.	0.000	416.16
Anchor layout and drilling, per inch of depth (1/2″ diameter, 2″ deep)	4.000	Ea.	0.640	0.28
1/2″ anchors (for attachment of wood framing to building)	4.000	Ea.	0.400	10.48
Totals			**56.140**	**1,795.93**

Total per linear foot including general contractor's overhead and profit **$116**

Total per each 60-foot ramp including general contractor's overhead and profit **$6,983**

Install new concrete dog-leg ramp

Description	Quantity	Unit	Labor-Hours	Material
Excavation	90.000	C.Y.	13.334	0.00
Concrete forms, footings	272.000	SFCA	17.947	195.84
Concrete forms, walls	1824.000	SFCA	69.494	930.24
Reinforcing (@ 50 lbs./C.Y.)	0.830	Ton	8.853	448.20
Place concrete, direct chute	33.000	C.Y.	17.600	0.00
Concrete, material only	33.000	C.Y.	0.000	2,392.50
Backfilling	67.000	C.Y.	9.926	0.00
Gravel under slab	5.000	C.Y.	0.233	26.50
Slab for ramp, 4″ thick	5.000	C.Y.	4.783	425.00
Aluminum pipe rail (2 rail)	136.000	L.F.	27.200	2,264.40
Totals			**169.370**	**6,682.68**

Total per linear foot including general contractor's overhead and profit **$428**

Total per each 60-foot ramp including general contractor's overhead and profit **$25,674**

Notes

11

Construct New Ramp: Below-Grade

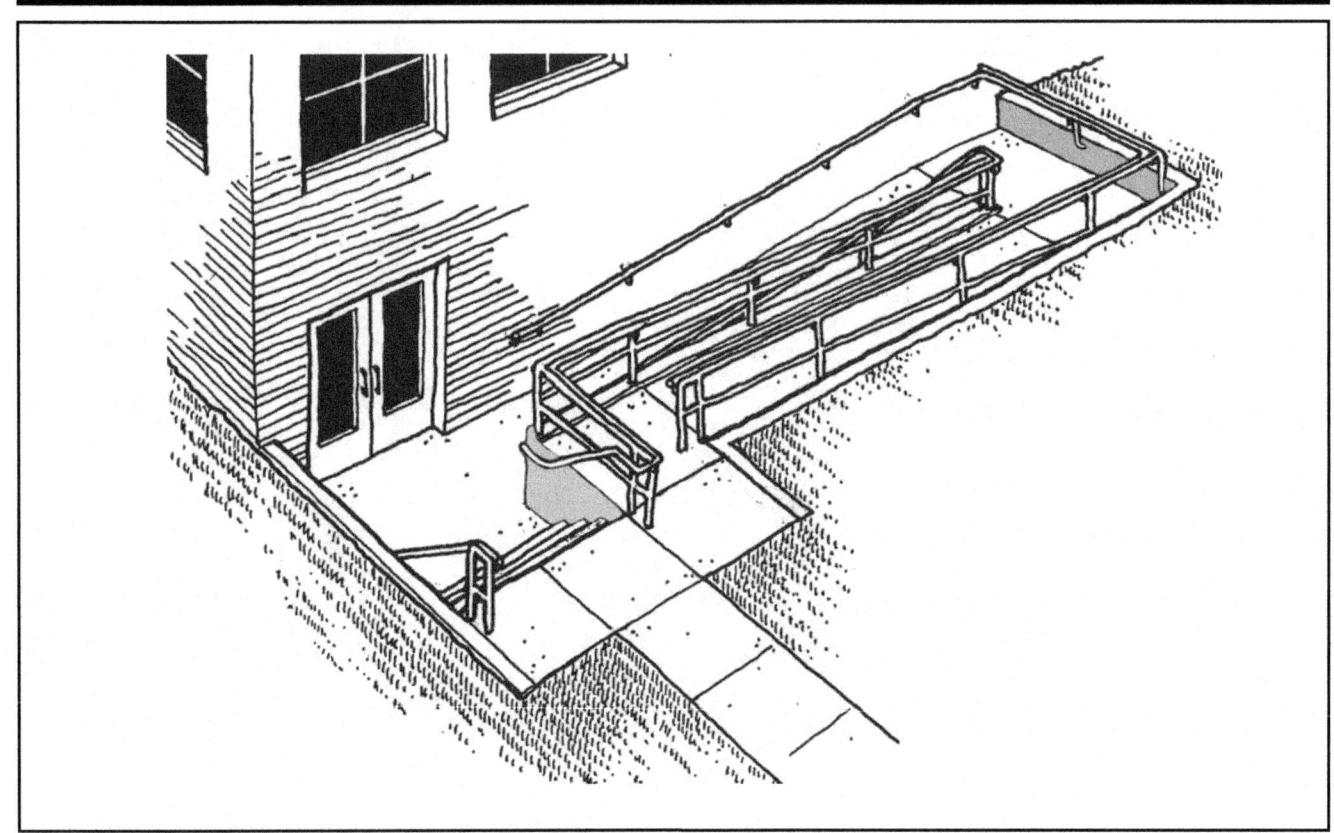

Many existing buildings were constructed with the first floor almost a full story above grade. For some of these buildings, it may be easier to make an accessible entry on the below-grade floor. Still other buildings have an entrance below grade. With proper drainage, these can be made accessible by a ramp connecting the accessible route to the new or existing entrance.

ADAAG References
206.4, Entrances
207, Accessible Means of Egress
302, Floor or Ground Surfaces
302.3, Openings (for drainage grates)
405, Ramps
505, Handrails
(Consult ADAAG 202.5 for special allowances for historic facilities and alterations.)

Where Applicable
Public entrances and routes of travel where steps or different levels exist, and a ramp is installed below existing grade level.

Design Requirements
• If there is a dog-leg, the landing must be at least 60" minimum in both directions.

- See Project 8 for other design requirements.

Design Suggestions

Below-grade ramps must comply with all the design requirements and suggestions applicable to other ramps. If the existing entrance is not large enough to accommodate a ramp, excavation and landscaping may be needed, along with retaining walls and drainage to keep water from collecting at the base of the ramp. Even if excavation is necessary, a ramp should be considered rather than a lift, which would require maintenance and would run the risk of being locked or otherwise not readily available when needed.

When a building's main entrance is above or below grade, a secondary entrance (but not a service entrance or a loading dock) can be used if it is along an accessible route, leads to an accessible route, and is clearly marked by signage. (Note: If this entrance is *not* the main entrance, it will still have to remain unlocked whenever the main entrance is unlocked. This may compromise the security of the building. A doorbell is generally an unacceptable alternative.)

See Project 8 for other design suggestions.

Key Items
- Ramp material: concrete.
- Earthwork and excavation.
- Rails: pipe rails, metal rails, wood banisters or rails, uprights, attachments.
- Possibly will require slip-resistant applied materials: sand paint, sandpaper strips.
- Will require drainage.

Level of Difficulty

High. Excavation machinery may be required. Concrete requires foundation work. Possible landscaping work involved. Design drawings and building permits may be required.

Estimates

Install new concrete below-grade switch-back ramp

Description	Quantity	Unit	Labor-Hours	Material
Excavation	115.000	C.Y.	17.037	0.00
Remove concrete wall at existing stairs	64.000	S.F.	21.333	0.00
Concrete forms, footings	240.000	SFCA	15.835	172.80
Concrete forms, walls	1440.000	SFCA	54.864	734.40
Reinforcing (@ 50 lbs./C.Y.)	0.600	Ton	6.400	324.00
Place concrete, direct chute	24.000	C.Y.	12.800	0.00
Concrete, material only	24.000	C.Y.	0.000	1,740.00
2 coat bituminous dampproofing	360.000	S.F.	5.760	36.00
Backfilling	31.000	C.Y.	4.593	0.00
Gravel under slab	6.000	C.Y.	0.280	31.80
Slab for ramp, 4" thick	6.000	C.Y.	5.739	510.00
Aluminum pipe rail (2 rail)	90.000	L.F.	18.000	1,498.50
Aluminum wall pipe railing	150.000	L.F.	22.535	1,320.00
Haul excess material	60.000	C.Y.	6.857	0.00
Totals			192.033	6,367.50

Total per linear foot including general contractor's overhead and profit **$472**

Total per each 60-foot ramp including general contractor's overhead and profit **$28,321**

12
Modify Existing Ramp

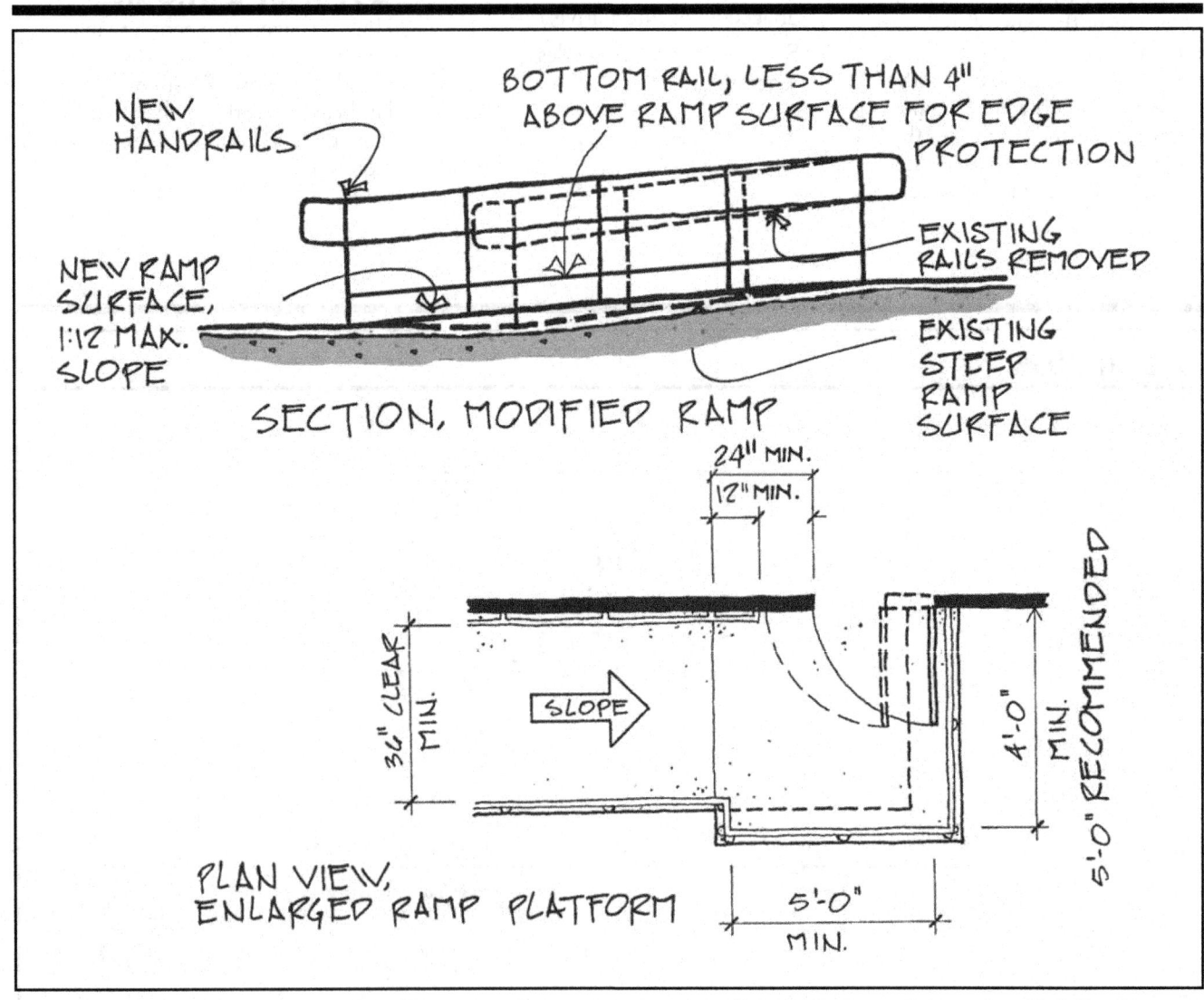

NEW HANDRAILS

BOTTOM RAIL, LESS THAN 4" ABOVE RAMP SURFACE FOR EDGE PROTECTION

EXISTING RAILS REMOVED

NEW RAMP SURFACE, 1:12 MAX. SLOPE

EXISTING STEEP RAMP SURFACE

SECTION, MODIFIED RAMP

24" MIN.

12" MIN.

36" CLEAR MIN.

SLOPE

4'-0" MIN.

5'-0" RECOMMENDED

5'-0" MIN.

PLAN VIEW, ENLARGED RAMP PLATFORM

Some modifications to a ramp can be expensive, and still not create a fully accessible ramp. Before assuming that the existing ramp should be altered, assess the cost of modifications versus replacing the ramp with a new one (or determine if another accessible means of vertical circulation is necessary or possible).

ADAAG Reference
See Project 8 for references.

Where Applicable
All existing ramps at public entrances or routes of travel that do not meet ADAAG because of slope or width.

Design Requirements
• If there is a dog-leg, the landing must be at least 60" deep in both directions, i.e., 60" by 60", minimum.

See Project 8 for other design requirements.

Design Suggestions
Ramps steeper than 1:12 were often built because of architectural features impinging on one leg or the other. For this reason, it may be necessary to correct the slope by adding a dog-leg or switch-back segment.

See Project 8 for additional design suggestions.

Key Items
• For applying materials to surface: special coatings, such as sand paint or sandpaper strips.
• For modifying the slope by adding surface materials: concrete, asphalt, wood to match existing.
• For changing rails: pipe rails, metal rails, wood banisters, uprights, attachments.
• For increasing the size of the landing: ramp materials, and new rails to match.

Level of Difficulty
Variable. Low for creating a slip-resistant surface. Moderate for changing handrails. Moderate to high for widening the platform and for decreasing the slope.

Estimates

Repave concrete ramp to shallower slope, including 1-1/2″ pipe handrails

Description	Quantity	Unit	Labor-Hours	Material
Torch cut handrail supports	8.000	Ea.	0.305	0.00
Loading and hauling rails for disposal	1.000	C.Y.	0.667	0.00
4″ sidewalk, broom finish	36.000	S.F.	1.440	43.56
Aluminum pipe rail (2 rail)	24.000	L.F.	4.800	399.60
Totals			7.212	443.16

Total per linear foot including general contractor's overhead and profit **$117**

Total per 12 linear feet including general contractor's overhead and profit **$1,400**

Repave asphalt ramp to shallower slope, including 1-1/2″ pipe handrails

Description	Quantity	Unit	Labor-Hours	Material
Torch cut handrail supports	8.000	Ea.	0.305	0.00
Loading and hauling rails for disposal	1.000	C.Y.	0.667	0.00
Install asphalt sidewalk, 2-1/2″ thick	5.340	S.Y.	0.388	22.64
Aluminum pipe rail (2 rail)	24.000	L.F.	4.800	399.60
Totals			6.160	422.24

Total per linear foot including general contractor's overhead and profit **$108**

Total per 12 linear feet including general contractor's overhead and profit **$1,294**

12. Modify Existing Ramp *(continued)*

Widen switch-back concrete ramp (including handrails)

Description	Quantity	Unit	Labor-Hours	Material
Excavation	35.000	C.Y.	5.185	0.00
Torch cut handrail supports	20.000	Ea.	0.762	0.00
Loading and hauling rails for disposal	6.000	C.Y.	4.000	0.00
Concrete forms, footings	78.000	SFCA	5.146	56.16
Concrete forms, walls	546.000	SFCA	20.803	278.46
Reinforcing (@ 50 lbs./C.Y.)	0.300	Ton	3.200	162.00
Place concrete, direct chute	12.000	C.Y.	6.400	0.00
Concrete, material only	12.000	C.Y.	0.000	870.00
Backfilling	23.000	C.Y.	3.407	0.00
Aluminum pipe rail (2 rail)	78.000	L.F.	15.600	1,298.70
Totals			**64.503**	**2,665.32**

Total per linear foot including general contractor's overhead and profit **$170**

Total per each 60-foot ramp including general contractor's overhead and profit **$10,199**

Resurface existing ramp (sand paint)

Description	Quantity	Unit	Labor-Hours	Material
Prepare surface	1.000	S.Y.	0.003	0.00
Nonskid pavement renewal	1.000	S.Y.	0.023	0.64
Totals			**0.026**	**0.64**

Total per square yard including general contractor's overhead and profit **$3**

Resurface existing ramp (sandpaper strips)

Description	Quantity	Unit	Labor-Hours	Material
Anti-skid sandpaper strips (6" x 24"), one per linear foot	1.000	S.F.	0.041	4.60
Totals			**0.041**	**4.60**

Total per linear foot including general contractor's overhead and profit **$11**

Enlarge pressure-treated wood ramp platform

Description	Quantity	Unit	Labor-Hours	Material
Remove board decking	16.000	S.F.	0.582	0.00
Remove handrails	32.000	L.F.	0.320	0.00
Remove framing	28.000	L.F.	0.477	0.00
Remove 4" x 4" posts	16.000	L.F.	0.320	0.00
Hand excavating for post footings	0.500	C.Y.	1.000	0.00
Concrete forms, 12" diameter tubes (2 forms, 4' deep)	8.000	L.F.	1.707	16.80
Hand backfilling around post footings	0.250	C.Y.	0.182	0.00
Place concrete for footings, direct chute	0.250	C.Y.	0.218	0.00
Concrete, material only	0.250	C.Y.	0.000	18.13
Install two-piece galvanized steel post foot	2.000	Ea.	0.123	10.80
4" x 4" post framing, w/pressure-treated lumber	0.020	M.B.F.	0.000	19.30
2" x 8" joist framing, w/pressure-treated lumber	0.010	M.B.F.	0.000	7.15
1" x 8" board decking	25.000	SF Flr.	0.457	20.00
2" x 6" railing enclosure	44.000	L.F.	0.000	33.66
Handrail stock	44.000	L.F.	4.400	53.68
Drilling bolt holes	66.000	Inch	1.173	0.00
Bolts, nuts & washers	33.000	Ea.	1.886	9.90
Handrail bracket	11.000	Ea.	1.833	49.61
Anchor layout and drilling, per inch of depth (1/2" diameter, 2" deep)	8.000	Ea.	1.280	0.56
1/2" anchors (for attachment of wood framing to building)	4.000	Ea.	0.400	10.48
Totals			**16.358**	**250.07**

Total per square foot including general contractor's overhead and profit	**$63**
Total per each 25-square-foot landing including general contractor's overhead and profit	**$1,578**

Widen concrete ramp platform (including handrails)

Description	Quantity	Unit	Labor-Hours	Material
Excavation	4.000	C.Y.	0.593	0.00
Torch cut handrail supports	3.000	Ea.	0.114	0.00
Loading and hauling rails for disposal	1.000	C.Y.	0.667	0.00
Concrete forms, footings	12.000	SFCA	0.792	8.64
Concrete forms, walls	36.000	SFCA	1.372	18.36
Reinforcing (@ 50 lbs./C.Y.)	0.050	Ton	0.533	27.00
Place concrete, direct chute	2.000	C.Y.	1.067	0.00
Concrete, material only	2.000	C.Y.	0.000	145.00
Backfilling	2.000	C.Y.	0.296	0.00
Aluminum pipe rail (2 rail)	12.000	L.F.	2.400	199.80
Totals			**7.834**	**398.80**

Total per square foot including general contractor's overhead and profit	**$56**
Total per each 25-square-foot platform including general contractor's overhad and profit	**$1,388**

Copyright Reed Construction Data, 2004

13
Install or Modify Ramp Handrails

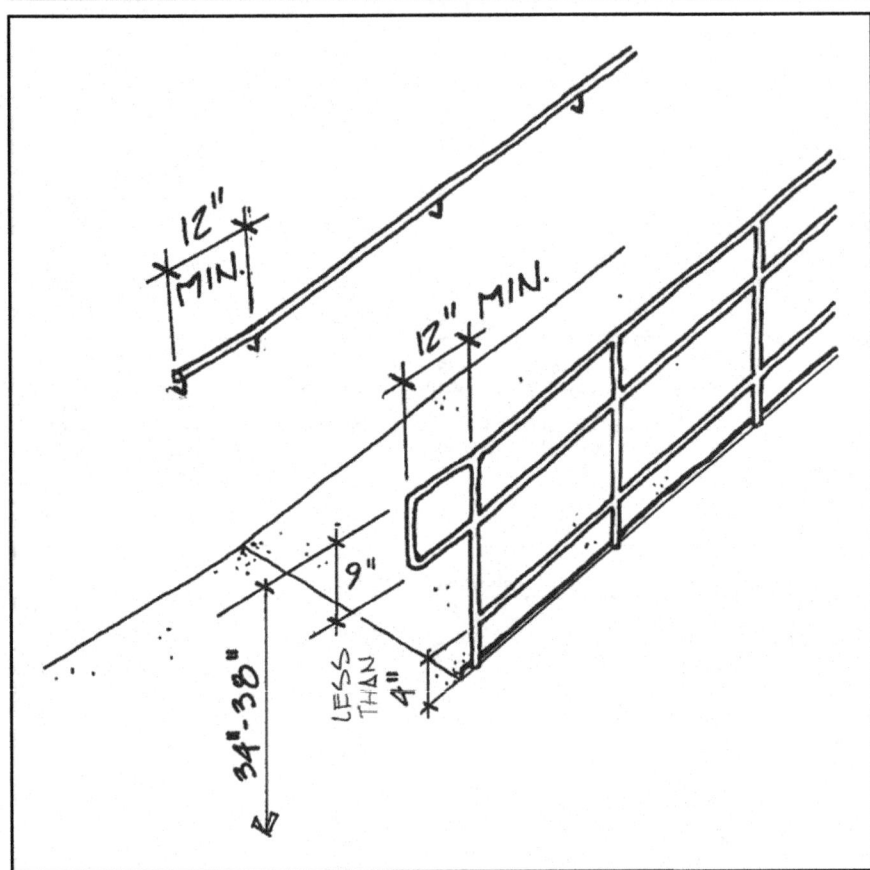

The design of a handrail is just as vital to a ramp's usability as the slope and surface. Rails provide an essential gripping surface that enables people with a range of abilities to use a ramp. Without proper rails, ramps, especially in exterior situations, are just pitched walkways that can be very difficult to use. Adding or replacing railings on a ramp is an essential modification that increases compliance and accessibility.

ADAAG References
405.5, Clear Width
405.8, Handrails
405.9, Edge Protection
505, Handrails

Where Applicable
Both sides of all ramps with a vertical rise greater than 6".

Design Requirements
- Rails on both sides of ramp, 36" minimum clearance between.
- 34" to 38" above ramp surface, 1-1/2" from wall exactly, with no rotation in fittings.
- 1-1/4" to 2" diameter if round, or, if not round, the perimeter between 4" and 6-1/4" and the maximum width 2-1/4".
- Continuous rail without interruption by posts or other construction. (Brackets attached to bottom are acceptable if they drop 1-1/2" or more before returning to the wall.)
- 12" horizontal extensions at the top and bottom; return to wall, post, or floor/ground.
- No handrail is needed if the vertical rise is 6" or less.
- If no shoulder or edge protection, bottom rail 4" above ramp surface, maximum.
- Edge protection: curb at 4" minimum, rail less than 4" above ramp surface, or the adjacent surface following the slope of the ramp at least 1' beyond the handrails.

Design Suggestions

Railings should be attached in such a way that a person's grasp is uninterrupted. Brackets under the rails work best, with the rails separated from the uprights instead of resting on them. This allows someone to use the rail without letting go. Also, especially on older buildings, the design should incorporate the rail into the existing aesthetic, using color, ornament, and so forth, as necessary.

Consider installing a rail a minimum of 9" below the top rail if the ramp will be used by children. If the adjacent wall is brick, consider installing a board behind the rail so persons will not scrape their knuckles.

When specifying rails, check nominal dimensions against actual requirements. Pipe rails specified with a 1-1/2" diameter can have an actual diameter of 1-7/8" or more. Metal rails are durable, but exterior rails can become quite hot in summer and dangerously cold in winter. Schedule 40 plastic pipe, 1-1/4" I.D., can be an alternative, as can be rails with an insulated coating or sleeve. These should be installed where wood rails are inappropriate.

Key Items

Pipe rails, metal rails, plastic pipe rails, wood banisters or rails, uprights, attachments.

Level of Difficulty

Low to moderate. Attaching rails to masonry can require a contractor to set the bolts. Welding extensions to existing pipe requires a skilled welder.

Estimates

Bolt pipe rail to wood surface

Description	Quantity	Unit	Labor-Hours	Material
Aluminum pipe rail (2 rail)	1.000	L.F.	0.200	16.65
Strap brackets (2 per upright)	0.500	Ea.	0.061	1.75
Drilling bolt holes	1.500	Inch	0.027	0.00
Bolts, nuts & washers	1.000	Ea.	0.057	0.30
Totals			0.345	18.70

Total per linear foot including general contractor's overhead and profit ___ **$63**

Set pipe rail in concrete

Description	Quantity	Unit	Labor-Hours	Material
Pipe rail set in concrete (2 rail)	1.000	L.F.	0.200	16.65
Totals			0.200	16.65

Total per linear foot including general contractor's overhead and profit ___ **$48**

Set hand-forged wrought iron railing in concrete

Description	Quantity	Unit	Labor-Hours	Material
Wrought iron railing	1.000	L.F.	0.667	71.00
Totals			0.667	71.00

Total per linear foot including general contractor's overhead and profit ___ **$185**

13. Install or Modify Ramp Handrails (continued)

Mount wood dowel railing on brass brackets (wood uprights existing)

Description	Quantity	Unit	Labor-Hours	Material
Handrail stock	1.000	L.F.	0.100	1.22
Drilling bolt holes	3.000	Inch	0.053	0.00
Bolts, nuts & washers	0.750	Ea.	0.043	0.23
Handrail bracket	0.250	Ea.	0.042	1.13
Totals			**0.238**	**2.58**

Total per linear foot including general contractor's overhead and profit **$21**

Mount wood dowel railing on brass brackets (pipe rail uprights)

Description	Quantity	Unit	Labor-Hours	Material
Handrail stock	1.000	L.F.	0.100	1.22
Drilling bolt holes	3.000	Inch	0.053	0.00
Bolts, nuts & washers	0.750	Ea.	0.043	0.23
Handrail bracket	0.250	Ea.	0.042	1.13
Totals			**0.238**	**2.58**

Total per linear foot including general contractor's overhead and profit **$21**

Attach pipe railing to brick wall

Description	Quantity	Unit	Labor-Hours	Material
Pipe rail attached to brick wall	1.000	L.F.	0.150	8.80
Totals			**0.150**	**8.80**

Total per linear foot including general contractor's overhead and profit **$30**

Attach wood dowel railing to brick wall

Description	Quantity	Unit	Labor-Hours	Material
Handrail stock	1.000	L.F.	0.100	1.22
Drilling bolt holes, per inch of depth	2.250	Ea.	0.360	0.16
Expansion bolts & shields	0.750	Ea.	0.075	1.97
Handrail bracket	0.250	Ea.	0.042	1.13
Totals			**0.577**	**4.48**

Total per linear foot including general contractor's overhead and profit **$50**

Weld handrail extensions to existing pipe

Description	Quantity	Unit	Labor-Hours	Material
Fabricate 1-1/2" diameter pipe extensions	1.000	Ea.	0.200	11.45
Install pipe extensions on site	1.000	L.F.	0.000	6.35
Totals			**0.200**	**17.80**

Total per each including general contractor's overhead and profit **$51**

Attach handrail extensions to end posts (wood)

Description	Quantity	Unit	Labor-Hours	Material
Handrail stock	2.000	L.F.	0.200	2.44
Drilling bolt holes	24.000	Inch	0.427	0.00
Bolts, nuts & washers	6.000	Ea.	0.343	1.80
Handrail bracket	2.000	Ea.	0.333	9.02
Additional wood posts	0.010	M.B.F.	0.308	8.75
Totals			**1.611**	**22.01**

Total per each including general contractor's overhead and profit **$156**

14
Install Vertical Platform Lift

Vertical platform lifts are useful for creating access where a ramp would not fit or would be too long, an elevator would be too expensive, or a stair lift would block an egress stair. Platform lifts can be installed inside or outside and can be open or enclosed. The maximum height of a vertical lift is usually less than 8', although 12' heights are available.

ADAAG References
206.7, Platform Lifts
207.2, Platform Lifts, Means of Egress
302, Floor or Ground Surfaces
303, Changes in Level
305, Clear Floor or Ground Space
309, Operable Parts
404.2, Doors, Doorways, and Gates
404.3, Automatic Doors and Gates
410, Platform Lifts
ASME A18.1-2003
IBC 1007 (2003 Edition)

Where Applicable
Allowed in existing buildings to create accessible vertical circulation between levels where installation of an elevator or ramp is not possible. Allowed in certain specific locations in new construction, such as performance areas, speakers' platforms, areas in courtrooms, and incidental spaces that are not public and are used by five or fewer people. They can also be used in assembly areas to achieve wheelchair space dispersion or line of sight requirements and where exterior site restraints make use of a ramp or elevator not feasible. They are allowed in certain situations in transient guest rooms, amusement rides, play areas, team seating areas, and boating facilities and fishing piers.

Design Requirements

- Accessible route to and off lift, top and bottom.
- Maneuvering space at swinging gate or door meeting ADAAG 404.2.4, Maneuvering Clearances.
- Stable, firm, slip-resistant surface at approaches, top and bottom.
- 80" head clearance along path of travel.
- Cannot be attendant-operated; must allow unassisted entry and exit. Gates must be power operated if serving more than two landings.
- Controls within reach range (48" high, maximum) and operable with a closed fist.
- To fit in a lift, you will probably need to rough-in an area about 54" by 54" for a straight-through design for a nominal 36" by 48" platform, and 60" by 72" for the smallest 90° turn configuration.
- In compliance with ASME A18.1 and *International Building Code* (IBC) Section 1007.
- Standby power required when lift is part of a means of egress.

Design Suggestions

While lifts can be used as a means of achieving vertical access, they require servicing, periodic testing, staff training, and protection against vandalism. They often end up locked (usually against regulations) and take time to use. Accordingly, they should be avoided as a means of making a heavily-used route accessible. Consider the possibility of using a ramp, or even an elevator, as an alternative, even though these will be more expensive. Frequently, lifts are limited to wheelchair users by local safety codes (but not by ADAAG).

Because of the wide gate to the lift, check to see if a person can get around the gate to open it. Maneuverability is often the largest impediment to using a lift.

Needless to say, the difficulty of fitting a lift into the existing aesthetic increases as the height increases. Unless it cannot be avoided, use a straight-line configuration where the on/off direction of travel is a straight line (making a turn is very difficult). The best design allows users to enter from one end and exit from the opposite end.

Lifts must be usable independently (i.e., unassisted). Where the lift must be kept locked, explore other options, because a locked lift complies with the regulations only if an attendant is readily available *at all times*. Consider installing a call button within reach, even when the lift is not locked, in case assistance is needed. ASME A18.1, referenced in ADAAG, sets strict technical requirements for the length of travel and methods of operation for lifts. Most lifts are standard items and meet these requirements, and are so labeled. Accordingly, the actual specifications must be reviewed only for quality based on a facility's particular need, not for regulatory compliance. Local codes, however, may have additional requirements, and should be consulted prior to any installation. A manufacturer's representative can assist you in making a choice.

Key Items

Accessible route to lift and maneuvering room at entrance/exit. Lift equipment. Concrete pad for exterior applications. Enclosure, if desired, prefab or custom (important in older building applications). Call button recommended, even if not necessary.

Level of Difficulty

Moderate to high. Installation is usually done by lift manufacturer's representatives. Requires preparatory electrical wiring. Possible concrete work for pad and carpentry if custom enclosure.

Estimates

Install exterior unenclosed lift with concrete pad

Description	Quantity	Unit	Labor-Hours	Material
Excavation	0.600	C.Y.	0.089	0.00
Concrete formwork	32.000	SFCA	4.388	16.64
15" thick concrete pad	16.000	S.F.	0.460	58.24
Unenclosed lift	1.000	Ea.	16.000	5,375.00
Totals			20.937	5,449.88

Total per each including general contractor's overhead and profit **$11,071**

14. Install Vertical Platform Lift *(continued)*

Install exterior enclosed lift with concrete pad

Description	Quantity	Unit	Labor-Hours	Material
Excavation	0.600	C.Y.	0.089	0.00
Concrete formwork	32.000	SFCA	4.388	16.64
15″ thick concrete pad	16.000	S.F.	0.460	58.24
Enclosed lift	1.000	Ea.	32.000	12,700.00
Totals			**36.937**	**12,774.88**

Total per each including general contractor's overhead and profit   **$24,932**

Install interior unenclosed lift

Description	Quantity	Unit	Labor-Hours	Material
Interior unenclosed lift	1.000	Ea.	16.000	5,375.00
Totals			**16.000**	**5,375.00**

Total per each including general contractor's overhead and profit   **$10,592**

Notes

15
Install Stairway Chairlift

Many older buildings have no elevator, floor-to-floor distances too great to ramp, or no room for a vertical lift. Even where the vertical distance between levels is not excessive, there is often no room to install a ramp. In such situations, installing a lift on a stair can create access between levels with minimum modification to the building.

Stair lifts need not be confined to single flights; some are capable of traveling up to six flights of stairs continuously, but such installations are expensive. The major drawback of stair lifts is that they use 40" to 50" of the stair width.

ADAAG References
See references for Project 14.

Where Applicable
ADAAG includes stairway chairlifts in the same category as platform lifts. They are allowed in existing buildings to create accessible vertical circulation between levels where installation of an elevator or ramp is not possible and the stairs are wide enough. Also see Project 14.

Design Requirements
- Accessible route to and off stairway chairlift, top and bottom.
- Maneuvering space, top and bottom.
- Stable, firm, slip-resistant surface at approaches, top and bottom.
- 80" head clearance along path of travel.
- Cannot be attendant-operated; must allow unassisted entry and exit.
- Standby power required if lift is part of an accessible means of egress.
- Stair wide enough to accommodate lift in down position without intruding into required clear fire egress widths. (A straight lift meeting the regulations is about 42" wide.)

- Controls within reach range (48" high, maximum) and operable with a closed fist.

Design Suggestions

Make sure that other options (such as elevators or ramps) are not feasible. As with vertical platform lifts, inclined lifts are usually restricted by local building codes for use by people in wheelchairs. Fire egress widths are calculated with the lift in the operating (down) position (about 42"; folded up, the lifts are about a foot wide). Fire marshals assume that the lift will be in use during an emergency. This usually prevents the use of inclined lifts on fire egress stairs.

As with vertical lifts, stairway chairlifts need servicing and periodic testing, often end up locked (usually against regulations), and take time to use. They are often a nuisance when installed to make a heavily-used route accessible. Lifts should be usable independently (unassisted); where the lift must be kept locked, an attendant must be readily available *at all times*. In case a person needs assistance, consider installing a call button within reach, even when the lift is not locked.

Most lifts are standard items and meet ASME A18.1, and are so labeled. Accordingly, actual specifications only need to be reviewed for the quality and the facility's particular need, not for regulatory compliance. Local building codes, however, may have additional requirements, and should be consulted prior to installation. A manufacturer's representative can assist you in making a choice.

Key Items

Lift, operating mechanism, call button (recommended even if not necessary).

Level of Difficulty

Moderate to high. Usually installed by manufacturer's representative. Preliminary electrical work required.

Estimates

Install stairway lift, straight run

Description	Quantity	Unit	Labor-Hours	Material
Straight run stair lift	1.000	Ea.	16.000	10,600.00
Totals			16.000	10,600.00

Total per each including general contractor's overhead and profit **$19,477**

Install stairway lift, one turn

Description	Quantity	Unit	Labor-Hours	Material
One turn stair lift	1.000	Ea.	80.000	16,800.00
Totals			80.000	16,800.00

Total per each including general contractor's overhead and profit **$35,684**

Copyright Reed Construction Data, 2004

16
Install New Stairs

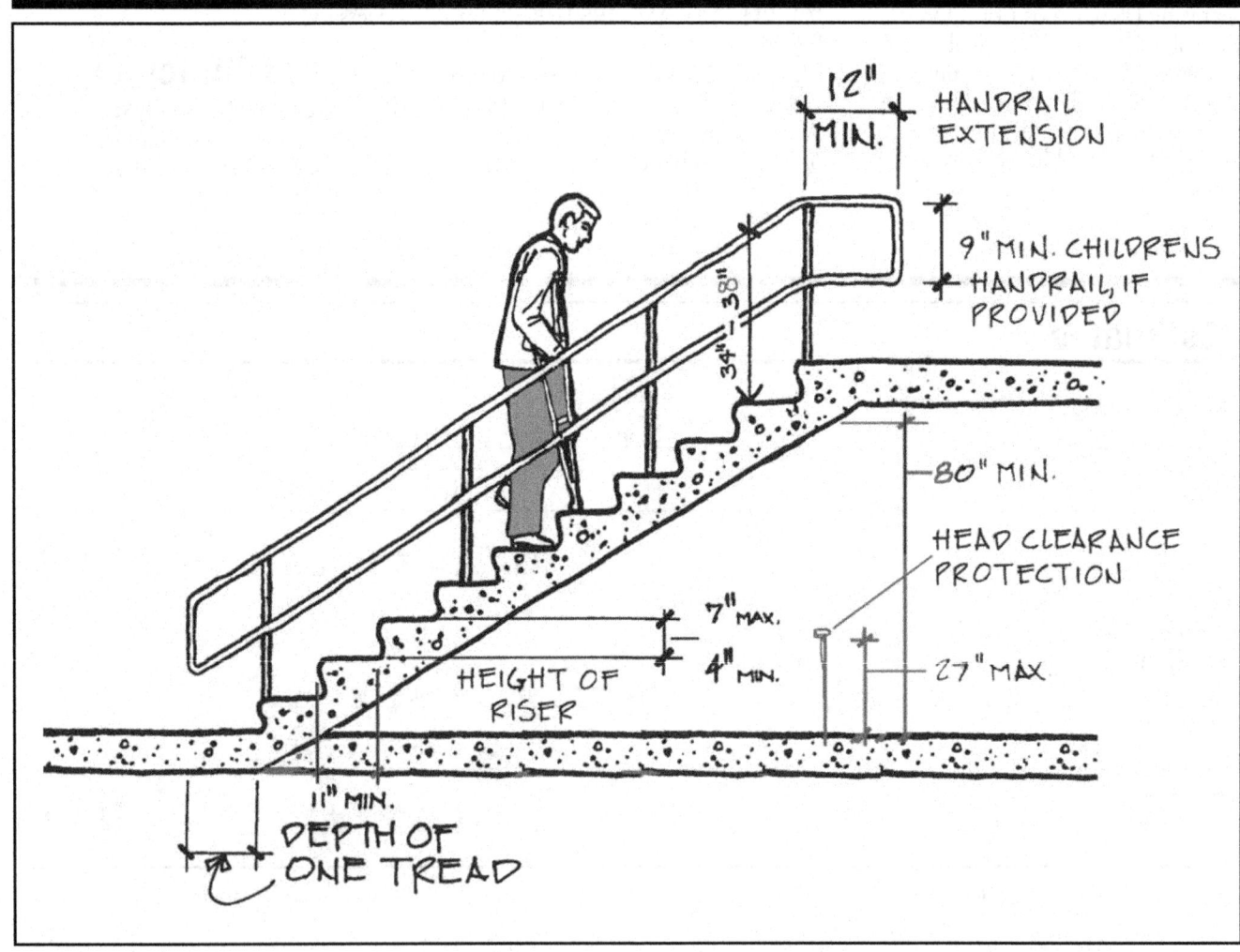

12" MIN.

HANDRAIL EXTENSION

9" MIN. CHILDRENS HANDRAIL, IF PROVIDED

34" - 38"

80" MIN.

HEAD CLEARANCE PROTECTION

27" MAX

7" MAX.

4" MIN.

HEIGHT OF RISER

11" MIN.

DEPTH OF ONE TREAD

Stairs are sometimes overlooked in an access survey, since they are not typically used by people who use wheelchairs. However, stairs are such a heavily used means of vertical circulation that attention to design is critical in creating an accessible facility. An inaccessible stair that would be difficult to modify may, in some cases, be better replaced with an accessible stair that can greatly increase a facility's usability and safety.

ADAAG References

210, Stairways
216.4, Signs at Egress Stairs
302, Floor or Ground Surfaces
307, Protruding Objects
504, Stairways
505, Handrails

Where Applicable

All stairs that are part of a means of egress. Local building codes, however, may require all stairs to meet ADAAG requirements. Also, all stairs must meet the handrail requirements.

Design Requirements

- Protection against bumping head below stairs for 80" height or less. If protection is a rail, maximum height of rail is 27".
- Minimum tread depth of 11"; minimum riser height of 4"; maximum riser of 7".
- Slip-resistant surface.
- A slope of 2% is allowed for drainage. No pooling water allowed on exterior stairs and landings.
- No protrusion greater than 1-1/2" over the tread below. Underside of

nosing sloped back at a 30° angle or less with the riser.
- Radius of the leading edge 1/2" or less.
- No open risers.
- Rails on both sides, 34" to 38" above nosing, 1-1/2" exactly from wall, with no rotation in fittings.
- Rails 1-1/4" to 2" diameter round, or, if not round, the perimeter between 4" and 6-1/4" and the maximum width 2-1/4".
- Continuous inside rail without interruption by newel posts or other construction. Brackets attached to bottom are acceptable if they drop 1-1/2" or more before returning to the wall.
- Rail extension: 12" minimum extension at the top parallel to the floor. Length of one tread, minimum, at the bottom, following the slope of the stair. Return to wall, post, or floor.

Design Suggestions

While ADAAG covers only egress stairs, it is strongly recommended that all stairs conform to ADAAG regulations, since stairs are a heavily

used element by people with a wide range of abilities, and accidents on stairs are frequent and disastrous.

Consider making the leading edges of treads a contrasting color. Also consider installing a rail for children that is a minimum of 9" below the standard handrail.

There are some reasons to consider complete replacement of a stair, such as if it is too steep, has open risers, or has especially slippery tread surfaces.

Key Items

Foundation/footings, stairs (wood, concrete, metal pan), rails (wood, pipe, metal).

Level of Difficulty

Moderate to high, depending on the materials used. Installation of wood stairs requires skilled carpenters, concrete stairs require foundation and formwork, and metal stairs require welding. This project may require design drawings, and will require a building permit.

Estimates

Install painted wood stairs

Description	Quantity	Unit	Labor-Hours	Material
Stair stringers, 2" x 12"	0.140	M.B.F.	8.615	102.90
Pine risers, 3/4" x 7-1/2"	64.000	L.F.	7.757	195.20
Aluminum pipe rail (2 rail)	18.000	L.F.	3.600	299.70
Strap brackets (2 per upright)	10.000	Ea.	1.221	35.00
Drilling bolt holes	30.000	Inch	0.533	0.00
Bolts, nuts & washers	20.000	Ea.	1.143	6.00
Pipe rail attached to brick wall	18.000	L.F.	2.704	158.40
Painting, treads, risers, and stringers	160.000	L.F.	3.939	24.00
Totals			29.512	821.20

Total per riser including general contractor's overhead and profit	**$231**
Total per 16 risers including general contractor's overhead and profit	**$3,696**

16. Install New Stairs *(continued)*

Install pressure-treated wood stairs

Description	Quantity	Unit	Labor-Hours	Material
Stair stringers, 2" x 12"	0.140	M.B.F.	8.615	102.90
Pine risers, 3/4" x 7-1/2"	64.000	L.F.	7.757	195.20
2" x 6" stair treads	0.120	M.B.F.	2.560	91.80
Aluminum pipe rail (2 rail)	18.000	L.F.	3.600	299.70
Strap brackets (2 per upright)	10.000	Ea.	1.221	35.00
Drilling bolt holes	30.000	Inch	0.533	0.00
Bolts, nuts & washers	20.000	Ea.	1.143	6.00
Pipe rail attached to brick wall	18.000	L.F.	2.704	158.40
Painting, risers	60.000	L.F.	1.477	9.00
Totals			**29.610**	**898.00**

Total per riser including general contractor's overhead and profit **$241**

Total per 16 risers including general contractor's overhead and profit **$3,860**

Install concrete stairs

Description	Quantity	Unit	Labor-Hours	Material
Freestanding concrete stairs	64.000	LF Nose	37.012	441.60
Safety treads	64.000	L.F.	4.452	700.80
Aluminum pipe rail (2 rail)	18.000	L.F.	3.600	299.70
Aluminum wall railing	18.000	L.F.	2.704	158.40
Totals			**47.768**	**1,600.50**

Total per riser including general contractor's overhead and profit **$399**

Total per 16 risers including general contractor's overhead and profit **$6,383**

Install metal pan stairs, concrete fill

Description	Quantity	Unit	Labor-Hours	Material
Metal pan stairs with railings	16.000	Riser	17.067	3,024.00
Non-slip concrete fill for treads	64.000	S.F.	5.120	245.12
Totals			**22.187**	**3,269.12**

Total per riser including general contractor's overhead and profit **$476**

Total per 16 risers including general contractor's overhead and profit **$7,613**

Copyright Reed Construction Data, 2004

Notes

17
Modify Existing Stair Risers/Treads

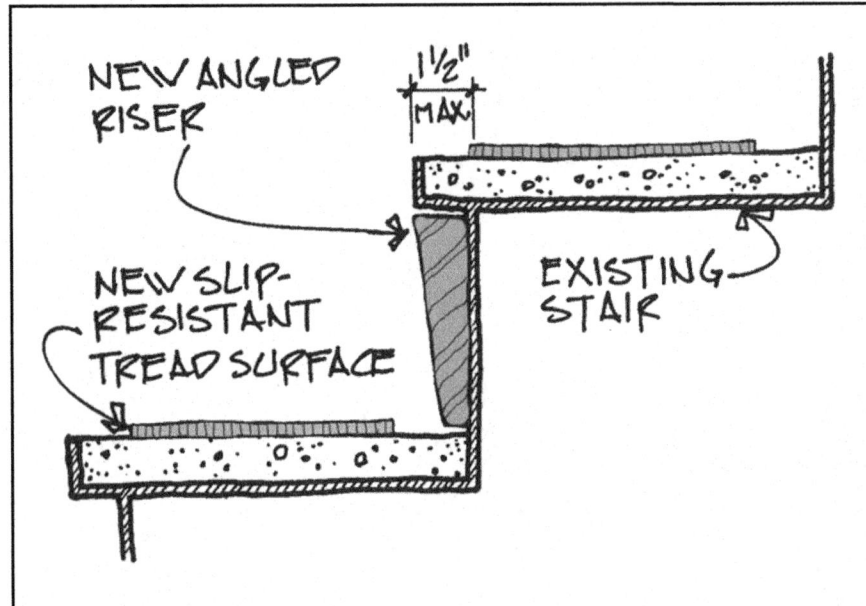

NEW ANGLED RISER

NEW SLIP-RESISTANT TREAD SURFACE

1½" MAX.

EXISTING STAIR

Inaccessible stairs can make travel between levels extremely difficult for people with balance, grasping, or visual impairments. This situation can be dangerous, especially when stairs are used for emergency evacuation. Protruding nosing on which a person can catch their toes or slippery surfaces can, in fact, make the stair awkward and unsafe for all users. Accessible modifications to a flight of stairs result in a much more accessible and safe building for all users.

ADAAG References
210, **Stairways**
302, **Floor or Ground Surfaces**
504, **Stairways**

Where Applicable
All stairs that are part of a means of egress, except stairs in existing buildings that connect levels accessible by elevators, ramps, or lifts. Local building codes, however, may require all stairs that are part of any means of egress to meet ADAAG.

Design Requirements
- Minimum tread depth of 11"; minimum riser height of 4"; maximum riser of 7".
- Slip-resistant surface.
- A slope of 2% is allowed for drainage. No pooling water allowed on exterior stairs and landings.
- No protruding nosing greater than 1-1/2"; underside of nosing sloped back at a 30° angle or less with the riser. Radius of the leading edge l/2" or less.
- No open risers.

Design Suggestions
For existing stairs, ADAAG covers only egress stairs that connect levels not connected by an accessible means of vertical access. It *does*, however, require the handrails on all egress stairs to meet ADAAG. In other words, an existing stair with open

risers or nosings not meeting ADAAG can remain unmodified, as long as the handrails are fixed. However, it is strongly recommended that all stairs conform to ADAAG regulations, since they are a heavily used element by people with a wide range of abilities, and accidents on stairs are frequent and disastrous.

For adding slip-resistant surfaces, use a material that will stand up to serious wear, such as carpet for interior installations, or rubber treads for exterior applications. Applied stair treatments must be carefully installed to resist detaching. It is possible to meet the slip-resistance criteria in ADAAG with coatings, such as sand paint, but these will require frequent maintenance and are not as durable as applied tread materials.

Continuous carpet runners are often used to cover the blemishes where one has filled in open risers or blocked out to eliminate the nosings. It, however, has several drawbacks that should be of concern. Carpeting can be slippery, may stretch and become loose, and people may find it difficult to visually distinguish one tread from the next.

Key Items
Wood risers or bevels for nosings; wood or metal risers; slip-resistant surfaces (sandpaper strips, carpet).

Level of Difficulty
Low to moderate. Modifying metal stairs requires welding or bolting; modifications to wood stairs require carpentry. Applying sandpaper strips or carpet usually does not require contracted labor.

Estimates

Fill in open stair riser (wood stairs)

Description	Quantity	Unit	Labor-Hours	Material
3/4" x 7-1/2" stair risers	1.000	L.F.	0.121	3.05
Painting	1.000	S.F.	0.012	0.05
Totals			0.133	3.10

Total per linear foot of riser including general contractor's overhead and profit **$16**

Fill in open stair riser (metal pan stairs)

Description	Quantity	Unit	Labor-Hours	Material
Plate steel for risers (8" wide x 4' long, 6.81 lbs./L.F.)	0.270	Cwt.	0.000	17.28
Welding	0.200	Hr.	0.200	0.60
Painting, both sides	8.000	L.F.	0.200	0.40
Totals			0.400	18.28

Total per linear foot of riser including general contractor's overhead and profit **$67**

Bevel stair nosing (wood)

Description	Quantity	Unit	Labor-Hours	Material
3/4" x 7-1/2" stair risers	1.000	L.F.	0.121	3.05
Painting	1.000	L.F.	0.012	0.05
Totals			0.133	3.10

Total per linear foot of riser including general contractor's overhead and profit **$16**

17. Modify Existing Stair Risers/Treads *(continued)*

Bevel stair nosing (concrete)

Description	Quantity	Unit	Labor-Hours	Material
Remove concrete nosing	0.120	C.F.	0.074	0.00
Drill 1/2" holes for reinforcing pockets (per inch of depth)	8.000	Ea.	1.280	0.56
Wedge anchors for reinforcing support	2.000	Ea.	0.128	2.86
Deformed reinforcing dowels	2.000	Ea.	0.067	0.84
Steel angle edging for stair nosing	4.000	L.F.	0.388	14.24
Form board (4 L.F.)	4.000	S.F.	1.011	4.64
2" nonshrink grout	4.000	S.F.	1.280	73.40
Totals			**4.228**	**96.54**

Total per linear foot of riser including general contractor's overhead and profit **$118**

Total per 4-foot-wide riser including general contractor's overhead and profit **$472**

Bevel stair nosing (metal pan)

Description	Quantity	Unit	Labor-Hours	Material
Plate steel for risers (8" wide x 4' long, 6.81 lbs./L.F.)	0.270	Cwt.	0.000	17.28
Welding	0.200	Hr.	0.200	0.60
Painting, both sides	8.000	L.F.	0.200	0.40
Totals			**0.400**	**18.28**

Total per linear foot of riser including general contractor's overhead and profit **$17**

Total per 4-foot-wide riser including general contractor's overhead and profit **$67**

Add non-slip surface to stair tread (sandpaper strips)

Description	Quantity	Unit	Labor-Hours	Material
Anti-skid sandpaper strips (6" x 24"), one per riser	1.000	S.F.	0.041	4.60
Totals			**0.041**	**4.60**

Total per riser including general contractor's overhead and profit **$11**

Notes

18
Install or Modify Stair Handrails

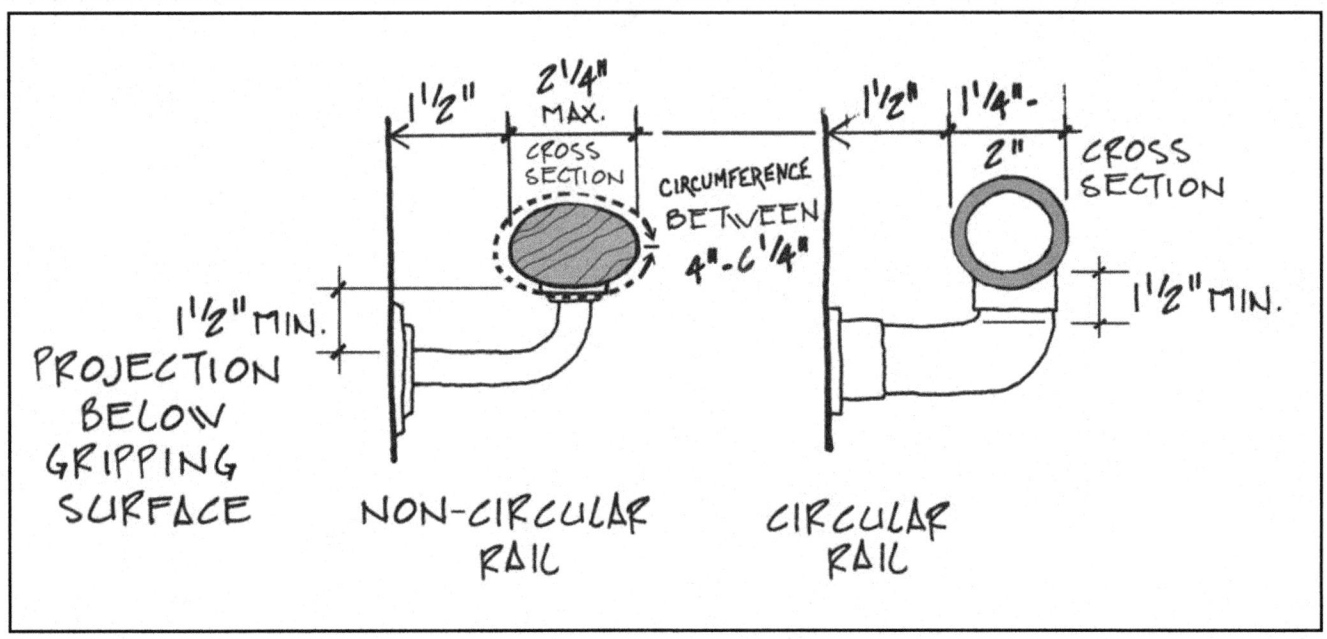

Adding or replacing handrails on stairs is often relatively easy to do, and delivers an enormous benefit. It is strongly recommended that all stairs have handrails meeting ADAAG, whether or not they are required.

ADAAG References

210, Stairways
504, Stairways
505, Handrails

Where Applicable

All egress stairs between levels, except stairs in alterations where the levels are connected by an accessible route.

Design Requirements

- On both sides, 34" to 38" above nosings, 1-1/2" from wall, with no rotation in fittings.
- 1-1/4" to 2" diameter round, or, if not round, the perimeter between 4" and 6-1/4" and the maximum width 2-1/4".
- Continuous inside rail without interruption by newel posts or other construction. Brackets attached to bottom are acceptable if they drop 1-1/2" or more before returning to the wall.

- Rail ends: 12" extensions at the top parallel to the floor. Length of one tread extension at the bottom, following slope of the stair. Return to wall, post, or floor.

Design Suggestions

On dog-leg or switch-back stairs, outside rail should also be continuous at a landing if possible. Under-rail brackets work best, with the rails separated from the uprights. This allows someone to use the rail without letting go. Check nominal dimensions against actual requirements when specifying rails. Metal or Schedule 40 rails specified at 1-1/2" (nominal dimension) will have an actual diameter closer to 1-7/8".

Rails are a highly visible element. Especially on older buildings, it is important to choose design options that will incorporate handrails into the existing aesthetic, such as posts with the same design as stair newel posts. Wall space for extensions sometimes is a problem, but many older landings are oversized, and one can find a clever way to have the extension without it looking "institutional."

Key Items

Rails (wood, pipe, Schedule 40, metal), uprights, anchors.

Level of Difficulty

Moderate to high, depending on the material used. Installing wood rails requires a skillful carpenter; pipe or metal rail installation requires welding or bolting, possibly to a masonry surface.

Estimates

Add new wall-mounted pipe rail

Description	Quantity	Unit	Labor-Hours	Material
1-1/2" wall-mounted aluminum pipe rail	1.000	L.F.	0.150	8.80
Totals			0.150	8.80

Total per linear foot including general contractor's overhead and profit **$30**

Add new free-standing pipe rail

Description	Quantity	Unit	Labor-Hours	Material
1-1/2" freestanding aluminum pipe rail (2 rail)	1.000	L.F.	0.200	16.65
Totals			0.200	16.65

Total per linear foot including general contractor's overhead and profit **$48**

Add new wall-mounted wood dowel rail

Description	Quantity	Unit	Labor-Hours	Material
Handrail stock	1.000	L.F.	0.100	1.22
Drilling bolt holes	1.500	Inch	0.027	0.00
Bolts, nuts & washers	0.750	Ea.	0.043	0.23
Handrail bracket	0.250	Ea.	0.042	1.13
Totals			0.212	2.58

Total per linear foot including general contractor's overhead and profit **$20**

18. Install or Modify Stair Handrails *(continued)*

Add new wood dowel handrail including pipe rail uprights bolted to wood

Description	Quantity	Unit	Labor-Hours	Material
Aluminum pipe railing	0.750	L.F.	0.150	12.49
Handrail stock	1.000	L.F.	0.100	1.22
Handrail bracket	0.250	Ea.	0.042	1.13
Drilling bolt holes for bracket	1.500	Inch	0.027	0.00
Bolts, nuts & washers	0.750	Ea.	0.043	0.23
Strap brackets (2 per upright)	0.500	Ea.	0.061	1.75
Drilling bolt holes	1.500	Inch	0.027	0.00
Bolts, nuts & washers	1.000	Ea.	0.057	0.30
Totals			**0.507**	**17.12**

Total per linear foot including general contractor's overhead and profit — **$70**

Add new wall-mounted metal rail

Description	Quantity	Unit	Labor-Hours	Material
1-1/2" wall-mounted steel railing	1.000	L.F.	0.182	6.50
Totals			**0.182**	**6.50**

Total per linear foot including general contractor's overhead and profit — **$30**

Add new free-standing metal railing

Description	Quantity	Unit	Labor-Hours	Material
1-1/2" freestanding pipe rail (2 rail)	1.000	L.F.	0.200	11.45
Totals			**0.200**	**11.45**

Total per linear foot including general contractor's overhead and profit — **$40**

Add 12" extension to existing wall-mounted pipe railing

Description	Quantity	Unit	Labor-Hours	Material
Fabricate and install 1-1/2" diameter pipe extensions	1.000	Ea.	0.200	11.45
Drilling bolt holes, per inch of depth	6.000	Inch	0.960	0.42
Expansion bolts & shields	3.000	Ea.	0.300	7.86
Totals			**1.460**	**19.73**

Total per each including general contractor's overhead and profit — **$146**

Add 12″ extension to existing free-standing pipe railing

Description	Quantity	Unit	Labor-Hours	Material
Fabricate 1-1/2″ diameter pipe extensions	1.000	Ea.	1.000	57.25
Install pipe extensions on site	1.000	Ea.	0.200	11.45
Totals			**1.200**	**68.70**

Total per each including general contractor's overhead and profit **$395**

Add 12″ extension to existing wall-mounted wood railing

Description	Quantity	Unit	Labor-Hours	Material
Handrail stock	1.000	L.F.	0.100	1.22
Drilling bolt holes	12.000	Inch	0.213	0.00
Bolts, nuts & washers	6.000	Ea.	0.343	1.80
Handrail bracket	2.000	Ea.	0.333	9.02
Totals			**0.989**	**12.04**

Total per each including general contractor's overhead and profit **$92**

Add 12″ extension to existing free-standing wood railing

Description	Quantity	Unit	Labor-Hours	Material
Handrail stock	2.000	L.F.	0.200	2.44
Drilling bolt holes	24.000	Inch	0.427	0.00
Bolts, nuts & washers	6.000	Ea.	0.343	1.80
Handrail bracket	2.000	Ea.	0.333	9.02
Additional wood posts	0.010	M.B.F.	0.308	8.75
Totals			**1.611**	**22.01**

Total per each including general contractor's overhead and profit **$156**

Add 12″ extension to existing wall-mounted metal railing

Description	Quantity	Unit	Labor-Hours	Material
Fabricate 1-1/2″ diameter pipe extensions	1.000	Ea.	1.000	57.25
Install pipe extensions on site	1.000	Ea.	0.182	6.50
Drilling bolt holes, per inch of depth	6.000	Inch	0.960	0.42
Expansion bolts & shields	3.000	Ea.	0.300	7.86
Totals			**2.442**	**72.03**

Total per each including general contractor's overhead and profit **$333**

Add 12″ extension to existing free-standing metal railing

Description	Quantity	Unit	Labor-Hours	Material
Fabricate 1-1/2″ diameter pipe extensions	1.000	Ea.	1.000	57.25
Install pipe extensions on site	1.000	Ea.	0.200	11.45
Totals			**1.200**	**68.70**

Total per each including general contractor's overhead and profit **$395**

19
Install Under-Stair Barrier

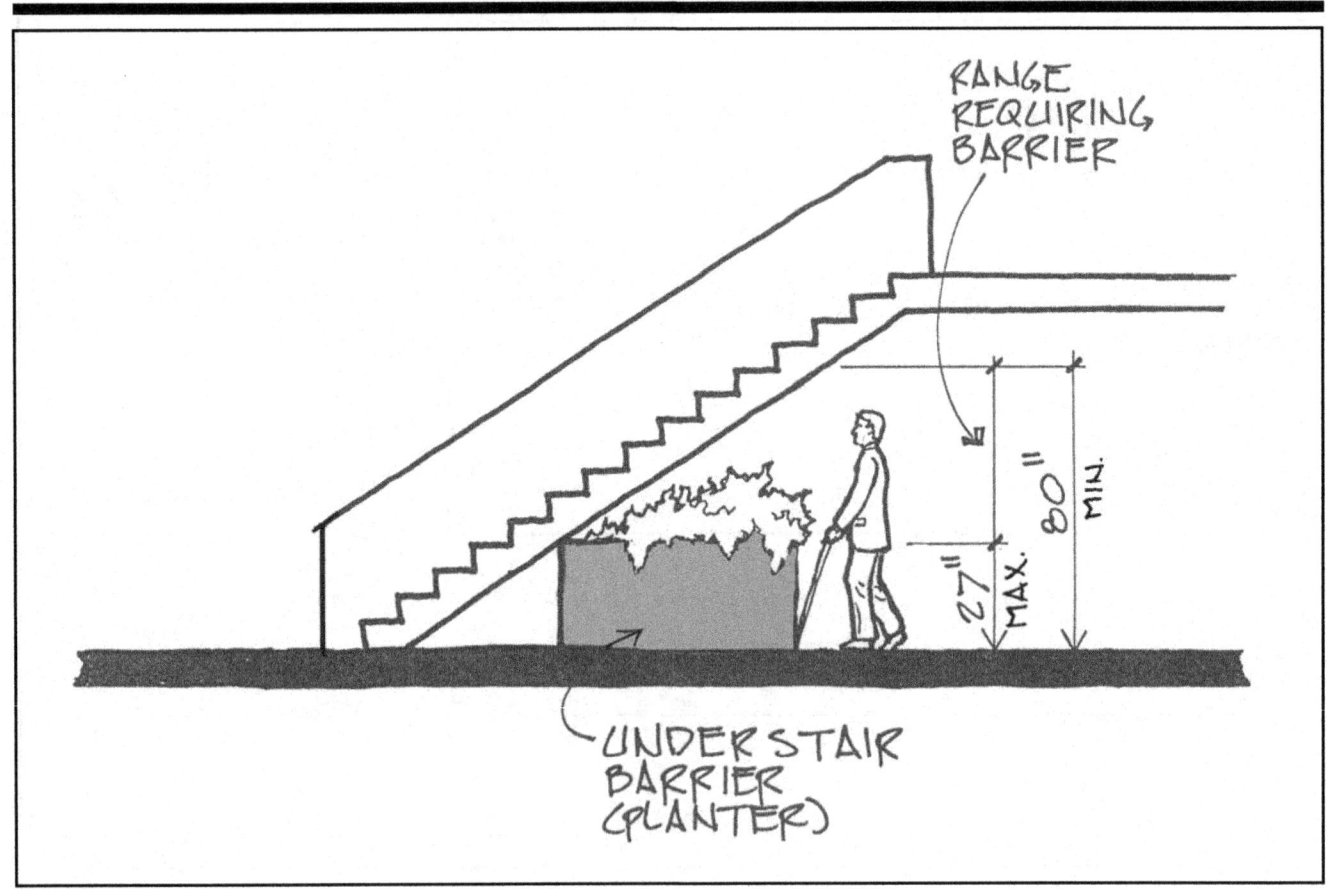

RANGE REQUIRING BARRIER

UNDER STAIR BARRIER (PLANTER)

80" MIN.

27" MAX.

A lack of cane-detectable barriers under free-standing stairs can be extremely dangerous for people with visual impairments. When someone sweeps a cane in front of them, his or her head may hit a sloping object below 80" before the cane touches the bottom of it. Installing a barrier under a stair is vital to creating a safe, accessible facility.

ADAAG Reference
307.4, Vertical Clearance

Where Applicable
Under free-standing staircases, in stair halls with switch-back stairs, and any area along a public route of travel (not just accessible routes) where the headroom is reduced below 80".

Design Requirements
• Barrier located 27" a.f.f. or lower where headroom is less than 80" a.f.f.

Design Suggestions
The requirements for protection from overhead objects apply to all circulation paths, not just an

accessible route. It should not be assumed that an area is not within a route of travel just because it is not the intended route. Install barriers under all free-standing stairs and similar overhanging hazards. The barrier could be functional, such as seating, a planter, or a built-in trash receptacle. If space permits, the area under the stairs could be enclosed with partitions and used for storage. Movable objects that are not anchored in place of fixed barriers are generally not acceptable. Check local building code requirements for under-stair use.

Key Items

A barrier or barriers under each low-headroom hazard stair. Pipe rails, masonry, wood/drywall partitions.

Level of Difficulty

Low.

Estimates

Install barrier under stairs

Description	Quantity	Unit	Labor-Hours	Material
Square wood planter, 48" x 48" x 24"	1.000	Ea.	1.067	820.00
Totals			1.067	820.00

Total per each including general contractor's overhead and profit **$1,457**

20
Modify Existing Door

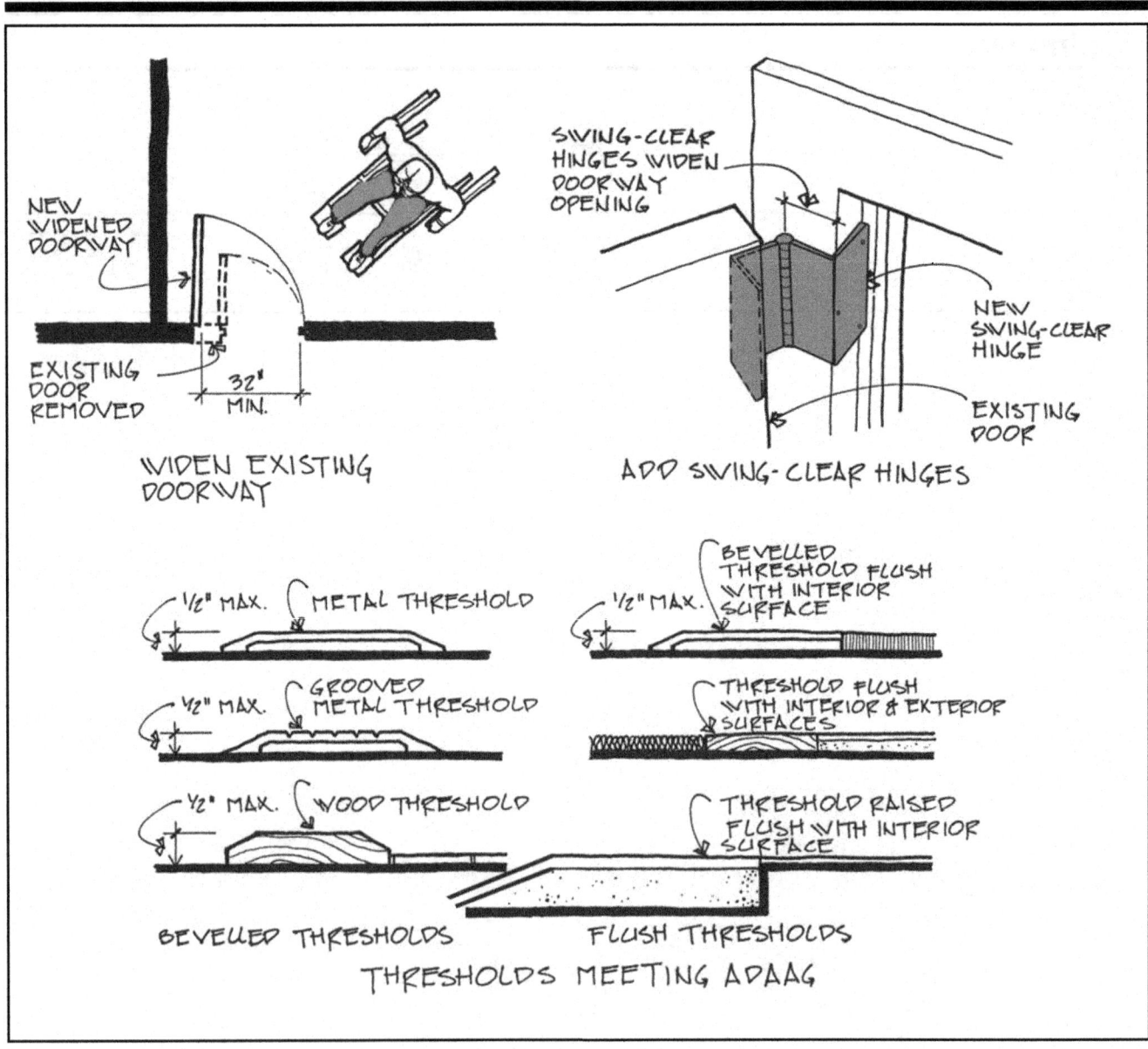

NEW WIDENED DOORWAY

EXISTING DOOR REMOVED

32" MIN.

WIDEN EXISTING DOORWAY

SWING-CLEAR HINGES WIDEN DOORWAY OPENING

NEW SWING-CLEAR HINGE

EXISTING DOOR

ADD SWING-CLEAR HINGES

1/2" MAX. METAL THRESHOLD

1/2" MAX. GROOVED METAL THRESHOLD

1/2" MAX. WOOD THRESHOLD

BEVELLED THRESHOLDS

1/2" MAX. BEVELLED THRESHOLD FLUSH WITH INTERIOR SURFACE

THRESHOLD FLUSH WITH INTERIOR & EXTERIOR SURFACES

THRESHOLD RAISED FLUSH WITH INTERIOR SURFACE

FLUSH THRESHOLDS

THRESHOLDS MEETING ADAAG

Some modifications are relatively simple on any door. Examples include replacing or modifying existing hardware, beveling an existing threshold, or removing an existing threshold. Reversing a door swing can sometimes solve maneuvering clearance problems, but can be difficult with metal doors because hinges and hardware have to be changed, and the frame sometimes has to be reversed in the opening. Swing-clear hinges free the space normally occupied by the door (when the door is open), although they can also be difficult to install on metal doors.

ADAAG References

206.5, Doors, Doorways, and Gates
303.3, Beveled Changes in Level
309, Operable Parts
404, Doors, Doorways, and Gates

Where Applicable

Doors along an accessible route that do not meet ADAAG in one or more ways.

Design Requirements

- 32" minimum clear width measured between the door stop and the face of the door open to 90°.

- Typical maneuvering dimensions: 18" minimum clearance adjacent to the latch on the pull side of the door. If the door has closers, 12" minimum on the push side. See 404.2.4 for clearances when a turn is involved.
- Level maneuvering space on both sides of the door, depending on approach. Between 42" and 60", minimum, required in front of door, depending on approach and whether door has a closer. (See ADAAG Tables 404.2.4.1 & 404.2.4.2.) Surface of required maneuvering space must be level (2% maximum slope allowed for drainage).
- Threshold: 1/2" maximum high; beveled at a 1:2 slope, (vertical to horizontal) maximum.
- Five pounds maximum pull or push weight on interior doors (no ADAAG standard for exterior doors) and other fire doors.
- Accessible hardware (acceptable if operable with a closed fist).
- With door closer, five seconds minimum closing time to an angle of 12°. With spring hinges, 1.5 seconds minimum from 70° to closed position.

- Door surfaces: bottom 10" on push side smooth, extending full width of door.
- Vision panels: 43" maximum, to at least one vision panel.
- See Project 23 if automatic opener is contemplated.

Design Suggestions

To address opening force, if there is a closer on the door, the closing speed can usually be adjusted. If there are hinges, oiling them is another simple way to make a door easier to open. It is almost always possible (but sometimes expensive) to install an automatic opener to compensate for insufficient latchside clearance or heavy door opening weights. However, this will require stand-by power. Some building inspectors will not allow this on egress doors.

Key Items

Door, door hardware, hinges, thresholds, automatic opener.

Level of Difficulty

Low to moderate. Some finish work required for threshold removal. Automatic openers require preliminary electrical work, and can be expensive.

20. Modify Existing Door (continued)

Reverse door swing, wood door

Description	Quantity	Unit	Labor-Hours	Material
Remove door	1.000	Ea.	0.400	0.00
Remove door frame & trim	1.000	Ea.	0.500	0.00
Interior door frame	18.000	L.F.	0.768	71.82
Door trim set, 1 head, 2 sides, 2-1/2" pine	2.000	Opng.	2.712	28.90
Remove door hinge	3.000	Ea.	0.500	0.00
Install door hinge	3.000	Ea.	0.500	0.00
Re-install door	1.000	Job	2.000	0.00
Remove lockset	1.000	Ea.	0.400	0.00
Re-install lockset	1.000	Ea.	0.800	68.50
Paint door frame & trim	18.000	L.F.	0.543	2.88
Minimum labor/equipment charge	0.500	Job	1.000	0.00
Totals			**10.123**	**172.10**

Total per each including general contractor's overhead and profit **$1,095**

Replace existing hinges with swing-clear hinges

Description	Quantity	Unit	Labor-Hours	Material
Remove interior door	1.000	Ea.	0.400	0.00
Remove door hinge	3.000	Ea.	0.500	0.00
Swing clear hinges	1.500	Pr.	0.000	193.50
Install door hinge	3.000	Ea.	0.500	0.00
Re-install door	1.000	Ea.	1.143	305.00
Totals			**2.543**	**498.50**

Total per set of hinges including general contractor's overhead and profit **$1,029**

Replace existing lockset with lever-handled lockset

Description	Quantity	Unit	Labor-Hours	Material
Remove lockset	1.000	Ea.	0.400	0.00
Lever-handled lockset	1.000	Ea.	0.800	121.00
Totals			**1.200**	**121.00**

Total per each including general contractor's overhead and profit **$294**

Install push plates on wood door

Description	Quantity	Unit	Labor-Hours	Material
Aluminum push plates, both sides of door	1.000	Ea.	0.571	13.35
Totals			**0.571**	**13.35**

Total per each including general contractor's overhead and profit **$65**

Install push plates and panic bar

Description	Quantity	Unit	Labor-Hours	Material
Aluminum push plates, both sides of door	1.000	Ea.	0.571	13.35
Panic bar and verticle rod, for exit only	1.000	Ea.	1.600	490.00
Totals			2.171	503.35

Total per each including general contractor's overhead and profit **$1,018**

Remove threshold, fill floor flush with existing flooring (vinyl tile)

Description	Quantity	Unit	Labor-Hours	Material
Remove threshold	1.000	Ea.	0.021	0.00
Vinyl tile	3.000	S.F.	0.048	9.87
Minimum labor to patch subfloor and lay new tile	1.000	Job	2.000	0.00
Totals			2.069	9.87

Total per each including general contractor's overhead and profit **$152**

Bevel 2 existing wood thresholds

Description	Quantity	Unit	Labor-Hours	Material
Minimum labor to bevel 2 wood thresholds	1.000	Job	2.000	0.00
Totals			2.000	0.00

Total per each including general contractor's overhead and profit **$73**

Total per 2 thresholds including general contractor's overhead and profit **$146**

Remove 2 existing thresholds

Description	Quantity	Unit	Labor-Hours	Material
Minimum labor to remove 2 existing thresholds (no patching included)	1.000	Job	2.000	0.00
Totals			2.000	0.00

Total per each including general contractor's overhead and profit **$73**

Total per 2 thresholds including general contractor's overhead and profit **$146**

20. Modify Existing Door *(continued)*

Adjust door-closing speed (5 doors)

Description	Quantity	Unit	Labor-Hours	Material
Minimum labor to adjust 5 door closers	1.000	Job	2.000	0.00
Totals			**2.000**	**0.00**

Total per each including general contractor's overhead and profit	**$29**
Total per 5 closers including general contractor's overhead and profit	**$146**

Install automatic door opener

Description	Quantity	Unit	Labor-Hours	Material
Cutout demolition of partition	1.000	Ea.	0.333	0.00
Conductor fished to nearby junction box	0.100	C.L.F.	0.333	2.15
Junction box	1.000	Ea.	0.400	2.10
Automatic opener, button operation	1.000	Ea.	5.333	355.00
Totals			**6.399**	**359.25**

Total per each including general contractor's overhead and profit	**$1,083**

Copyright Reed Construction Data, 2004

Notes

21

Install New Door: Drywall Partition

NEW DOOR OPENING

32" CLEAR MIN.

Installing a new door is one of the easiest ways of getting access to a room that is not accessible because of restrictions in the accessible route.

ADAAG References

206.5, Doors, Doorways, and Gates
303.3, Beveled Changes in Level
309, Operable Parts
404, Doors, Doorways, and Gates

Where Applicable

All rooms along accessible routes that have walls constructed of studs and wallboard.

Design Requirements

- 32" minimum clear width measured between the door stop and the face of the door open to 90°.
- Typical maneuvering dimensions: 18" minimum clearance adjacent to the latch on the pull side of the door. If the door has closers, 12" minimum on the push side. See 404.2.4 for clearances when a turn is involved.
- Level maneuvering space on both sides of the door, depending on approach. Between 42" and 60", minimum, required in front of door, depending on approach and whether door has a closer. (See ADAAG Tables 404.2.4.1 & 404.2.4.2.) Surface of required maneuvering space must be level (2% maximum slope allowed for drainage).
- Threshold: 1/2" maximum high; beveled at a 1:2 slope, (vertical to horizontal) maximum.
- Five pounds maximum pull or push weight on interior doors (no ADAAG standard for exterior doors) and other than fire doors.
- Accessible hardware (acceptable if operable with a closed fist).
- With door closer, five seconds minimum closing time to an angle of 12°. With spring hinges, 1.5 seconds minimum from 70° to closed position.

- Door surfaces: bottom 10" on push side smooth, extending full width of door.
- Vision panels: 43" maximum, to at least one vision panel.
- See Project 24 if automatic opener is desired.

Design Suggestions

Because the 32" clear opening is measured from the face of the door in a 90° open position to the stop on the opposite jamb, the door itself must be wider. Usually a 36" door is used. Doors sized 2'- 10" can be used, but they are not as readily available.

There are several accessible hardware options: a loop pull (allow at least 1-1/2" between inside of loop and face of door), lever handles, push plate, or panic bar. Where opening force is necessarily high or where adequate maneuvering space cannot be provided, installation of an automatic opener may be a solution (See Project 24).

The maneuvering space in Tables 404.2.4.1 and .2 to make a turn from a corridor through a door are minimum, and wider dimensions are recommended to avoid banged-up walls. For instance, the new door cannot be installed if one must make a 90° turn from a 36" wide corridor. Maneuvering space on each side of the door is determined by how it is approached. A 60" by 60" minimum clear space with the door off-set to the latch side is recommended, since it complies with all approaches cited in ADAAG. Where only a straight-on approach is available, a 60" deep space measured from the face of the door is required on the pull side, and a 48" deep space is required on the push side. Where only a side approach is available, the required depth of the clear area in front of the door varies from 42" to 60", and the width is affected by the latch edge clearances and presence or absence of a door opener.

Key Items

Door (may need to be fire-coded), new structural members for opening, wall/floor finishes to match existing, hardware.

Level of Difficulty

Moderate to high. Higher for load-bearing walls. Involves demolition, structural framing, and finish work. May require a building permit and possibly design drawings.

Estimates

Install solid core wood door in metal stud/drywall partition

Description	Quantity	Unit	Labor-Hours	Material
Remove metal studs/gypsum board	9.000	S.F.	0.415	0.00
Steel door frame, 3'-0" x 6'-8"	1.000	Ea.	1.000	75.00
Solid core wood door, flush birch face	1.000	Ea.	1.143	167.00
Lever-handled lockset	2.000	Ea.	1.600	242.00
Hinges	1.500	Pr.	0.000	66.75
Threshold	1.000	Ea.	0.400	30.00
Paint door	1.000	Ea.	1.600	10.30
Painting	76.000	S.F.	1.930	16.72
Totals			8.088	607.77

Total per each including general contractor's overhead and profit   **$1,584**

22
Install New Door: Masonry

A new accessible door can often prevent the need for a long, circuitous route of travel to an existing accessible entrance. In some cases, an existing window opening can be enlarged to create an accessible entrance to a facility or space.

ADAAG References
See references for Project 20.

Where Applicable
All rooms along accessible routes that have walls constructed of brick, block, or stone.

Design Requirements
• If door is recessed, the recess cannot be greater than 8" without recessing the maneuvering spaces required by 404.2.4.3.

See Project 20 for additional design requirements.

Design Suggestions
If the masonry wall is over 12" wide, care must be taken in choosing where in the width to locate the door. The maximum recess from the face of the door to the wall surface is 8". If the door surface exceeds 8", the opening must be extended to 60" wide to meet the required 18" on the latch side of the door.

See Project 20 for additional design suggestions.

Key Items
Door (may need to be fire-coded), masonry materials, possibly new structural members for opening, wall/floor finishes to match existing, hardware.

Level of Difficulty
Moderate to high. Higher for load-bearing walls. Involves demolition, bracing of new opening, structural framing, and finish work. Requires a building permit and possibly design drawings.

32" CLEAR OPENING REQUIRED-- MASONRY OPENING DETERMINED BY FIELD CONDITIONS

Estimates

Install solid core wood door in 12″ thick brick exterior wall

Description	Quantity	Unit	Labor-Hours	Material
Saw cutting brick wall, per inch of depth	288.000	L.F.	15.359	92.16
Brick wall demolition	34.000	C.F.	6.182	0.00
Remove window	1.000	Ea.	0.615	0.00
Labor minimum for masonry repairs	1.000	Job	8.000	0.00
Steel lintel, 4″ x 3-1/2″ x 1/4″, 5′-0″ long	1.000	Ea.	0.381	15.80
Steel door frame, 3′-0″ x 6′-8″	1.000	Ea.	1.000	75.00
Solid core wood door, flush birch face	1.000	Ea.	1.143	167.00
Door closer	1.000	Ea.	1.333	125.00
Lever-handled lockset	1.000	Ea.	0.800	121.00
Hinges	1.500	Pr.	0.000	66.75
Threshold	1.000	Ea.	0.400	30.00
Paint door	1.000	Ea.	1.600	10.30
Silicone sealant	17.000	L.F.	0.579	3.74
Totals			**37.392**	**706.75**

Total per each including general contractor's overhead and profit **$4,261**

Install solid core wood door in masonry veneer exterior wall

Description	Quantity	Unit	Labor-Hours	Material
Saw cutting brick wall, per inch of depth	96.000	L.F.	5.120	30.72
Brick wall demolition	12.000	C.F.	2.182	0.00
Remove window	1.000	Ea.	0.615	0.00
Remove metal studs, interior & exterior gypsum board	34.000	S.F.	1.569	0.00
Labor minimum for masonry repairs	1.000	Job	8.000	0.00
Steel lintel, 4″ x 3-1/2″ x 1/4″, 5′-0″ long	1.000	Ea.	0.381	15.80
Steel door frame, 3′-0″ x 6′-8″	1.000	Ea.	1.000	75.00
Solid core wood door, flush birch face	1.000	Ea.	1.143	167.00
Door closer	1.000	Ea.	1.333	125.00
Lever-handled lockset	1.000	Ea.	0.800	121.00
Hinges	1.500	Pr.	0.000	66.75
Threshold	1.000	Ea.	0.400	30.00
Paint door	1.000	Ea.	1.600	10.30
Silicone sealant	17.000	L.F.	0.579	3.74
Totals			**24.722**	**645.31**

Total per each including general contractor's overhead and profit **$2,936**

Copyright Reed Construction Data, 2004

Install solid core wood door in 8″ block exterior wall

Description	Quantity	Unit	Labor-Hours	Material
Saw cutting block wall, per inch of depth	192.000	L.F.	12.288	63.36
Concrete block demolition	34.000	S.F.	1.679	0.00
Remove window	1.000	Ea.	0.615	0.00
Labor minimum for masonry repairs	1.000	Job	8.000	0.00
Steel lintel, 4″ x 3-1/2″ x 1/4″, 5′-0″ long	1.000	Ea.	0.381	15.80
Steel door frame, 3′-0″ x 6′-8″	1.000	Ea.	1.000	75.00
Solid core wood door, flush birch face	1.000	Ea.	1.143	167.00
Door closer	1.000	Ea.	1.333	125.00
Lever-handled lockset	1.000	Ea.	0.800	121.00
Hinges	1.500	Pr.	0.000	66.75
Threshold	1.000	Ea.	0.400	30.00
Paint door	1.000	Ea.	1.600	10.30
Silicone sealant	17.000	L.F.	0.579	3.74
Totals			**29.818**	**677.95**

Total per each including general contractor's overhead and profit

$3,614

Install hollow metal door in 12″ thick brick exterior wall

Description	Quantity	Unit	Labor-Hours	Material
Saw cutting brick wall, per inch of depth	288.000	L.F.	15.359	92.16
Brick wall demolition	34.000	C.F.	6.182	0.00
Remove window	1.000	Ea.	0.615	0.00
Labor minimum for masonry repairs	1.000	Job	8.000	0.00
Steel lintel, 4″ x 3-1/2″ x 1/4″, 5′-0″ long	1.000	Ea.	0.381	15.80
Steel door frame, 3′-0″ x 6′-8″	1.000	Ea.	1.000	75.00
Hollow metal door	1.000	Ea.	0.941	166.00
Door closer	1.000	Ea.	1.333	125.00
Lever-handled lockset	1.000	Ea.	0.800	121.00
Hinges	1.500	Pr.	0.000	66.75
Threshold	1.000	Ea.	0.400	30.00
Paint door	1.000	Ea.	1.600	10.30
Silicone sealant	17.000	L.F.	0.579	3.74
Totals			**37.190**	**705.75**

Total per each including general contractor's overhead and profit

$4,245

Copyright Reed Construction Data, 2004

Install hollow metal door in masonry veneer exterior wall

Description	Quantity	Unit	Labor-Hours	Material
Saw cutting brick wall, per inch of depth	96.000	L.F.	5.120	30.72
Brick wall demolition	12.000	C.F.	2.182	0.00
Remove window	1.000	Ea.	0.615	0.00
Remove metal studs, interior & exterior gypsum board	34.000	S.F.	1.569	0.00
Labor minimum for masonry repairs	1.000	Job	8.000	0.00
Steel lintel, 4" x 3-1/2" x 1/4", 5'-0" long	1.000	Ea.	0.381	15.80
Steel door frame, 3'-0" x 6'-8"	1.000	Ea.	1.000	75.00
Hollow metal door	1.000	Ea.	0.941	166.00
Door closer	1.000	Ea.	1.333	125.00
Lever-handled lockset	1.000	Ea.	0.800	121.00
Hinges	1.500	Pr.	0.000	66.75
Threshold	1.000	Ea.	0.400	30.00
Paint door	1.000	Ea.	1.600	10.30
Silicone sealant	17.000	L.F.	0.579	3.74
Totals			**24.520**	**644.31**

Total per each including general contractor's overhead and profit **$2,920**

Install hollow metal door in 8" block exterior wall

Description	Quantity	Unit	Labor-Hours	Material
Saw cutting block wall, per inch of depth	192.000	L.F.	12.288	63.36
Concrete block demolition	34.000	S.F.	1.679	0.00
Remove window	1.000	Ea.	0.615	0.00
Labor minimum for masonry repairs	1.000	Job	8.000	0.00
Steel lintel, 4" x 3-1/2" x 1/4", 5'-0" long	1.000	Ea.	0.381	15.80
Steel door frame, 3'-0" x 6'-8"	1.000	Ea.	1.000	75.00
Hollow metal door	1.000	Ea.	0.941	166.00
Door closer	1.000	Ea.	1.333	125.00
Lever-handled lockset	1.000	Ea.	0.800	121.00
Hinges	1.500	Pr.	0.000	66.75
Threshold	1.000	Ea.	0.400	30.00
Paint door	1.000	Ea.	1.600	10.30
Silicone sealant	17.000	L.F.	0.579	3.74
Totals			**29.616**	**676.95**

Total per each including general contractor's overhead and profit **$3,598**

23
Install New Door: Glass Storefront

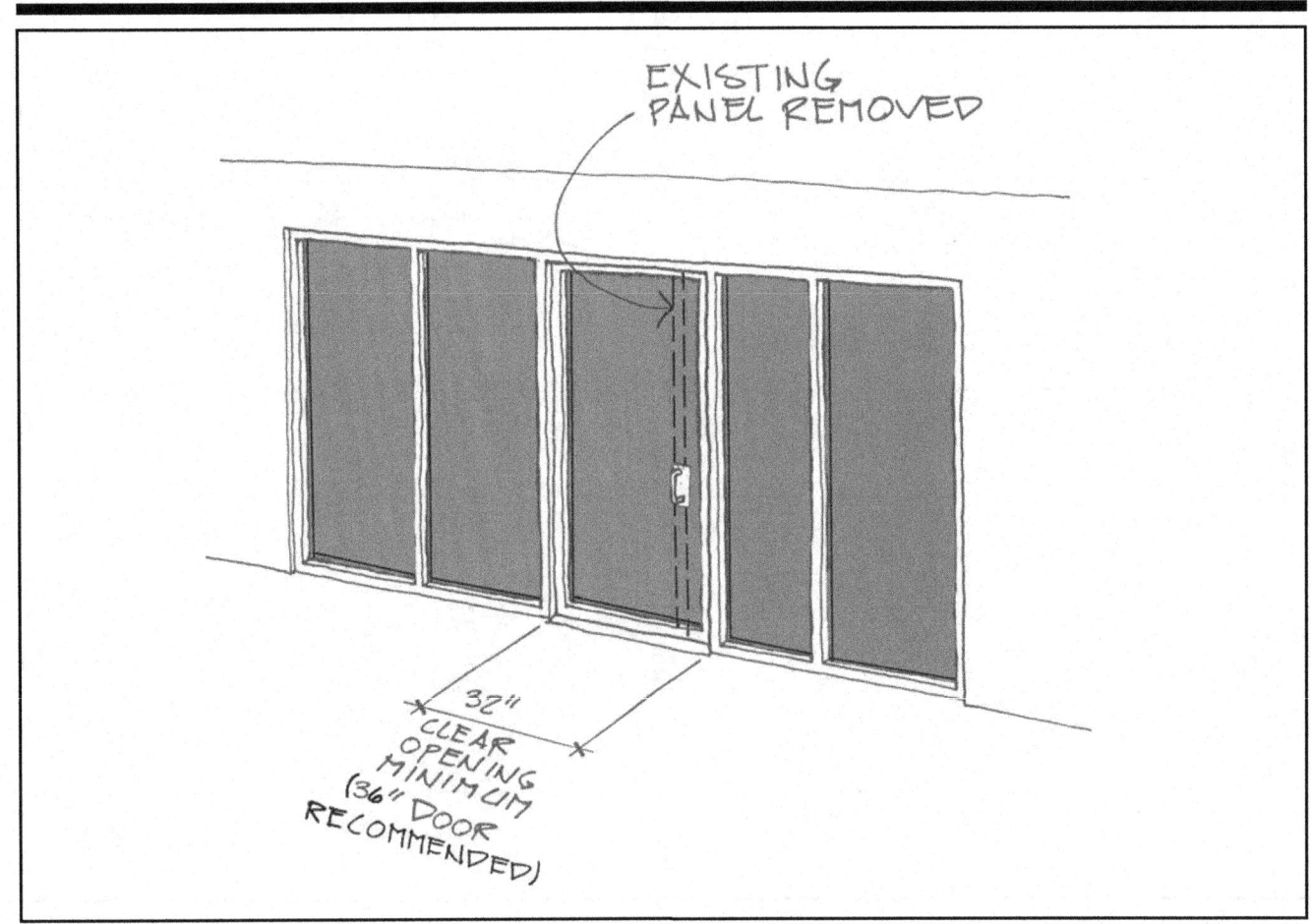

EXISTING PANEL REMOVED

32" CLEAR OPENING MINIMUM (36" DOOR RECOMMENDED)

Many public facilities have glass storefront façades. Although systems and materials vary widely, many storefronts are made of modules or partitions that can be removed and replaced relatively simply. This allows for the installation of an accessible door in an existing building while maintaining the building design.

ADAAG References
303.3, Beveled Changes in Level
309.4, Operable Parts
404, Doors, Doorways, and Gates

Where Applicable

Areas on accessible routes of travel that have walls constructed of metal framed glass.

Design Requirements

See Project 20 for design requirements.

Design Suggestions

See Project 20 for design suggestions.

Key Items

Door, new structural members for opening, wall/floor finishes to match existing. Installation of an accessible door in an existing metal and glass storefront wall involves removal of adjacent panels and replacing with narrower panels to accommodate the wider door.

Level of Difficulty

Moderate to high. Involves demolition, bracing of new opening, and finish work. If storefront is one integral unit, installing or widening a door can require removal of the wall. Requires a building permit and, possibly, design drawings.

Estimates

Install new storefront door

Description	Quantity	Unit	Labor-Hours	Material
Remove glass	60.000	S.F.	3.200	0.00
Cut back tube frame sills	0.125	Job	1.000	0.00
New tube frame jambs	20.000	L.F.	4.571	269.00
Install new tube frame header	3.000	L.F.	0.615	36.45
Door stop (snap in)	23.000	L.F.	3.200	165.60
Install new insulated glass	39.000	S.F.	8.320	936.00
Install new storefront door	1.000	Ea.	2.286	231.00
Totals			23.192	1,638.05

Total per each including general contractor's overhead and profit **$4,419**

Remove glass panels

Description	Quantity	Unit	Labor-Hours	Material
Remove glass, 3'-0" x 8'-0"	24.000	S.F.	1.280	0.00
Totals			1.280	0.00

Total per each including general contractor's overhead and profit **$74**

Copyright Reed Construction Data, 2004

24
Install Automatic Door Opener

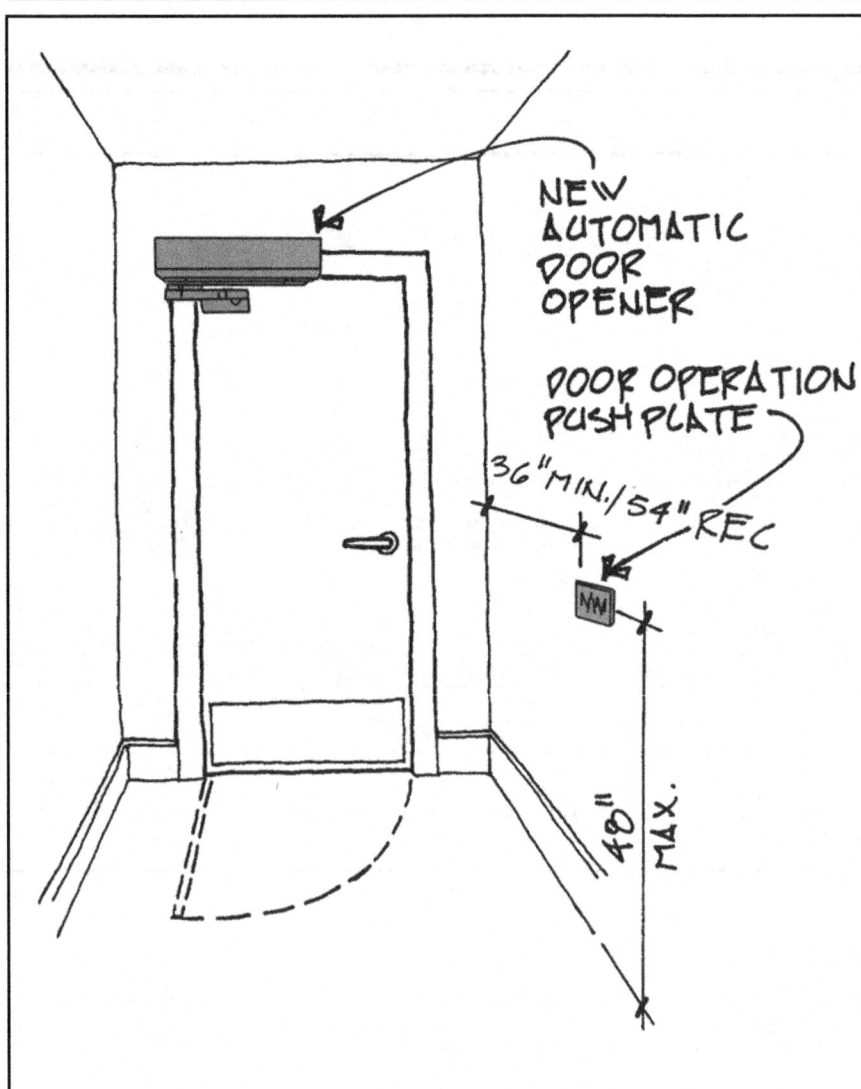

NEW AUTOMATIC DOOR OPENER

DOOR OPERATION PUSH PLATE

36" MIN. / 54" REC

48" MAX.

Automatic door openers are not required in new construction but in older buildings, automatic openers are useful for making doors accessible where the door opening pressure is excessive or there is insufficient maneuvering clearance on one or both sides of the door.

ADAAG References
309, Operable Parts
404.2.4, Maneuvering Clearances
404.2.6, Doors in Series and Gates in Series
404.3, Automatic Doors and Gates
703, Signs
ANSI/BHMA A156.10
ANSI/BHMA A156.19

Where Applicable
Doors on accessible routes of travel that meet the width requirement but are either heavy to open or do not have sufficient space in front of, or adjacent to, the door to allow wheelchair access. Doors in a series where the vestibule is too small.

Design Requirements
- Level maneuvering space on both sides of the door, subject to building inspector.
- With door closer, 5 seconds minimum closing time to an angle of 12°. With spring hinges, 1.5 seconds minimum from 70° to closed position.
- Stand by power if automatic opener is used because other ADAAG requirements cannot be met.

Design Suggestions
Some doors with automatic openers open by a nearby button, others have an arm that rests against the door allowing the option of automatic or manual use, and still others have an electric eye, which is preferable. If the operating device is a button on a wall or post, it must be far enough in front of the door (60" recommended) to allow for the door swing.

In theory, an automatic opener obviates the level space requirement and changes the requirements for maneuvering space, but maneuvering space is still required if there is a push/pull bar on the door or to reach the button. Also, when the approach is parallel to the wall, the maneuvering space is still required. The recommended maximum opening force for exterior doors is eight pounds of pressure; if the operating force exceeds eight pounds for exterior doors, consider installing an automatic door opener or a power assisted door.

Key Items

Door opener, electrical supply, possible wall demo and finish work.

Level of Difficulty

Low to moderate, depending on amount of electrical work and type of opener. Higher when mechanism is buried in load-bearing walls. May involve demolition, bracing of new opening, structural framing, and finish work.

Estimates

Install automatic exterior door opener

Description	Quantity	Unit	Labor-Hours	Material
Cutout demolition of exterior wall (12" thick brick wall)	1.000	Ea.	4.000	0.00
Conductor	0.100	C.L.F.	0.296	1.23
Install junction box	1.000	Ea.	0.400	7.75
Install outlet	1.000	Ea.	0.296	7.05
Install plate	1.000	Ea.	0.100	1.76
Automatic opener, button operation	1.000	Ea.	5.333	355.00
Totals			**10.425**	**372.79**

Total per each including general contractor's overhead and profit **$1,376**

Install automatic interior door opener

Description	Quantity	Unit	Labor-Hours	Material
Cutout demolition of partition	1.000	Ea.	0.333	0.00
Conductor	0.100	C.L.F.	0.296	1.23
Install junction box	1.000	Ea.	0.400	7.75
Install outlet	1.000	Ea.	0.296	7.05
Install plate	1.000	Ea.	0.100	1.76
Repair gypsum board	1.000	Job	4.000	0.00
Paint gypsum board – minimum	1.000	Job	2.000	0.00
Automatic opener, button operation	1.000	Ea.	5.333	355.00
Totals			**12.758**	**372.79**

Total per each including general contractor's overhead and profit **$1,555**

24. Install Automatic Door Opener *(continued)*

Install automatic door opener with infrared activator

Description	Quantity	Unit	Labor-Hours	Material
Single swinging door automatic opener	1.000	Ea.	20.000	3,550.00
Infrared detector	1.000	Ea.	3.478	254.00
Totals			**23.478**	**3,804.00**

Total per each including general contractor's overhead and profit **$8,260**

Install power assist interior door opener/closer

Description	Quantity	Unit	Labor-Hours	Material
Cast iron power assist door opener/closer	1.000	Ea.	20.000	3,200.00
Totals			**20.000**	**3,200.00**

Total per each including general contractor's overhead and profit **$6,966**

Notes

25
Modify Existing Double-Leaf Doors

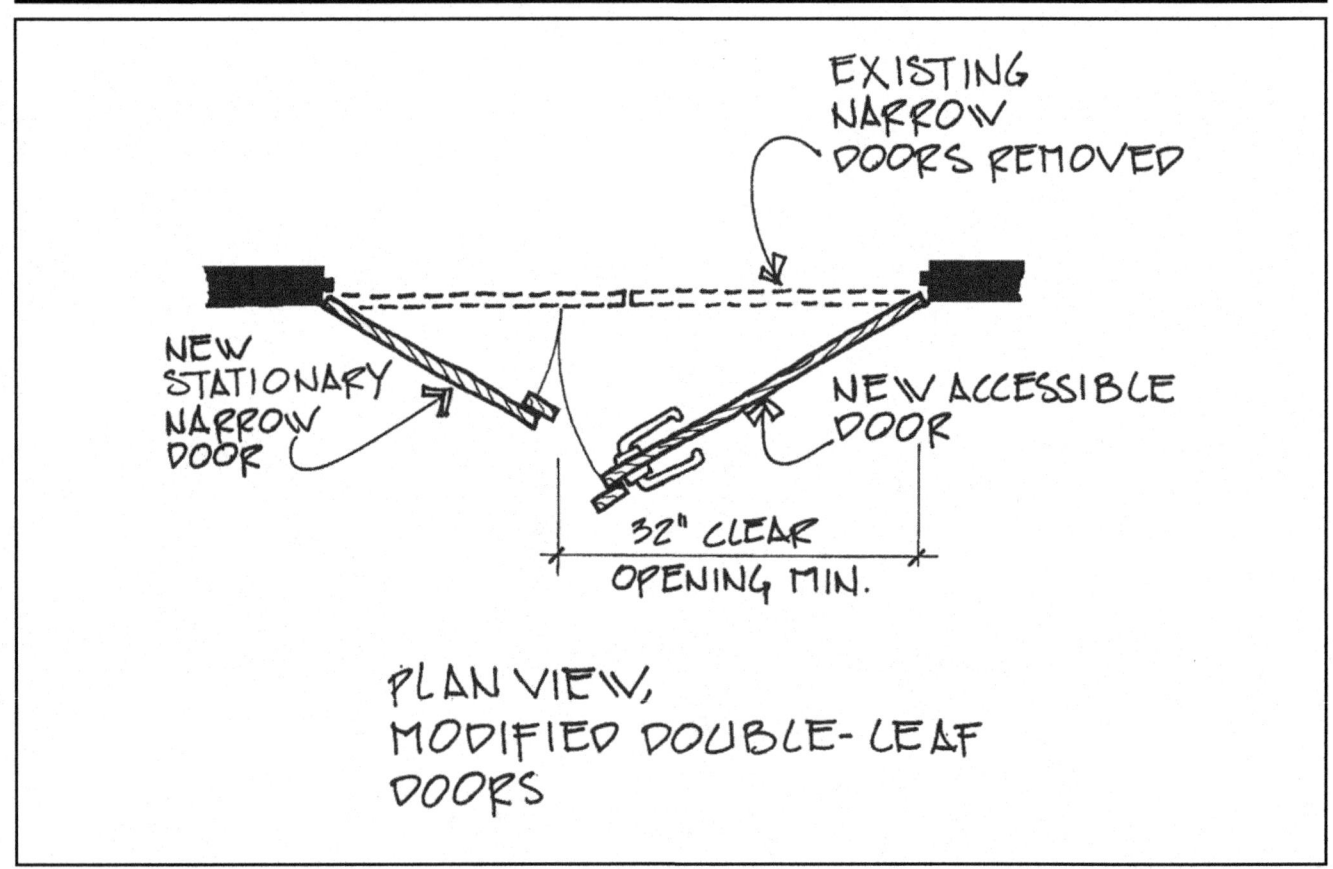

EXISTING
NARROW
DOORS REMOVED

NEW
STATIONARY
NARROW
DOOR

NEW ACCESSIBLE
DOOR

32" CLEAR
OPENING MIN.

PLAN VIEW,
MODIFIED DOUBLE-LEAF
DOORS

The entire opening of a double-leaf doorway is almost always sufficient for complying with the necessary clear width for an accessible door, so modifying the doors to comply can be a useful, and often simple, access modification.

ADAAG Reference
404.2.3, Doors, Clear Width

See Project 20 for additional requirements.

Where Applicable
All double doorways along accessible routes of travel where at least one leaf does not have a 32" clear opening. (A 32" clear opening usually means that one leaf is at least 34" wide.)

Design Requirements
- At least one leaf of a pair with 32" *clear* opening.
- All other door requirements must be met, such as hardware, maneuvering space, and so forth.
See Project 20.

Design Suggestions

The most frequent situation is a 5' wide opening with two 30" doors. For this situation, consider replacing the doors with uneven doors, one 36" wide and the other 24". However, depending on existing conditions, there may be several other options.

- Option 1: Widen doorway to allow two 34" doors (very difficult if doors span the width of a hallway). As with all doors, 34" is a minimum; 36" is recommended.
- Option 2: Install one 34" door (minimum) and fixed sidelight.
- Option 3: Install automatic door openers (both leaves). When the power goes out, as is the case in many emergencies, the door will not meet regulations unless there is stand-by power. For this reason, many building inspectors will not allow this solution.
- Option 4: Remove doors, if possible, without sacrificing privacy, security, or fire safety.
- Option 5: Keep doors open during public operating hours (if permitted). This solution is often not allowed by building inspectors.

Key Items

Replacement doors and hardware as required, finish materials, possibly structural bracing. Button-operated or infrared automatic opener, hold-open device.

Level of Difficulty

- Option 1: High; requires structural and finish work.
- Options 2: Moderate to high; more difficult for metal doors.
- Option 3: Varies; some automatic openers can be installed by maintenance staff, while others require electrical and finish work.
- Options 4 & 5: Low.

Estimates

Widen existing brick/block opening & replace w/ solid wood door

Description	Quantity	Unit	Labor-Hours	Material
Remove interior door	1.000	Ea.	0.400	0.00
Remove door frame	1.000	Ea.	0.500	0.00
Saw cutting block wall, per inch of depth	64.000	L.F.	4.096	21.12
Remove masonry partition	9.000	S.F.	0.444	0.00
Labor minimum for masonry repairs	1.000	Job	8.000	0.00
Steel lintel, 4″ x 3-1/2″ x 1/4″, 5'-0″ long	1.000	Ea.	0.381	15.80
Interior door frame	18.000	L.F.	0.768	71.82
Interior wood door, birch face, 3'-0″ x 6'-8″	1.000	Ea.	1.143	87.50
Add for solid core door	1.000	Ea.	0.000	29.50
Hinges	1.500	Pr.	0.000	66.75
Lever-handled lockset	1.000	Ea.	0.800	121.00
Threshold	1.000	Ea.	0.400	30.00
Paint door	1.000	Ea.	1.600	10.30
Painting	68.000	S.F.	1.727	14.96
Totals			**20.259**	**468.75**

Total per each including general contractor's overhead and profit _____ **$2,300**

25. Modify Existing Double-Leaf Doors (continued)

Widen existing brick/block opening & replace w/ hollow wood door

Description	Quantity	Unit	Labor-Hours	Material
Remove interior door	1.000	Ea.	0.400	0.00
Remove door frame	1.000	Ea.	0.500	0.00
Saw cutting block wall, per inch of depth	64.000	L.F.	4.096	21.12
Remove masonry partition	9.000	S.F.	0.444	0.00
Labor minimum for masonry repairs	1.000	Job	8.000	0.00
Steel lintel, 4" x 3-1/2" x 1/4", 5'-0" long	1.000	Ea.	0.381	15.80
Interior door frame	18.000	L.F.	0.768	71.82
Interior wood door, birch face, 3'-0" x 6'-8"	1.000	Ea.	1.143	87.50
Hinges	1.500	Pr.	0.000	66.75
Lever-handled lockset	1.000	Ea.	0.800	121.00
Threshold	1.000	Ea.	0.400	30.00
Paint door	1.000	Ea.	1.600	10.30
Painting	68.000	S.F.	1.727	14.96
Totals			**20.259**	**439.25**

Total per each including general contractor's overhead and profit — **$2,249**

Widen existing brick/block opening & replace w/ metal door

Description	Quantity	Unit	Labor-Hours	Material
Remove interior door	1.000	Ea.	0.400	0.00
Remove door frame	1.000	Ea.	0.500	0.00
Saw cutting block wall, per inch of depth	64.000	L.F.	4.096	21.12
Remove masonry partition	9.000	S.F.	0.444	0.00
Labor minimum for masonry repairs	1.000	Job	8.000	0.00
Steel lintel, 4" x 3-1/2" x 1/4", 5'-0" long	1.000	Ea.	0.381	15.80
Interior door frame	1.000	Ea.	1.000	75.00
Hollow metal flush door, 3'-0" x 6'-8"	1.000	Ea.	0.941	166.00
Hinges	1.500	Pr.	0.000	66.75
Lever-handled lockset	1.000	Ea.	0.800	121.00
Threshold	1.000	Ea.	0.400	30.00
Paint door	1.000	Ea.	1.600	10.30
Painting	68.000	S.F.	1.727	14.96
Totals			**20.289**	**520.93**

Total per each including general contractor's overhead and profit — **$2,391**

Widen existing brick/block opening & replace w/ double solid wood door

Description	Quantity	Unit	Labor-Hours	Material
Remove interior door	1.000	Ea.	0.400	0.00
Remove door frame	1.000	Ea.	0.500	0.00
Saw cutting block wall, per inch of depth	80.000	L.F.	5.120	26.40
Remove masonry partition	23.000	S.F.	1.136	0.00
Labor minimum for masonry repairs	1.000	Job	8.000	0.00
Steel lintel, 4" x 3-1/2" x 1/4", 9'-0" long	1.000	Ea.	0.667	28.50
Interior door frame	20.000	L.F.	0.853	79.80
Interior wood door, birch face, 3'-0" x 6'-8"	1.000	Ea.	1.143	87.50
Interior wood door, birch face, 2'-6" x 6'-8"	1.000	Ea.	1.067	80.00
Add for solid core door	2.000	Ea.	0.000	59.00
Hinges	3.000	Pr.	0.000	133.50
Lever-handled lockset	1.000	Ea.	0.800	121.00
Threshold	2.000	Ea.	0.800	60.00
Paint door	2.000	Ea.	3.200	20.60
Painting	80.000	S.F.	2.032	17.60
Totals			**25.718**	**713.90**

Total per each including general contractor's overhead and profit **$3,123**

Widen existing brick/block opening & replace w/ double hollow wood door

Description	Quantity	Unit	Labor-Hours	Material
Remove interior door	1.000	Ea.	0.400	0.00
Remove door frame	1.000	Ea.	0.500	0.00
Saw cutting block wall, per inch of depth	80.000	L.F.	5.120	26.40
Remove masonry partition	23.000	S.F.	1.136	0.00
Labor minimum for masonry repairs	1.000	Job	8.000	0.00
Steel lintel, 4" x 3-1/2" x 1/4", 9'-0" long	1.000	Ea.	0.667	28.50
Interior door frame	20.000	L.F.	0.853	79.80
Interior wood door, birch face, 3'-0" x 6'-8"	1.000	Ea.	1.143	87.50
Interior wood door, birch face, 2'-6" x 6'-8"	1.000	Ea.	1.067	80.00
Hinges	3.000	Pr.	0.000	133.50
Lever-handled lockset	1.000	Ea.	0.800	121.00
Threshold	2.000	Ea.	0.800	60.00
Paint door	2.000	Ea.	3.200	20.60
Painting	80.000	S.F.	2.032	17.60
Totals			**25.718**	**654.90**

Total per each including general contractor's overhead and profit **$3,023**

25. Modify Existing Double-Leaf Doors *(continued)*

Widen existing brick/block opening & replace w/ double hollow metal door

Description	Quantity	Unit	Labor-Hours	Material
Remove interior door	1.000	Ea.	0.400	0.00
Remove door frame	1.000	Ea.	0.500	0.00
Saw cutting block wall, per inch of depth	80.000	L.F.	5.120	26.40
Remove masonry partition	23.000	S.F.	1.136	0.00
Labor minimum for masonry repairs	1.000	Job	8.000	0.00
Steel lintel, 4" x 3-1/2" x 1/4", 9'-0" long	1.000	Ea.	0.667	28.50
Interior door frame	1.000	Ea.	1.143	90.50
Hollow metal flush door, 3'-0" x 6'-8"	1.000	Ea.	0.941	166.00
Hollow metal flush door, 2'-0" x 6'-8"	1.000	Ea.	0.800	159.00
Hinges	3.000	Pr.	0.000	133.50
Lever-handled lockset	1.000	Ea.	0.800	121.00
Threshold	2.000	Ea.	0.800	60.00
Paint door	2.000	Ea.	3.200	20.60
Painting	80.000	S.F.	2.032	17.60
Totals			**25.539**	**823.10**

Total per each including general contractor's overhead and profit **$3,295**

Widen existing stud wall opening & replace w/ solid wood door

Description	Quantity	Unit	Labor-Hours	Material
Remove interior door	1.000	Ea.	0.400	0.00
Remove door frame	1.000	Ea.	0.500	0.00
Remove stud wall partition	7.000	S.F.	0.323	0.00
Interior door frame	18.000	L.F.	0.768	71.82
Interior wood door, birch face, 3'-0" x 6'-8"	1.000	Ea.	1.143	87.50
Add for solid core door	1.000	Ea.	0.000	29.50
Hinges	1.500	Pr.	0.000	66.75
Lever-handled lockset	1.000	Ea.	0.800	121.00
Threshold	1.000	Ea.	0.400	30.00
Paint door	1.000	Ea.	1.600	10.30
Painting	68.000	S.F.	1.727	14.96
Totals			**7.661**	**431.83**

Total per each including general contractor's overhead and profit **$1,254**

Widen existing stud wall opening & replace w/ hollow wood door

Description	Quantity	Unit	Labor-Hours	Material
Remove interior door	1.000	Ea.	0.400	0.00
Remove door frame	1.000	Ea.	0.500	0.00
Remove stud wall partition	7.000	S.F.	0.323	0.00
Interior door frame	18.000	L.F.	0.768	71.82
Interior wood door, birch face, 3'-0" x 6'-8"	1.000	Ea.	1.143	87.50
Hinges	1.500	Pr.	0.000	66.75
Lever-handled lockset	1.000	Ea.	0.800	121.00
Threshold	1.000	Ea.	0.400	30.00
Paint door	1.000	Ea.	1.600	10.30
Painting	68.000	S.F.	1.727	14.96
Totals			**7.661**	**402.33**

Total per each including general contractor's overhead and profit **$1,203**

Widen existing stud wall opening & replace w/ metal door

Description	Quantity	Unit	Labor-Hours	Material
Remove interior door	1.000	Ea.	0.400	0.00
Remove door frame	1.000	Ea.	0.500	0.00
Remove stud wall partition	7.000	S.F.	0.323	0.00
Interior door frame	1.000	Ea.	1.000	75.00
Hollow metal flush door, 3'-0" x 6'-8"	1.000	Ea.	0.941	166.00
Hinges	1.500	Pr.	0.000	66.75
Lever-handled lockset	1.000	Ea.	0.800	121.00
Threshold	1.000	Ea.	0.400	30.00
Paint door	1.000	Ea.	1.600	10.30
Painting	68.000	S.F.	1.727	14.96
Totals			**7.691**	**484.01**

Total per each including general contractor's overhead and profit **$1,345**

Widen existing stud wall opening & replace w/ double solid wood door

Description	Quantity	Unit	Labor-Hours	Material
Remove interior door	1.000	Ea.	0.400	0.00
Remove door frame	1.000	Ea.	0.500	0.00
Remove stud wall partition	21.000	S.F.	0.969	0.00
Interior door frame	20.000	L.F.	0.853	79.80
Interior wood door, birch face, 3'-0" x 6'-8"	1.000	Ea.	1.143	87.50
Interior wood door, birch face, 2'-6" x 6'-8"	1.000	Ea.	1.067	80.00
Add for solid core door	2.000	Ea.	0.000	59.00
Hinges	3.000	Pr.	0.000	133.50
Lever-handled lockset	1.000	Ea.	0.800	121.00
Threshold	2.000	Ea.	0.800	60.00
Paint door	2.000	Ea.	3.200	20.60
Painting	80.000	S.F.	2.032	17.60
Totals			**11.764**	**659.00**

Total per each including general contractor's overhead and profit **$1,914**

25. Modify Existing Double-Leaf Doors *(continued)*

Widen existing stud wall opening & replace w/ double hollow wood door

Description	Quantity	Unit	Labor-Hours	Material
Remove interior door	1.000	Ea.	0.400	0.00
Remove door frame	1.000	Ea.	0.500	0.00
Remove stud wall partition	21.000	S.F.	0.969	0.00
Interior door frame	20.000	L.F.	0.853	79.80
Interior wood door, birch face, 3'-0" x 6'-8"	1.000	Ea.	1.143	87.50
Interior wood door, birch face, 2'-6" x 6'-8"	1.000	Ea.	1.067	80.00
Hinges	3.000	Pr.	0.000	133.50
Lever-handled lockset	1.000	Ea.	0.800	121.00
Threshold	2.000	Ea.	0.800	60.00
Paint door	2.000	Ea.	3.200	20.60
Painting	80.000	S.F.	2.032	17.60
Totals			**11.764**	**600.00**

Total per each including general contractor's overhead and profit   **$1,813**

Widen existing stud wall opening & replace w/ double hollow metal door

Description	Quantity	Unit	Labor-Hours	Material
Remove interior door	1.000	Ea.	0.400	0.00
Remove door frame	1.000	Ea.	0.500	0.00
Remove stud wall partition	21.000	S.F.	0.969	0.00
Interior door frame	1.000	Ea.	1.143	90.50
Hollow metal flush door, 3'-0" x 6'-8"	1.000	Ea.	0.941	166.00
Hollow metal flush door, 2'-0" x 6'-8"	1.000	Ea.	0.800	159.00
Hinges	3.000	Pr.	0.000	133.50
Lever-handled lockset	1.000	Ea.	0.800	121.00
Threshold	2.000	Ea.	0.800	60.00
Paint door	2.000	Ea.	3.200	20.60
Painting	80.000	S.F.	2.032	17.60
Totals			**11.585**	**768.20**

Total per each including general contractor's overhead and profit   **$2,086**

Install automatic door opener

Description	Quantity	Unit	Labor-Hours	Material
Cutout demolition of partition	1.000	Ea.	0.333	0.00
Conductor fished to nearby junction box	0.100	C.L.F.	0.333	2.15
Junction box	1.000	Ea.	0.400	2.10
Automatic opener, button operation	1.000	Ea.	5.333	355.00
Totals			**6.399**	**359.25**

Total per each including general contractor's overhead and profit   **$1,083**

Remove doors

Description	Quantity	Unit	Labor-Hours	Material
Remove interior door	1.000	Ea.	0.400	0.00
Totals			**0.400**	**0.00**

Total per each including general
contractor's overhead and profit **$23**

Install magnetic hold-open devices

Description	Quantity	Unit	Labor-Hours	Material
Magnetic door holder	1.000	Ea.	2.000	77.50
Cutout demolition of partition	1.000	Ea.	0.333	0.00
Conductor fished to nearby junction box	0.100	C.L.F.	0.333	2.15
Junction box	1.000	Ea.	0.400	2.10
Totals			**3.066**	**81.75**

Total per each including general
contractor's overhead and profit **$387**

Copyright Reed Construction Data, 2004

26
Install Sliding Door

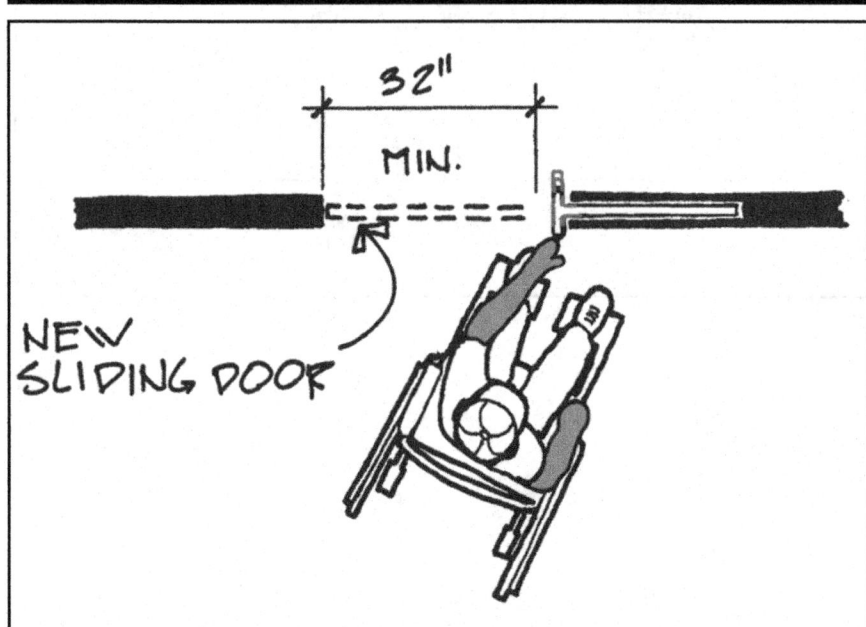

32"

MIN.

NEW
SLIDING DOOR

The installation of an accessible sliding door can create an accessible doorway where door swings might otherwise prevent access. Sliding doors can fit within the width of a standard stud wall, and can be a useful and creative method of creating an accessible route between two spaces when clearances are tight.

ADAAG References
404.2.4.2, Doorways without Doors or Gates, Sliding Doors, and Folding Doors
See Project 20 for References.

Where Applicable
Doors to areas on accessible routes of travel where swinging doors block required maneuvering space.

Design Requirements
See Project 20 for design requirements.

Design Suggestions
To meet the 32" clear opening requirement, expect to install a 36" door. No sliding door hardware exists that meets the regulations when the door is fully open. There are several ways to overcome this problem. One is to install blocking — either on the door or in the pocket — that keeps the door from opening all the way, leaving 2" to 3" exposed for projection hardware. Another is to apply a piece of trim the full length of the door, which serves as a ledge for a person to grab when the door is open, and, when closed, fits flush against the opposite jamb.

Key Items
Door, door frame, track hardware, new structural members for opening, and wall/floor finishes to match existing.

Level of Difficulty

Moderate to high. Involves demolition, bracing of new opening, structural framing, finish work.

Estimates

Install solid core wood pocket door in metal stud wall

Description	Quantity	Unit	Labor-Hours	Material
Remove interior door	1.000	Ea.	0.400	0.00
Remove door frame	1.000	Ea.	0.500	0.00
Remove stud wall partition	21.000	S.F.	0.969	0.00
Pocket door frame	1.000	Ea.	1.000	53.50
Gypsum board finish on pocket door frame	64.000	S.F.	1.061	14.72
Cased opening jamb and header	10.000	L.F.	0.400	9.70
Door trim set, 1 head, 2 sides, 2-1/2" pine	2.000	Opng.	2.712	28.90
Interior wood door, birch face, 3'-0" x 6'-8"	1.000	Ea.	1.143	87.50
Add for solid core door	1.000	Ea.	0.000	29.50
Lever-handled lockset	1.000	Ea.	0.800	121.00
Totals			**8.985**	**344.82**

Total per each including general contractor's overhead and profit **$1,225**

Install solid core wood pocket door in wood stud wall

Description	Quantity	Unit	Labor-Hours	Material
Remove interior door	1.000	Ea.	0.400	0.00
Remove door frame	1.000	Ea.	0.500	0.00
Remove stud wall partition	21.000	S.F.	0.969	0.00
Pocket door frame	1.000	Ea.	1.000	53.50
Gypsum board finish on pocket door frame	64.000	S.F.	1.061	14.72
Cased opening jamb and header	10.000	L.F.	0.400	9.70
Door trim set, 1 head, 2 sides, 2-1/2" pine	2.000	Opng.	2.712	28.90
Interior wood door, birch face, 3'-0" x 6'-8"	1.000	Ea.	1.143	87.50
Add for solid core door	1.000	Ea.	0.000	29.50
Lever-handled lockset	1.000	Ea.	0.800	121.00
Totals			**8.985**	**344.82**

Total per each including general contractor's overhead and profit **$1,225**

Install hollow core wood pocket door in metal stud wall

Description	Quantity	Unit	Labor-Hours	Material
Remove interior door	1.000	Ea.	0.400	0.00
Remove door frame	1.000	Ea.	0.500	0.00
Remove stud wall partition	21.000	S.F.	0.969	0.00
Pocket door frame	1.000	Ea.	1.000	53.50
Gypsum board finish on pocket door frame	64.000	S.F.	1.061	14.72
Cased opening jamb and header	10.000	L.F.	0.400	9.70
Door trim set, 1 head, 2 sides, 2-1/2" pine	2.000	Opng.	2.712	28.90
Interior wood door, birch face, 3'-0" x 6'-8"	1.000	Ea.	1.143	87.50
Lever-handled lockset	1.000	Ea.	0.800	121.00
Totals			**8.985**	**315.32**

Total per each including general contractor's overhead and profit **$1,174**

Install hollow core wood pocket door in wood stud wall

Description	Quantity	Unit	Labor-Hours	Material
Remove interior door	1.000	Ea.	0.400	0.00
Remove door frame	1.000	Ea.	0.500	0.00
Remove stud wall partition	21.000	S.F.	0.969	0.00
Pocket door frame	1.000	Ea.	1.000	53.50
Gypsum board finish on pocket door frame	64.000	S.F.	1.061	14.72
Cased opening jamb and header	10.000	L.F.	0.400	9.70
Door trim set, 1 head, 2 sides, 2-1/2" pine	2.000	Opng.	2.712	28.90
Interior wood door, birch face, 3'-0" x 6'-8"	1.000	Ea.	1.143	87.50
Lever-handled lockset	1.000	Ea.	0.800	121.00
Totals			**8.985**	**315.32**

Total per each including general contractor's overhead and profit **$1,174**

Notes

27
Enlarge Vestibule

Vestibules at building entrances serve the function of conserving energy, and those at restrooms create visual privacy. However, a small vestibule can create a barrier for people who need larger maneuvering space. A vestibule with maneuvering space between doors not only fulfills the energy conservation and visual functions of a smaller vestibule, but also makes the area accessible for all users.

ADAAG References

404.2.6, Two Doors in Series
See Project 20 for more references.

Where Applicable

All vestibules on accessible routes with insufficient maneuvering clearances.

Design Requirements

- 48" clear minimum between door swings.

See Project 20 for additional design requirements.

Design Suggestions

Vestibules are most often built at building entrances and toilet rooms. It might be possible to remove the inner set of doors of a vestibule to enlarge it. However, this is not recommended for a building entrance from an energy standpoint (and might possibly be prohibited by state or local building codes).

Reversing the door swings at either set of doors might solve wheelchair maneuvering clearance issues, but this is likely to be prohibited by local fire egress codes. Depending on the vestibule's configuration, it could be less expensive to extend it into the building, rather than adding to its existing face, which would require new foundation work.

Even less costly might be adding synchronized door openers, operable before one enters the vestibule. For

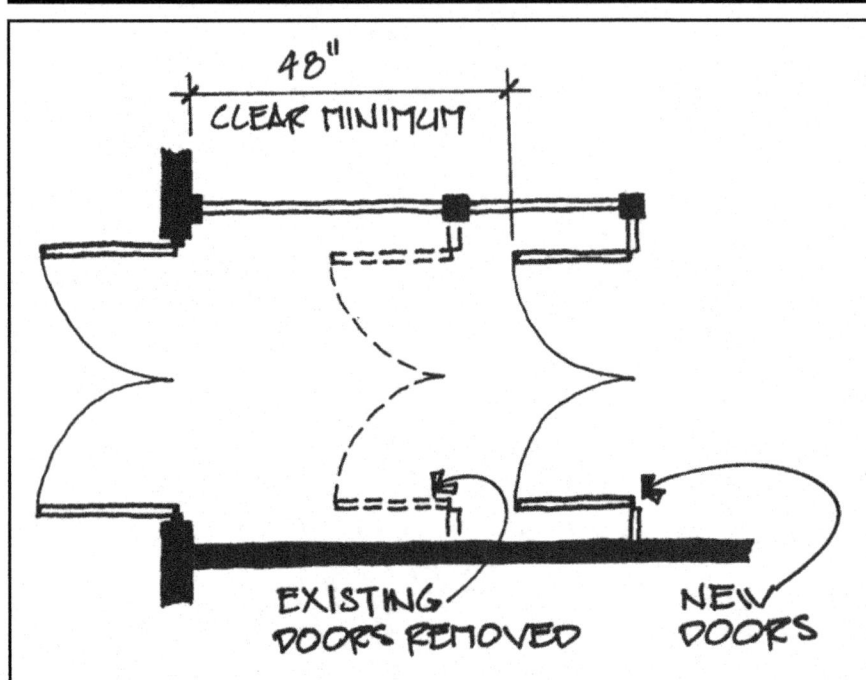

48"
CLEAR MINIMUM

EXISTING
DOORS REMOVED

NEW
DOORS

this option, stand-by power is required. Check first to see if the building inspector will allow it.

Key Items

New vestibule and doors, usually interior finish work on floor and wall surfaces, possibly new foundation work.

Level of Difficulty

High.

Estimates

Enlarge existing vestibule by adding on

Description	Quantity	Unit	Labor-Hours	Material
Excavation	8.000	C.Y.	1.185	0.00
Footing formwork	24.000	SFCA	1.584	17.28
Wall forms	72.000	SFCA	5.712	35.28
Reinforcing (@ 50 lbs./C.Y.)	0.100	Ton	1.067	54.00
Concrete, material only	4.000	C.Y.	0.000	290.00
Placing concrete	4.000	C.Y.	2.133	0.00
Backfill	4.000	C.Y.	2.286	0.00
Slab on grade, 4"	16.000	S.F.	0.312	15.04
Storefront system, commercial grade (with door)	96.000	S.F.	10.240	1,209.60
Flat roof framing	35.000	L.F.	0.448	18.20
Sheathing	25.000	S.F.	0.286	16.25
Fascia	18.000	L.F.	0.640	27.18
Cornice	18.000	L.F.	0.576	20.70
Drip edge, 8" girth aluminum	18.000	L.F.	0.360	5.40
Roll roofing, 1 layer 15# felt under mineral surfaced roofing	0.250	Sq.	0.519	9.75
Lead flashing	5.000	S.F.	0.296	14.00
Reglet	5.000	L.F.	0.178	4.70
Polyethylene backer rod	0.420	L.F.	0.730	3.84
Silicone sealant	42.000	L.F.	1.430	9.24
Totals			**29.982**	**1,750.46**

Total per square foot including general contractor's overhead and profit	**$321**
Total per each 16 square foot vestibule including general contractor's overhead and profit	**$5,129**

Remove vestibule doors

Description	Quantity	Unit	Labor-Hours	Material
Remove double doors	1.000	Ea.	0.500	0.00
Remove glazing	64.000	S.F.	3.413	0.00
Totals			**3.913**	**0.00**

Total per set including general contractor's overhead and profit	**$226**

Copyright Reed Construction Data, 2004

27. Enlarge Vestibule *(continued)*

Extend vestibule inside building

Description	Quantity	Unit	Labor-Hours	Material
Remove double doors	1.000	Ea.	0.500	0.00
Remove glazing	64.000	S.F.	3.413	0.00
Storefront system, commercial grade	24.000	S.F.	2.560	302.40
6' x 7' door with hardware	1.000	Ea.	12.308	795.00
Totals			**18.781**	**1,097.40**

Total per set including general contractor's overhead and profit   **$3,439**

Notes

28
Modify Buzzer or Intercom

Even when a facility's entrance has door widths and maneuvering spaces that meet ADAAG, if an individual is unable to reach the communication panel to gain entry, as far as they are concerned, the building is inaccessible. Locating the building's buzzers or intercoms at accessible heights, in accessible locations, and with accessible controls creates an entrance that is usable by a much larger group of people with a much wider range of abilities.

ADAAG References

230, Two-Way Communication Systems
305, Clear Floor or Ground Space
308, Reach Ranges
309, Operable Parts
708, Two-Way Communication Systems

Where Applicable

All accessible entrances with door opening controls, identification devices, doorbells, buzzers, or intercoms.

Design Requirements

- Intercom, buzzer, or control within reach range (48" a.f.f. maximum).
- Accessible controls (operable with a closed fist).
- 30" × 48" minimum clear floor space in front of device.
- No objects below the device that block access (plants, ashtrays, etc.).
- If there is a display, both visual and audible display required.
- TTY capabilities for residential dwelling unit communications systems.
- If handset provided, 29" cord, minimum.

Design Suggestions

If the communication panel is being lowered, it might be possible to make other alterations at the same time, such as adding larger buttons or characters. It may be possible to surface-mount controls to reduce costs. Large push-plates are easier to operate than buttons. If an intercom panel is being replaced or lowered, non-voice communication must be installed as well.

Consider installing devices that do not use handsets, and are easily used by more people.

Key Items

Electrical and finish work to match existing.

Level of Difficulty

Moderate to high, depending on the wall material and whether the panel is flush-mounted or installed on surface.

Estimates

Lower existing buzzer/intercom panel in 8″ block wall (painted)

Description	Quantity	Unit	Labor-Hours	Material
Remove panel and fixture	1.000	Ea.	0.031	0.00
Remove junction box	1.000	Ea.	0.100	0.00
Cutout demolition of 8″ block	1.000	Ea.	1.481	0.00
Re-install junction box	1.000	Ea.	0.444	2.10
Install outlet	1.000	Ea.	0.296	7.05
Install plate	1.000	Ea.	0.100	1.76
Miscellaneous materials for block repair and painting	1.000	Job	0.111	1.89
Repair block	1.000	Job	2.000	0.00
Paint block - minimum	1.000	Job	2.000	0.00
Totals			6.563	12.80

Total per each including general contractor's overhead and profit ___ **$448**

28. Modify Buzzer or Intercom *(continued)*

Lower existing buzzer/intercom panel in gypsum board/metal stud wall

Description	Quantity	Unit	Labor-Hours	Material
Remove panel and fixture	1.000	Ea.	0.031	0.00
Remove junction box	1.000	Ea.	0.100	0.00
Cutout demolition of partition	1.000	Ea.	0.333	0.00
Re-install junction box	1.000	Ea.	0.444	2.10
Install outlet	1.000	Ea.	0.296	7.05
Install plate	1.000	Ea.	0.100	1.76
Miscellaneous materials for gypsum board painting and repair	1.000	Job	0.800	0.14
Repair gypsum board	1.000	Job	4.000	0.00
Paint gypsum board – minimum	1.000	Job	2.000	0.00
Totals			**8.104**	**11.05**

Total per each including general
contractor's overhead and profit **$597**

Notes

29
Remove or Relocate Partitions

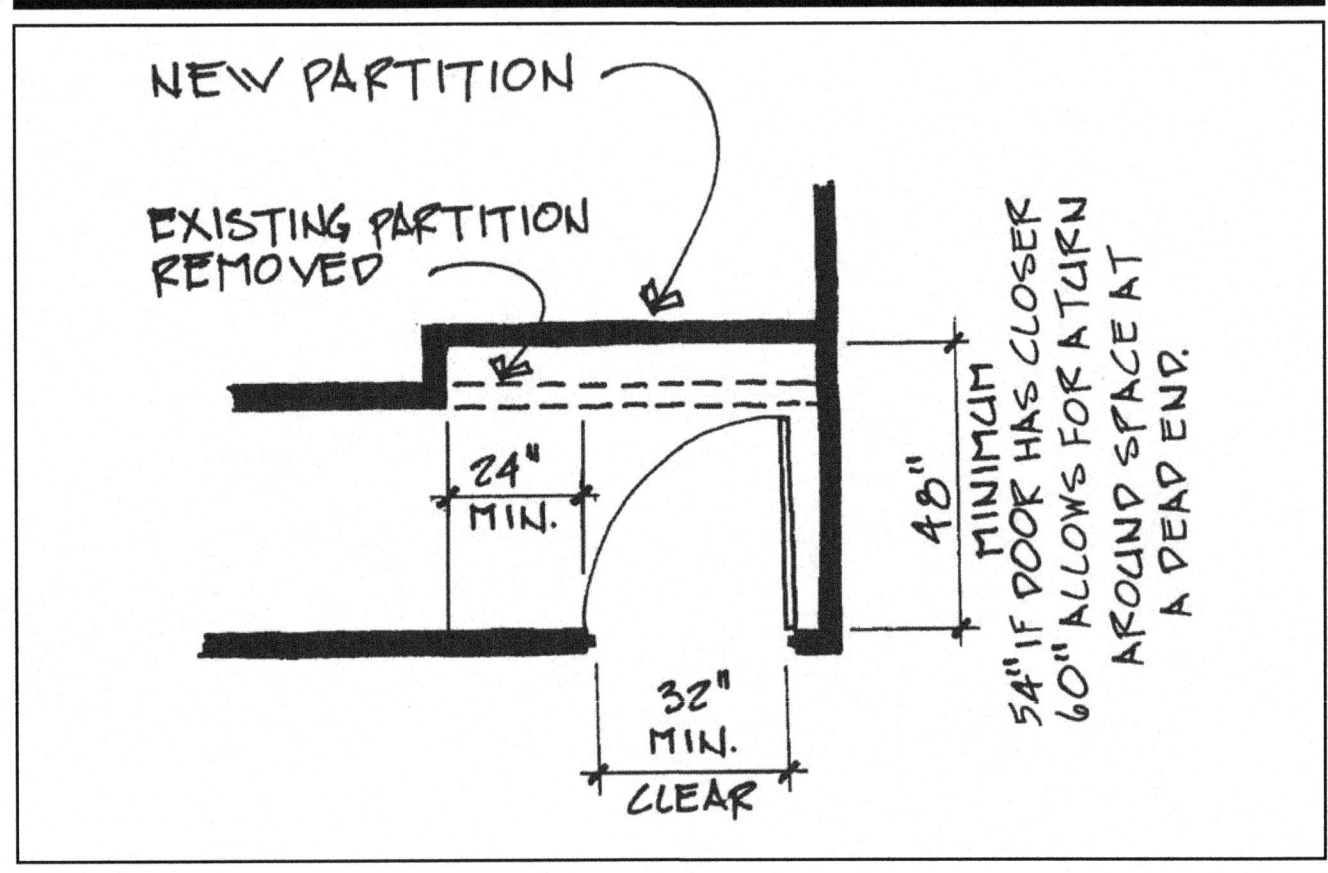

NEW PARTITION

EXISTING PARTITION REMOVED

24" MIN.

32" MIN. CLEAR

48" MINIMUM 54" IF DOOR HAS CLOSER 60" ALLOWS FOR A TURN AROUND SPACE AT A DEAD END.

Many accessibility problems can occur within an interior route of travel, such as changes in level, protruding objects, slippery surfaces, or insufficient lighting, but tight maneuvering spaces can make the route especially difficult to use. Removing or relocating a few linear feet of partition can often make all the difference, as well as reducing wear and tear on doors, trim, and walls.

ADAAG References
302, Floor or Ground Surfaces
303, Changes in Level

304, Turning Space
402, Accessible Routes
403.5, Clearances
404.2.4, Maneuvering Clearances at Doors
404.2.6, Two Doors in Series
405.7, Ramp Landings

Where Applicable

Interior routes of travel with insufficient maneuvering space.

Design Requirements

- 36" wide accessible route. Route may narrow to 32" for a distance of 24" or less, providing these narrow spots are at least 48" apart.
- Level maneuvering spaces on both sides of each door as required by 404.2.4 (between 42" and 60" from the door, depending on approach).

Design Suggestions

Although the width of an accessible route may be decreased to 32" through a doorway, level maneuvering space always must be provided. For ease of use and to protect wall surfaces, a 42" or more wide route of travel is recommended at 90° turns. Try to always have 60" by 60" turning space at the end of a hallway.

Key Items

New wall materials (studs, gypsum wallboard), new wall/ceiling finish materials to match existing.

Level of Difficulty

Moderate to high. Requires carpentry, electrical, and finish work.

Estimate

Remove 8-foot-high wood stud/gypsum board partition and replace (no ceiling work)

Description	Quantity	Unit	Labor-Hours	Material
Remove stud wall partition	1.000	S.F.	0.046	0.00
1/2" gypsum board on both sides of 2" x 4" wall	1.000	S.F.	0.052	0.80
Painting, 3 coats by roller	2.000	S.F.	0.025	0.28
Vinyl tile	0.130	S.F.	0.002	0.43
Totals			0.125	1.51

Total per square foot including general contractor's overhead and profit **$11**

30
Install Wing Walls

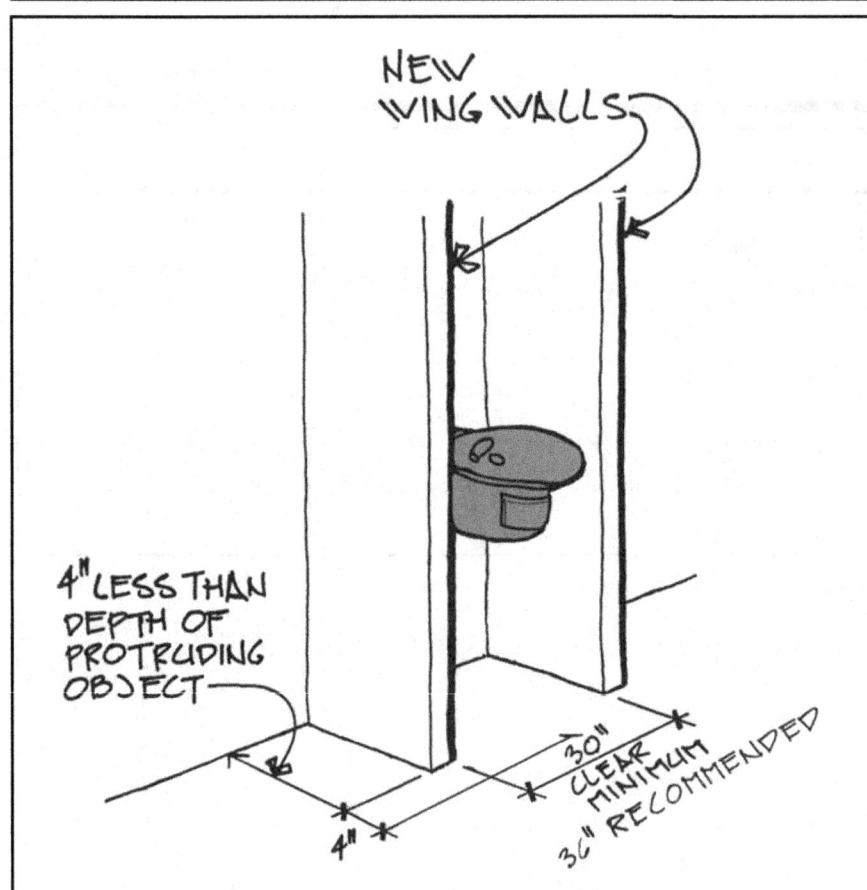

NEW WING WALLS

4" LESS THAN DEPTH OF PROTRUDING OBJECT

4"

30" CLEAR MINIMUM
36" RECOMMENDED

Where a building element such as a drinking fountain or public pay phone protrudes into a path of travel, it can present a hazard to people with visual impairments. Building wing walls on either side of it can be an effective and permanent solution to preventing danger without having to remove or relocate the object.

ADAAG References
305, Clear Floor or Ground Space
307, Protruding Objects

Where Applicable
Where an object protrudes more than 4" into the route of travel and is higher than 27" above the floor.

Design Requirements
• Wing walls must extend to within 4" of the edge of the protruding object.
• Wing walls that do not go to the floor must start below 27" a.f.f.

Design Suggestions
Wing walls can ensure that a protruding object complies with the ADA without replacing or removing the object. A 36" wide route of travel must be maintained. Ensure that wing walls do not intrude into 36" of clear dimension.

Wing walls need not start at the floor, and often it is more aesthetic to construct walls that start a foot or two above the floor and do not extend to the ceiling. If that is the case, they must start no higher than 27", although 20" is preferred. The minimum width between wing walls is 30", but if the remaining width of the accessible route is less than 60", it is quite likely that the wing walls will be damaged by turning wheelchairs.

Key Items
Studs, gypsum wallboard, finishes to match existing.

Level of Difficulty
Low to moderate.

Estimates

Install wing walls

Description	Quantity	Unit	Labor-Hours	Material
1/2" gypsum board on both sides of 2" x 4" wall	1.000	S.F.	0.052	0.80
Painting, 3 coats by roller	2.000	S.F.	0.025	0.28
Totals			0.077	1.08

Total per square foot including general contractor's overhead and profit _____ **$7**

31
Install Slip-Resistant Flooring Materials

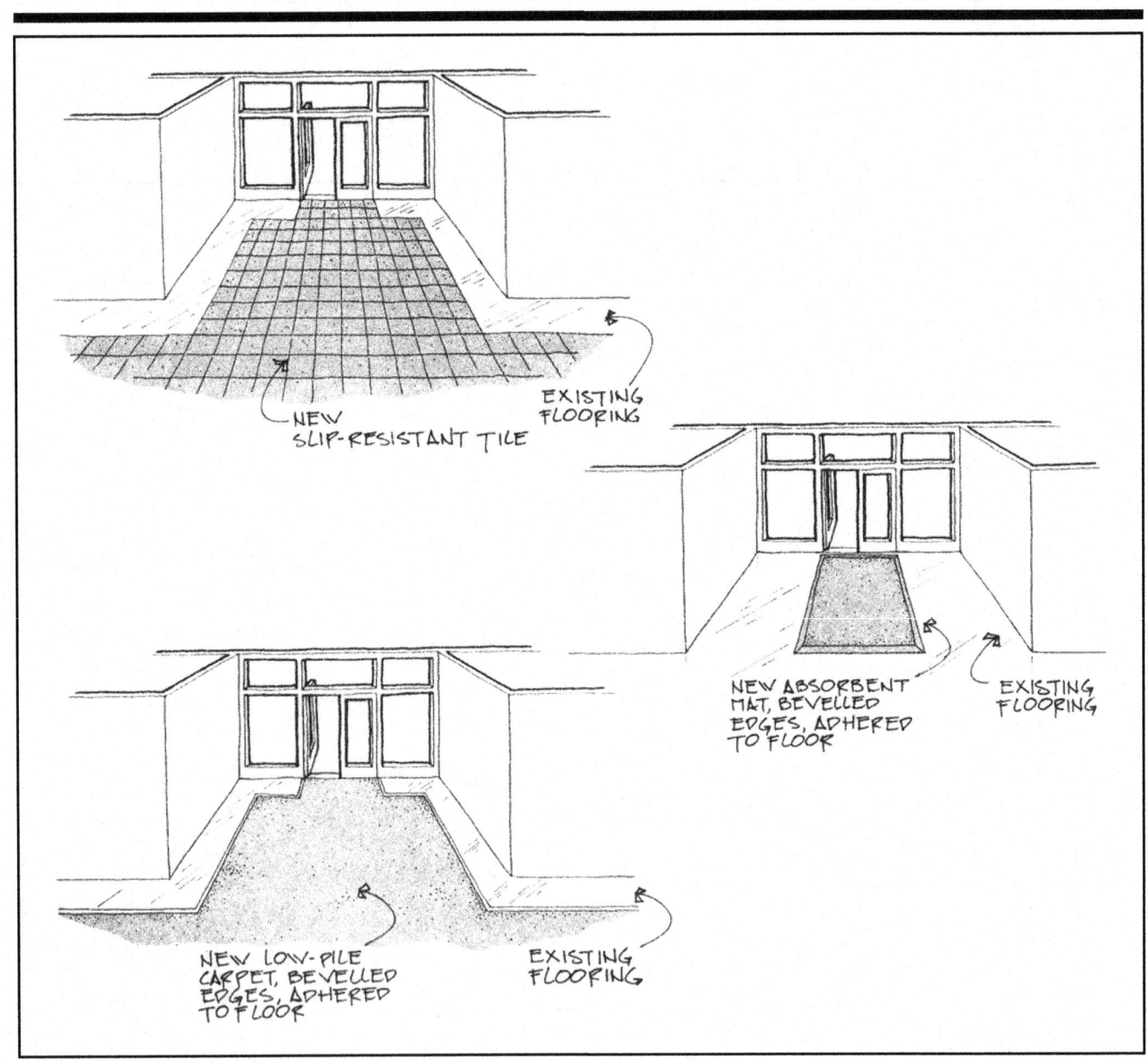

NEW
SLIP-RESISTANT TILE

EXISTING
FLOORING

NEW ABSORBENT
MAT, BEVELLED
EDGES, ADHERED
TO FLOOR

EXISTING
FLOORING

NEW LOW-PILE
CARPET, BEVELLED
EDGES, ADHERED
TO FLOOR

EXISTING
FLOORING

There is no real list of slip-resistant materials, since slip resistance is affected by many factors, such as the presence of water or slush on a floor surface and the slipperiness of the sole of a person's shoe. Some materials that are generally considered slippery are polished stone, polished terrazzo, and glazed tile with a glossy finish. Others, such as some untextured vinyl tiles designated as slip-resistant by the manufacturer, may be slippery when wet. Even carpets when used on ramps and stair nosings can sometimes be slippery, and should be checked.

ADAAG Reference

302, **Floor or Ground Surfaces**

Where Applicable

Routes of travel, particularly ramp surfaces and stair nosings and treads.

Design Requirements

- Flooring to be stable, firm, and slip-resistant. (Check manufacturer's specifications of surfacing materials for level surfaces prior to installation, if available.)
- Carpet securely attached, with a firm cushion or no pad. Maximum pile thickness 1/2".
- Floor mats securely attached.

Design Suggestions

There are two main ways to make a slippery surface more slip resistant: treating already-in-place surfaces, or installing new ones. If new flooring is being installed along a public route of travel, a material should be chosen that is designated by the manufacturer as slip-resistant. A static coefficient of friction of 0.6 is recommended for accessible routes and 0.8 for ramps. It is useful, however, to examine similar installations and obtain input from people with mobility impairments to help determine the actual slip resistance of a specific material in use.

There are at least four solutions:

1. Replace the flooring. This can be expensive, especially with stone or tile.
2. Cover the flooring along the entire route of travel. This is not as expensive as replacing the flooring, but can be difficult at doorways.
3. Cover the floor with carpets or mats at key areas, such as entrance doors (both inside and outside), main interior routes (to information desks or elevators),

elevator lobbies, and other highly used areas. Mats can be installed permanently, which is preferable, or only during inclement weather. The mats must be firmly attached to the floor, and must either have beveled edges or be lower than 1/4" in height.
4. Apply a slip-resistant coating to the floor. Terrazzo and vinyl tile can be coated with a non-glossy finish, but the actual slip-resistance of such coatings is unknown. They must also be applied on a regular basis. Manufacturer's specifications should be checked, and it is recommended that this solution be used in conjunction with slip-resistant mats or carpeting.

Key Items

New flooring material (such as rough stone, unglazed tile, or carpet). Coverings (carpet, carpet runners, or slip-resistant mats).

Level of Difficulty

Low, except for installing new stone or ceramic tile, which is moderate to high.

Estimates

Remove vinyl tile and replace with unglazed quarry tile

Description	Quantity	Unit	Labor-Hours	Material
Remove 12" x 12" tiles	1.000	S.F.	0.016	0.00
6" x 6" red quarry tile, mud set, 1/2" thick	1.000	S.F.	0.114	2.96
Totals			0.130	2.96

Total per square foot including general contractor's overhead and profit **$13**

31. Install Slip-Resistant Flooring Materials *(continued)*

Remove vinyl tile and replace with carpet

Description	Quantity	Unit	Labor-Hours	Material
Remove 12" x 12" tiles	9.000	S.F.	0.144	0.00
40 oz. nylon carpet	1.000	S.Y.	0.107	31.50
Totals			**0.251**	**31.50**

Total per square yard including general contractor's overhead and profit — **$68**

Install carpet runner with beveled edges

Description	Quantity	Unit	Labor-Hours	Material
Olefin carpet runner (3' x 60')	1.000	Ea.	2.133	196.00
Totals			**2.133**	**196.00**

Total per each including general contractor's overhead and profit — **$476**

Install absorbent mat

Description	Quantity	Unit	Labor-Hours	Material
Black rubber mat, with nosing, 3/8" thick	1.000	S.F.	0.052	14.00
Totals			**0.052**	**14.00**

Total per square foot including general contractor's overhead and profit — **$27**

Notes

32
Install New Elevator (Exterior Shaft)

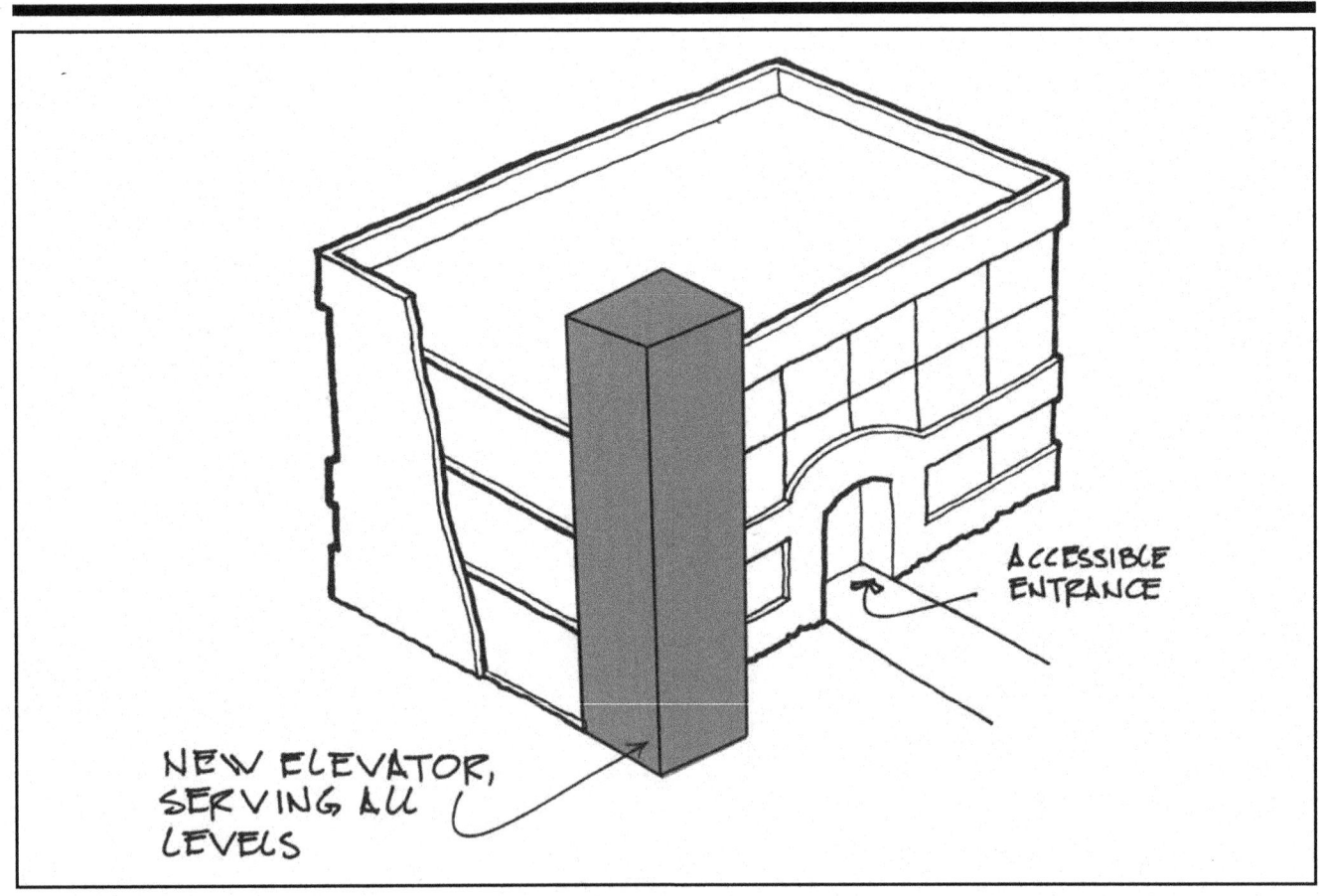

ACCESSIBLE ENTRANCE

NEW ELEVATOR, SERVING ALL LEVELS

No architectural modification for creating accessible vertical circulation is as effective as an elevator. Unlike a ramp, it is capable of bridging distances over several floors; unlike a wheelchair platform lift, it can be used by everyone. Installing an elevator against an existing building exterior wall allows for minimal disruption of the building's structure, may sometimes accommodate a larger shaft than an interior application, and can include an accessible exterior door at grade level. This is often an advantage when the first level of an existing building is a half level above grade. The disadvantage of an exterior elevator is that the elevator may be remote from the existing corridors.

ADAAG References

202, Existing Buildings
206.2.3, Accessible Routes, Multi-Story Buildings and Facilities
206.2.4, Accessible Routes, Spaces, and Elements
206.6, Elevators
216.7, Signs at Elevators
302, Floor or Ground Surfaces
304, Turning Space
305, Clear Floor or Ground Space
Table 404.2.4.2, Maneuvering Clearances
407, Elevators (in particular 407.2 Landings and 407.4, Car Requirements)
703, Signs

Where Applicable

Required in public accommodation and commercial facility new construction of buildings over two stories with at least one story exceeding 3,000 square feet and in additions to existing buildings conforming to the same conditions. Required in new construction of state and local government buildings, shopping centers, health care providers, transportation depots, and public buildings over one story. For Title II (state and local governments) facilities, elevators may be necessary to create program access.

Under Title III (public accommodations and commercial facilities), elevators may be required in renovations if an area of primary function is being modified, and alterations to the path of travel just to achieve access do not exceed 20% of total renovation costs.

Design Requirements

- Accessible route to the elevator and at all stops.
- Maneuvering space outside the elevator door for passing, reaching controls, and so forth.
- Cab dimensions ranging from 51" × 80" to 60" × 60", depending on door size and location.
- Compliance with all elevator cab signal, control, communication, and reach requirements (not cited here; consult ADAAG 407 for details). Elevator manufacturers will help you meet not only ADA, but local and state codes as well.
- ASME A17.1-2000 and strict building code requirements also apply.
- Where not all elevators in a bank comply, signs displaying the international access symbol are required clearly identifying those that are accessible.

Design Suggestions

Careful consideration must be given to elevator placement. Elevators should create accessible vertical circulation where it is most needed, but at the same time should serve all floors. If there are level changes within a building, the elevator should serve the most heavily used levels, and the possibility of creating accessible routes at level changes (such as ramps or lifts) should be examined.

Installation of an exterior elevator allows the building interior framing to remain intact, but will require bracing of the building façade and spandrel (edge) framing. The most efficient door configuration for a wheelchair user is a two-door cab, with the doors at opposite ends. This prevents the need for turns. A two-door cab with doors on adjacent sides works almost as well. Both of these options allow for half level access, often a necessity in existing buildings where the first floor is a half level above grade.

Key Items

Elevator shaft, cab, pit, mechanical room, excavation, foundations, electrical work. Accessible route to elevator and all stops.

Level of Difficulty

Very high. Requires major structural renovation, foundation and roof modifications, electrical preparation, and finish work. Necessitates design drawings, specifications, and shop drawings, approved by local building departments and/or elevator board.

Estimate

Add new elevator and shaft to building exterior

Description	Quantity	Unit	Labor-Hours	Material
Saw cutting exterior walls for elevator doorways (per inch of depth)	360.000	L.F.	19.199	115.20
Exterior wall demolition	54.000	S.F.	2.667	0.00
Excavation	15.000	C.Y.	2.222	0.00
Footing formwork	52.000	SFCA	3.431	37.44
Wall forms	156.000	SFCA	12.377	76.44
Reinforcing (@ 50 lbs./C.Y.)	0.150	Ton	1.600	81.00
Concrete, material only	6.000	C.Y.	0.000	435.00
Placing concrete	6.000	C.Y.	3.200	0.00
Backfill	5.000	C.Y.	2.857	0.00
Slab on grade, 4"	65.000	S.F.	1.267	61.10
Structural steel frame, plates, angles, and fasteners	3.000	Ton	23.762	3,900.00
Shaft wall, including light-gauge framing	1110.000	S.F.	80.730	987.90
Face brick, running bond	1110.000	S.F.	201.820	3,063.60
Scaffolding	11.100	C.S.F.	11.100	271.95
Elevator pit ladder	4.000	V.L.F.	1.506	104.00
3" deep, 22-gauge metal decking	65.000	S.F.	0.650	56.55
6" aluminum gravel stop	40.000	L.F.	2.370	145.60
2" thick polyisocyanurate roof insulation	65.000	S.F.	0.473	31.20
EPDM roofing with ballast	65.000	S.F.	0.510	44.20
Roofing minimum	1.000	Job	4.000	0.00
Hydraulic passenger elevator (3 stop)	1.000	Ea.	242.000	33,900.00
Totals			**617.741**	**43,311.18**

Total per each including general contractor's overhead and profit **$123,643**

Notes

33
Install New Elevator (Interior Shaft)

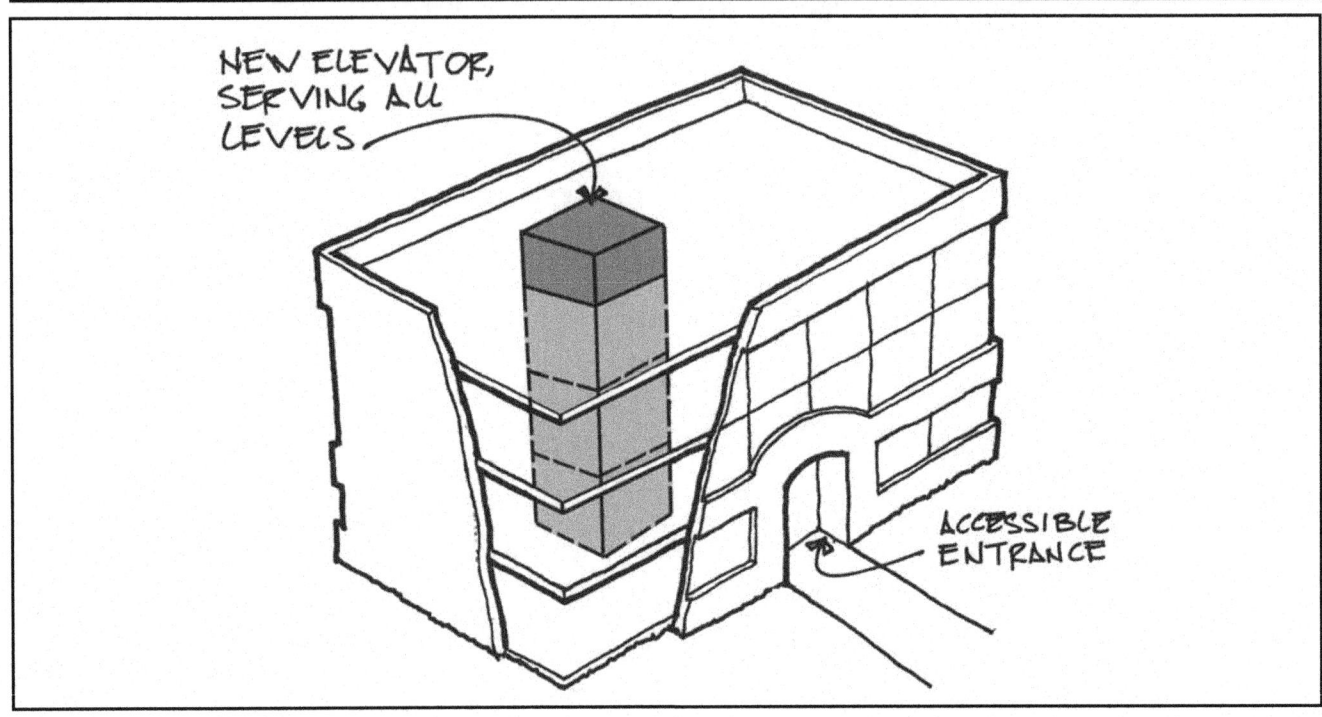

NEW ELEVATOR, SERVING ALL LEVELS

ACCESSIBLE ENTRANCE

The same benefits for installing an elevator on the exterior of a building apply to an interior installation: virtually unlimited height travel and unrestricted access use. Installation of an elevator in the interior of an existing building does not require a weatherproof shaft and does not block an exterior route of travel. Interior elevators can be monitored more easily if security is an issue. However, installation of an interior elevator may require extensive reinforcing of the building frame and nearly always requires some reconfiguration of the circulation inside the building.

ADAAG References

See Project 32 for References.

Where Applicable

Required in public accommodation and commercial facility new construction of buildings over two stories with at least one story exceeding 3,000 square feet and in additions to existing buildings conforming to the same conditions. Required in new construction of state and local government buildings, shopping centers, health care providers, transportation depots, and public buildings over one story. For Title II (state and local governments) facilities, elevators may be necessary to create program access.

Under Title III (public accommodations and commercial

facilities), elevators may be required in renovations if an area of primary function is being modified, and alterations to the path of travel just to achieve access do not exceed 20% of total renovation costs.

Design Requirements

See Project 32 for Design Requirements.

Design Suggestions

Careful consideration must be given to elevator placement so that vertical circulation is where it is most needed, although it is usually more important to access all floors. If there are level changes within a building, the elevator should serve the most heavily used ones, and the possibility of creating accessible routes at level changes (such as ramps or lifts) should be examined.

Often, a new elevator can be installed in a stairwell by removing the stairs. This has several advantages: it minimizes structural changes and plugs the elevator into the existing circulation system.

When the elevator is the minimum size, the most efficient door configuration for a wheelchair user is a two-door cab, with doors at opposite ends. This prevents the need for turns. A two-door cab with doors on adjacent walls works almost as well.

Key Items

Elevator shaft, cab, pit, mechanical room. Electrical and roofing work. Accessible route to elevator and all stops.

Level of Difficulty

Very high. Requires major structural renovation, foundation and roof modifications, electrical preparation, and finish work. Necessitates design drawings, specifications, and shop drawings, approved by local building departments and/or elevator board.

Estimate

Add new elevator and shaft to building interior

Description	Quantity	Unit	Labor-Hours	Material
Cut out demolition, 6" thick slab on grade	65.000	S.F.	24.762	0.00
Cut out demolition, floor slabs	195.000	S.F.	156.000	0.00
Shoring of floor openings	0.720	M.B.F.	15.709	529.20
Hand excavation	15.000	C.Y.	30.000	0.00
Footing formwork	26.000	SFCA	1.715	18.72
Wall forms	78.000	SFCA	6.189	38.22
Reinforcing (@ 50 lbs./C.Y.)	0.150	Ton	1.600	81.00
Concrete, material only	7.000	C.Y.	0.000	507.50
Placing concrete (pumped)	6.000	C.Y.	3.840	0.00
Backfill	5.000	C.Y.	2.857	0.00
Slab on grade, 4"	1.000	C.Y.	0.492	0.00
Structural steel frame, plates, angles, and fasteners	5.000	Ton	39.604	6,500.00
Shaft wall, including light-gauge framing	1320.000	S.F.	96.004	1,174.80
Face brick, running bond	330.000	S.F.	60.001	910.80
Scaffolding	0.330	C.S.F.	0.330	8.09
Elevator pit ladder	4.000	V.L.F.	1.506	104.00
3" deep, 22-gauge metal decking	65.000	S.F.	0.650	56.55
6" aluminum gravel stop	40.000	L.F.	2.370	145.60
2" thick polyisocyanurate roof insulation	65.000	S.F.	0.473	31.20
EPDM roofing with ballast	65.000	S.F.	0.510	44.20
Roofing labor minimum	1.000	Job	4.000	0.00
Painting, 3 coats by roller	990.000	S.F.	12.187	138.60
Electric passenger elevator	1.000	Ea.	320.000	65,500.00
Totals			**780.799**	**75,788.48**

Total per each including general contractor's overhead and profit **$190,542**

34
Modify Existing Elevator Hall Signals and Signs

Often a noisy elevator door is the only clue people with visual impairments have to tell them that an elevator has arrived. This, however, does not indicate which direction the elevator is traveling, and is of little use to someone with hearing and visual impairments. Both visual and audible hall signals are vital to let all users know where an elevator is and which direction it is going, especially at elevator banks with multiple cars.

ADAAG References
216.7, Elevators Signs
305, Clear Floor or Ground Space
308, Reach Ranges
309, Operable Parts
407, Elevators (in particular, 407.2, Landing Requirements)
703, Signs

Where Applicable
All passenger elevator lobbies.

Design Requirements
- Compliance with all elevator requirements (not cited here; consult ADAAG 407 for details). Elevator manufacturers will help you meet the ADA, and local and state building codes.
- Visible signals that register on the call button that a call has been made and turn off when answered.
- Visual and audible signals at each hoistway entrance to indicate which car is answering call, centered at 72" a.f.f., and at least 2-1/2" in the vertical dimension, visible from the call button area.
- Audible signals indicating car arrival, sounding once for up, twice for down, or else with a verbal message.
- Call buttons centered at 42" a.f.f.
- Call buttons at least 3/4" in diameter, raised or flush with the surrounding surface. No obstructions projecting more than 4" below call buttons.

- Hoistway entrances, both sides: floor designations in raised characters 2" high (ADAAG 703.2), and also in Braille (703.3); on the main entry level, tactile star on both jambs. Locate floor designations at 60" a.f.f.
- Where not all elevators in a bank are accessible, signs displaying the international access symbol are required identifying those that are.

Design Suggestions

In addition to visual elevator signals, ensure that lighting outside the elevator is sufficient to see all signage. Make the lobby as generous as possible. Plan for ashtrays, plants, and benches that will not impinge into the required clearances.

Key Items

Visual signal(s), audible signal(s), call buttons.

Level of Difficulty

Moderate. Signal and call button alterations must be done by elevator installer. Finish work to wall is also required.

Estimates

Add hallway visual signal

Description	Quantity	Unit	Labor-Hours	Material
Add visual signal	1.000	Ea.	Price includes an elevator contractor's materials and labor	
Totals				

Total per each including general contractor's overhead and profit **$725**

Add hallway audible signal

Description	Quantity	Unit	Labor-Hours	Material
Add audible signal	1.000	Ea.	Price includes an elevator contractor's materials and labor	
Totals				

Total per each including general contractor's overhead and profit **$725**

Lower hallway call buttons

Description	Quantity	Unit	Labor-Hours	Material
Remove panel and receptacle/switch/fixture	1.000	Ea.	0.031	0.00
Remove junction box	1.000	Ea.	0.100	0.00
Cutout demolition of partition	1.000	Ea.	0.333	0.00
Re-install junction box	1.000	Ea.	0.667	45.00
Re-install call button fixture	1.000	Job	1.000	0.00
Re-install call button panel	1.000	Job	0.200	0.00
Gypsum board repairs	1.000	Ea.	1.000	0.28
Paint gypsum board - minimum	1.000	Job	2.000	0.00
Totals			5.331	45.28

Total per each including general contractor's overhead and profit **$540**

34. Modify Existing Elevator Hall Signals *(continued)*

Replace hallway call buttons

Description	Quantity	Unit	Labor-Hours	Material
Remove panel and receptacle/switch/fixture	1.000	Ea.	0.031	0.00
Remove junction box	1.000	Ea.	0.100	0.00
Cutout demolition of partition	1.000	Ea.	0.333	0.00
Install new call button fixture	1.000	Ea.	2.000	320.00
Miscellaneous materials for gypsum board painting and repair	1.000	Ea.	1.000	0.28
Repair gypsum board	1.000	Job	4.000	0.00
Paint gypsum board - minimum	1.000	Job	2.000	0.00
Totals			**9.464**	**320.28**

Total per each including general contractor's overhead and profit   **$1,248**

Notes

35
Modify Existing Elevator Cab

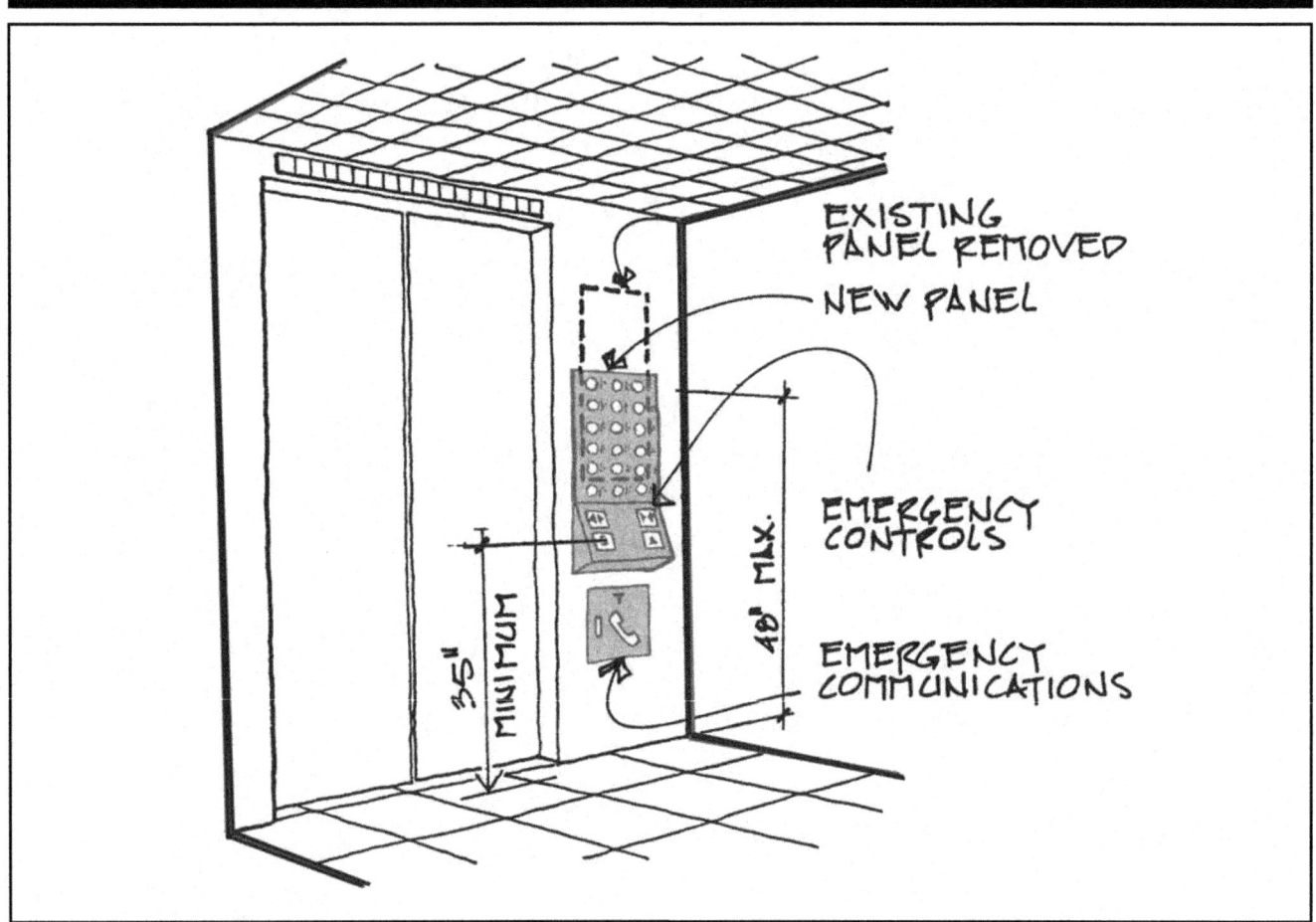

EXISTING PANEL REMOVED

NEW PANEL

EMERGENCY CONTROLS

EMERGENCY COMMUNICATIONS

48" MAX.

35" MINIMUM

If there were no numbers on the panel in an elevator, how would you know which button to push? Without some basic modifications, such as raised characters, Braille characters, or visual and audible signals, an elevator cab can be just as difficult to use for some people as if there were no numbers. These kinds of additions, and other simple modifications such as increased lighting, closing the gap at the door, and installing handrails, can greatly increase the usability of an elevator cab for all users.

ADAAG References
302, Floor and Grouted Surfaces
303, Changes in Level
309.4, Operable Parts

407, Elevators
(in particular 407.3, Door
Requirements and 407.4, Car
Requirements)
703, Signs

Where Applicable
All passenger elevators.

Design Requirements
- Compliance with all elevator cab, signal, control, communication, and reach requirements (not cited here; consult ADAAG 407 for details). Elevator manufacturers will help you meet not only ADAAG, but local and state building codes as well.
- Doors open for a minimum of three seconds when called; must reopen automatically if obstructed. Interval of time between the signal that the car has arrived and when the doors start to close is five seconds plus additional times determined by a formula found in ADAAG 407.3.4.
- 1-1/4" maximum platform to hoistway clearance.
- Audible signals indicating direction and passing floors.
- Visible signal indicating passing and arrival floors.
- Control panel at front or side of cab. Operating parts within reach

range, 48" maximum to top button. If over 16 floors, top button can be 54" a.f.f., maximum, if there is a parallel approach to the panel.
- Buttons raised or flush with panel, 3/4" diameter minimum. Visual indication of call registration. Raised characters and Braille next to each button. Emergency controls on bottom of panel.
- Emergency controls 35" a.f.f. minimum grouped at the bottom of the panel, operable with a closed fist.
- Slip-resistant flooring surface.
- Adequate lighting (5 foot-candles minimum).
- Where not all elevators in a bank meet ADAAG, signs displaying the international access symbol are required designating those that do.

Design Suggestions
It is recommended that elevator modifications be discussed with the manufacturer's representatives, as the requirements are complex. Existing elevators are exempt from many of the requirements; consult ADAAG before embarking on any up-grades. If the control panel is too high and the call buttons do not comply, consider adding a new panel and keeping the existing panel.

There are other design features not included in ADAAG that should be addressed in an up-grade, such as reducing reflective surfaces (they can make controls difficult to see and can be disorienting for people with low vision or cognitive impairments) and installing a handrail (helps users maintain balance and overcome vertigo). Increasing the lighting makes call button and characters easier to see for people with low vision; a lighting level twice the minimum is recommended.

If you are doing renovations in the lobby and are installing a new cab, consider increasing the door width.

Key Items
Signals, control panel, communications system, floor surfaces, tactile and Braille characters, possibly handrails.

Level of Difficulty
Low for installation of tactile and Braille numbers and floor surfacing, but high for all electrical work. Trades involved: electricians, possibly manufacturer's installers. Shop drawings may be necessary.

Estimates

Convert freight elevator to passenger elevator (5-stop, 3,000 lb. cap., 100 F.P.M.)

Description	Quantity	Unit	Labor-Hours	Material
Self-leveling device (hoist motor and controller change)	1.000	Ea.		
Lower existing control panel, add audible and visual signals	1.000	Ea.	Price includes an elevator	
Add railings to cab interior	20.000	L.F.	contractor's materials and labor	
Totals				
			Total per each including general contractor's overhead and profit	**$50,377**

Copyright Reed Construction Data, 2004

35. Modify Existing Elevator Cab *(continued)*

Replace self-leveler (5-stop, 3,500 lb. capacity, 100 F.P.M.)

Description	Quantity	Unit	Labor-Hours	Material
Self-leveling device (hoist motor and controller change)	1.000	Ea.	Price includes an elevator contractor's materials and labor	

Totals

Total per each including general contractor's overhead and profit	**$48,053**	

Install/replace non-compliant doors, including opening device (5-stop passenger elevator)

Description	Quantity	Unit	Labor-Hours	Material
Doors, frame and structure demolition	1.000	Job		
New cab door	1.000	Floor		
New lobby door	1.000	Ea.	Price includes an elevator	
New lobby frame	1.000	Ea.	contractor's materials and labor	

Totals

Total per floor including general contractor's overhead and profit	**$8,743**	

Add visual signal

Description	Quantity	Unit	Labor-Hours	Material
Add visual signal	1.000	Ea.	Price includes an elevator contractor's materials and labor	

Totals

Total per each including general contractor's overhead and profit	**$725**	

Add audible signal

Description	Quantity	Unit	Labor-Hours	Material
Add audible signal	1.000	Ea.	Price includes an elevator contractor's materials and labor	

Totals

Total per each including general contractor's overhead and profit	**$725**	

Lower existing elevator panel

Description	Quantity	Unit	Labor-Hours	Material
Existing panel relocation	1.000	Job	Price includes an elevator contractor's materials and labor	

Totals

Total per each including general contractor's overhead and profit	**$12,795**	

Replace existing elevator panel

Description	Quantity	Unit	Labor-Hours	Material
New control panel	1.000	Ea.	Price includes an elevator	
Removal of existing panel, installation of new panel and all controls	1.000	Job	contractor's materials and labor	
Totals				

Total per each including general contractor's overhead and profit **$18,802**

Add aluminum railing in cab (48″ long, one wall)

Description	Quantity	Unit	Labor-Hours	Material
1-1/2″ diameter aluminum railing	4.000	L.F.	0.601	35.20
Totals			0.601	35.20

Total per each including general contractor's overhead and profit **$121**

Add stainless steel railing in cab (48″ long, one wall)

Description	Quantity	Unit	Labor-Hours	Material
Stainless steel railing, 2″ x 3/8″	4.000	L.F.	1.196	120.00
Totals			1.196	120.00

Total per each including general contractor's overhead and profit **$323**

Add wood railing in cab (48″ long, one wall)

Description	Quantity	Unit	Labor-Hours	Material
Wood railing	4.000	L.F.	0.533	120.00
Totals			0.533	120.00

Total per each including general contractor's overhead and profit **$243**

Add raised character/Braille signage to control panel

Description	Quantity	Unit	Labor-Hours	Material
Adhesive-mounted control panel signage (for 9-stop elevator)	1.000	Ea.	2.500	159.00
Totals			2.500	159.00

Total per each including general contractor's overhead and profit **$415**

Replace emergency communications system

Description	Quantity	Unit	Labor-Hours	Material
Removal of existing and installation of new communication system	1.000	Job	Price includes an elevator	
Emergency system	1.000	Ea.	contractor's materials and labor	
Totals				

Total per each including general contractor's overhead and profit **$4,535**

36
Elevator Cab: Install Raised and Braille Characters

Installing raised and Braille characters is one of the easiest modifications to help make an elevator cab more accessible. Without these features, people who are blind or have severe vision impairments have no way of locating which button to push. Many of the standardized symbols, such as the star for the ground floor, can also be useful for others when the first floor, ground floor, or main floors are different.

ADAAG References
407, Elevators (in particular, 407.4, Elevator Car Requirements) 703, Signs

Where Applicable
All public elevators.

Design Requirements
- Raised character and Braille character 5/8" to 2" high adjacent to the left of all buttons in the cab, with a star indicating the first (ground) floor (regardless of other signaling devices). Raised symbol on emergency communication device door.
- Characters must have the size proportions and stroke widths as per ADAAG 703.

Design Suggestions
Except in those rare instances where an elevator has an audible voice system calling out every floor, raised and Braille characters on door jambs and call panels may be the only means that a blind person has of using an elevator unassisted. Also, the availability, easy installation, and low cost of adding raised and Braille characters usually makes this a readily achievable modification in a public accommodation.

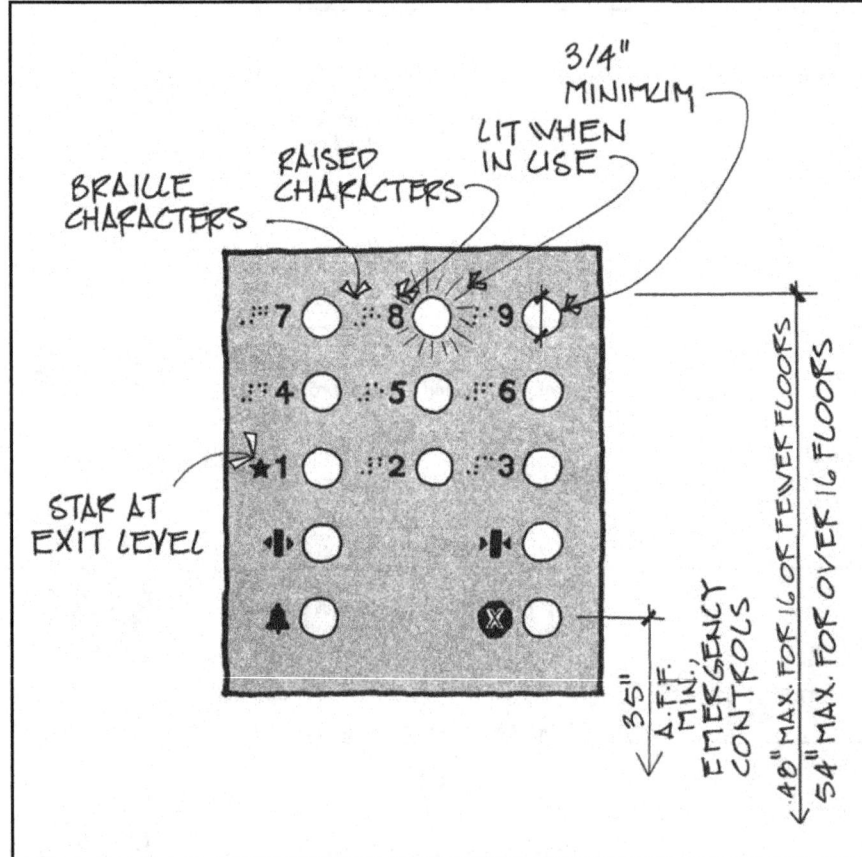

Key Items

Signage, either bolted or adhered to surface.

Level of Difficulty

Low. Installation may require only adhering new signage to existing panels.

Estimates

Add 5 adhesive-mounted floor designation signs

Description	Quantity	Unit	Labor-Hours	Material
Floor designation jamb signage	5.000	Ea.	1.250	117.50
Totals			1.250	117.50

Total per each sign including general contractor's overhead and profit **$55**

Total per 5 signs including general contractor's overhead and profit **$273**

Add adhesive-mounted metal elevator control panel signage

Description	Quantity	Unit	Labor-Hours	Material
Ahesive-mounted control panel signage (for 3-stop elevator)	1.000	Job	2.500	159.00
Totals			2.500	159.00

Total per each including general contractor's overhead and profit **$415**

37
Install or Modify Public Telephones for Access by Wheelchair Users

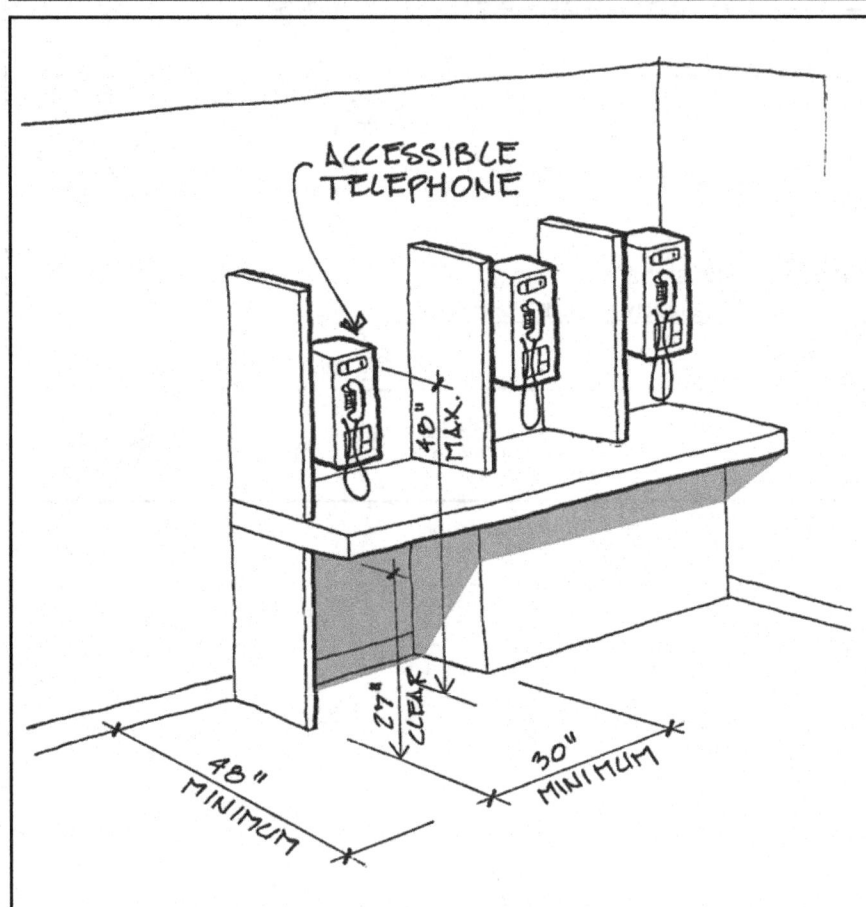

ACCESSIBLE TELEPHONE

48" MAX.

42" CLEAR

48" MINIMUM

30" MINIMUM

Telephones are one of the easiest building elements to make accessible. Although many people have cell phones, still many do not. Telephones still play a vital role in emergency situations, to find directions when an address is wrong, and so forth. Although not required, hospital public areas, transportation facilities, roadside stops, service plazas, and emergency stations should all have wheelchair-accessible phones.

Often, phones are located on accessible routes, but they are too high, or in narrow booths, or have no volume control. Simple modifications, such as removing a phone booth, lowering a phone, or installing a volume control phone can go a long way in making a facility more accessible for a wide range of users.

ADAAG References
216.9, Signs for TTYs
217, Telephones
305, Clear Floor or Ground Space
308, Reach Ranges
704, Telephones

Where Applicable
Coin-operated and coinless public pay phones. Public closed circuit telephones, courtesy phones.

Phones serving a primary function area that is being altered. For single phones, one wheelchair-accessible phone per floor/level or exterior site. For phone banks (two or more phones), at least one phone per bank.

Design Requirements
- All new and replaced phones must have a volume control.
- All banks of three or more phones must have a shelf and electrical outlet.
- Signage is required indicating the location of the nearest TTY, as well as signs indicating accessibility features.

144

- 30" x 48" clear floor space in front of phone.
- Positioned so that either a forward or parallel approach is possible.
- If in an alcove, deeper than 15" to the face of the phone, 60" is required for the parallel approach. A 36" wide space is required under the phone if the alcove is deeper than 24".
- Reach range: 48" maximum to the highest operable part. For a parallel approach, 10" from unit to front of counter; for a forward approach, 20" from unit to front of counter.

- 29" cord minimum.
- Push button controls.

Design Suggestions
Locate accessible phones adjacent to existing phones, in the most visible location. Do not locate phones in a corner unless absolutely necessary, and only if dimensions for a frontal approach are addressed. For wall- or post-mounted phones, always try to have two or more phones so that standing people can use them easily.

Key Items
Phone equipment, usually supplied by phone company. Signage as appropriate. Possible finish work if phones are installed or relocated on an existing surface.

Level of Difficulty
Most work is done by the telephone company, often at no charge. Installation of signage can be done by others. Trades involved: electrical, possible finish work.

Estimates

Lower public telephones

Description	Quantity	Unit	Labor-Hours	Material
Miscellaneous materials for gypsum board painting and repair	1.000	Ea.	1.000	0.28
Gypsum board patching labor minimum	1.000	Job	4.000	0.00
Paint gypsum board - minimum	1.000	Job	2.000	0.00
Totals			7.000	0.28

Total per each including general contractor's overhead and profit **$493**

Provide interior text telephone

Description	Quantity	Unit	Labor-Hours	Material
Telecommunications display device	1.000	Ea.	0.800	570.00
Totals			0.800	570.00

Total per each including general contractor's overhead and profit **$1,038**

Provide volume control

Description	Quantity	Unit	Labor-Hours	Material
Volume control handset	1.000	Ea.	1.000	63.00
Totals			1.000	63.00

Total per each including general contractor's overhead and profit **$191**

37. Install or Modify Public Telephones *(continued)*

Remove phone booth

Description	Quantity	Unit	Labor-Hours	Material
Demolition crew minimum	1.000	Job	7.000	0.00
Totals			**7.000**	**0.00**

Total per each including general
contractor's overhead and profit **$547**

Add signage to phone

Description	Quantity	Unit	Labor-Hours	Material
Plastic, adhesive-mounted signage	2.000	Ea.	0.500	26.70
Labor minimum	1.000	Job	2.000	0.00
Totals			**2.500**	**26.70**

Total per each including general
contractor's overhead and profit **$222**

Provide new accessible telephone

Description	Quantity	Unit	Labor-Hours	Material
Telephone company labor minimum	1.000	Job	2.286	0.00
Totals			**2.286**	**0.00**

Total per each including general
contractor's overhead and profit **$191**

Add counter and outlet for portable TDD

Description	Quantity	Unit	Labor-Hours	Material
Telephone enclosure	1.000	Ea.	3.200	940.00
Telephone receptacle w/20' of 4/C phone wire	1.000	Ea.	0.308	6.70
Totals			**3.508**	**946.70**

Total per each including general
contractor's overhead and profit **$1,885**

Notes

38
Install TTYs

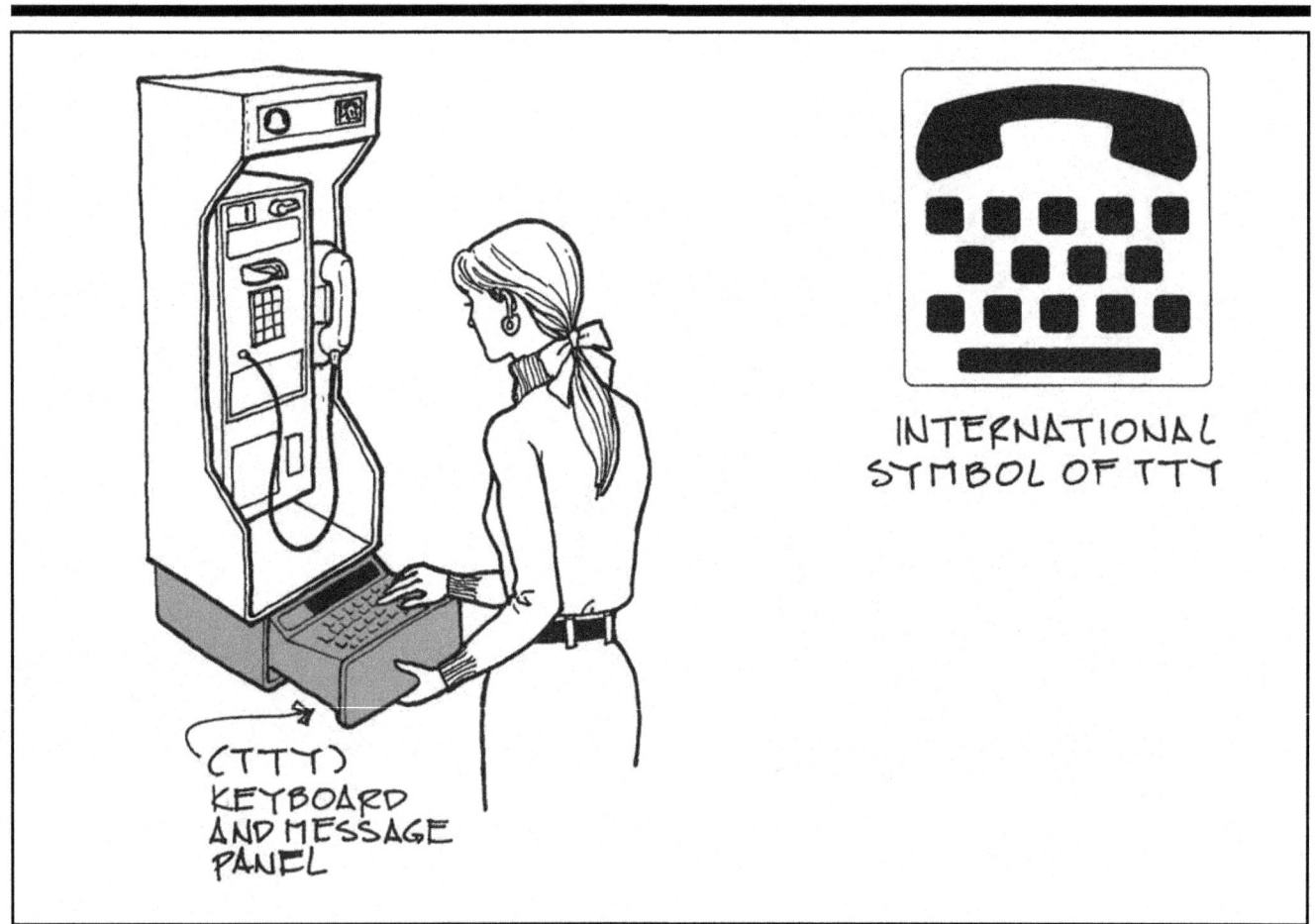

INTERNATIONAL SYMBOL OF TTY

(TTY) KEYBOARD AND MESSAGE PANEL

Accessibility is often thought of only in terms of people who use wheelchairs, but the TTY (sometimes called text telephone or TTD) is one of the most important advances in access technology. It enables someone with a hearing or speech impairment to communicate using the phone, something most people take for granted. A public telephone, with a TTY keyboard and display, is important not only as a building amenity, but can be vital in an emergency situation. It is an integral part of an accessible facility.

ADAAG References

216.9, Signs for TTYs
217.4, TTYs
217.5, Shelves for TTYs
305, Clear Floor or Ground Space
308, Reach Ranges
704.4, TTYs

Where Applicable

- Telephone banks. Where four or more public pay telephones, at least one public TTY.
- Public buildings. Where at least one public pay telephone on a floor, at least one public TTY on that floor. Where at least one public pay telephone in a public building, at least one public TTY in the building.
- Private buildings. Where four or more public pay telephones on a floor, at least one public TTY on that floor. Where four or more public pay telephones in a private building, at least one public TTY in the building.
- Exterior site requirement. Where four or more public pay telephones, at least one public TTY.
- Rest stops, emergency roadside stops, and service plazas. Where at least one public pay telephone, at least one public TTY.
- Hospitals. Where at least one public pay telephone in emergency room, hospital recovery room, or hospital waiting room, at least one public TTY at each location.
- Transportation facilities. In addition to the above, where at least one public pay telephone

serves a particular entrance, at least one public TTY to serve that entrance. In airports, in addition to the above, where four or more public pay telephones are in a terminal outside the security areas, a concourse within the security areas, or a baggage claim area in a terminal, at least one public TTY.
- Detention and correctional facilities. Where at least one pay telephone in a secured area used only by detainees or inmates and security personnel, at least one TTY.

Design Requirements

- Sign with international TTY symbol (see 703.7.2.2).
- All banks of three or more phones must have a TTY shelf and electrical outlet, unless a TTY is provided.
- 34" minimum above floor to key pad unless seats provided.
- Shelf and electrical outlet where portable TTYs required.

Design Suggestions

Try to locate TTY telephones adjacent to existing phones, in the most visible location. While TTYs usually cannot meet the criteria for wheelchair-accessible phones, they

nonetheless should be on an accessible route, meet maneuverability requirements, and be as close to the reach ranges for wheelchair-accessible phones as possible. Reading and typing at a TTY is more suited to sitting than standing. Provide seats where possible.

If possible, install both the TTY shelf and outlet, as well as the TTY phone.

Key Items

Phone equipment, usually supplied by the phone company. Signage as appropriate. Possible finish work if phones are installed or relocated on an existing surface.

Level of Difficulty

Most phone work is done by the phone company. Installation of signage can be done by others. Other trades involved: electrical, possible finish work.

Note: This book includes the cost of the public text telephone unit itself. In some instances, the text telephone is leased or supplied by the local phone company. In this case, construction costs would involve electrical work to supply the phone lines and outlet(s) and finish costs only.

Estimates

Install public text telephone

Description	Quantity	Unit	Labor-Hours	Material
Telecommunications display device in stainless steel enclosure	1.000	Ea.	4.000	570.00
Telephone company labor minimum	1.000	Job	2.286	0.00
Totals			6.286	570.00

Total per each including general contractor's overhead and profit **$1,498**

39
Install or Modify Drinking Fountains

As with telephones, inaccessible drinking fountains are often located on accessible routes. Replacing an inaccessible fountain can often use existing piping and electrical supply. Even in cases where this is not possible, installing an accessible fountain may be a relatively simple construction operation that adds an important accessible element to a facility.

ADAAG References

211, Drinking Fountains
305, Clear Floor or Ground Space (particularly 305.7.1, Forward Approach)
306, Knee and Toe Space
307, Protruding Objects
309, Operable Parts
602, Drinking Fountains

Where Applicable

Where drinking fountains are provided two are needed: one for children, short adults, and people in wheelchairs and the other for tall people and users who have problems bending over. If more than two are provided, 50% should be low and 50% high.

Design Requirements

- Low fountain, spout 36" a.f.f. maximum (32" a.f.f. is strongly recommended so the fountain is usable by children).
- 30" × 48" clear floor space minimum facing the fountain.
- Knee clearance below fountain: 27" a.f.f. minimum for a depth of 8" from the front apron.
- Toe clearance under fountain, 9" a.f.f. for a depth of 17".
- Lowest edge cannot intrude more than 4" into the circulation path, if higher than 27" a.f.f. and not otherwise protected (as with wing walls).
- High fountain, spout 38" to 43" a.f.f. (42" to 43" recommended for majority of the population).

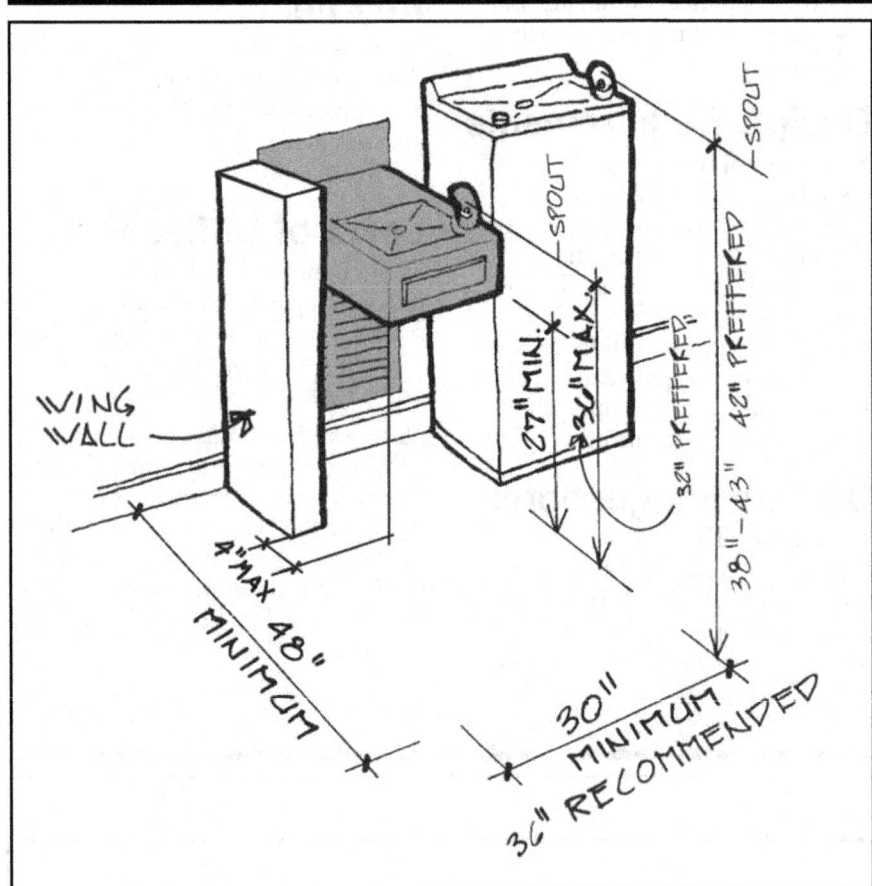

- Spout 5" maximum from the front of the fountain and 15" minimum from the wall.
- Water flow at least 4" high; angle of flow varies with spout location.

Design Suggestions

Drinking fountains are almost always along circulation paths and therefore frequently are protruding objects. To correct this, flange walls or permanent furnishings may be used.

All wheelchair-accessible fountains must be accessible via a front approach. Accordingly, a 30" alcove width is minimal and may result in scarred walls. A wider alcove is recommended.

Fountains with a push-bar control across the front are easier to use for a wider range of people.

Key Items

Drinking fountain, plumbing/water supply, electricity source, possible finish materials. Flange walls or other devices to meet protruding object regulations.

Level of Difficulty

Moderate to high; modifying or installing fountains requires plumbing and may require electrical, finish work, and sometimes structural reinforcement if the fountain is wall-mounted.

Estimates

Install new high/low fountain

Description	Quantity	Unit	Labor-Hours	Material
Remove existing freestanding water fountain	1.000	Ea.	1.000	0.00
Drinking fountain, wall-mounted, accessible type	1.000	Ea.	4.000	1,425.00
Water fountain supply, waste, and vent	1.000	Ea.	3.620	42.00
Totals			8.620	1,467.00

Total per each including general contractor's overhead and profit **$3,199**

Install new low, wall-mounted fountain

Description	Quantity	Unit	Labor-Hours	Material
Water fountain supply, waste, and vent	1.000	Ea.	3.620	42.00
Drinking fountain, wall-mounted, accessible type	1.000	Ea.	4.000	1,425.00
Totals			7.620	1,467.00

Total per each including general contractor's overhead and profit **$3,113**

Lower existing fountain

Description	Quantity	Unit	Labor-Hours	Material
Remove existing water fountain	1.000	Ea.	1.000	0.00
Water fountain supply, waste, and vent	1.000	Ea.	3.620	42.00
Re-install water fountain, labor minimum	1.000	Job	4.000	0.00
Totals			8.620	42.00

Total per each including general contractor's overhead and profit **$800**

40
Install or Modify Controls and Wall-Mounted Items

Controls and wall-mounted items are often vital for general safety. Examples include fire extinguishers, fire alarm pull boxes, and emergency assistance boxes. Placing such public items in accessible locations or making them accessible not only makes them usable by people with mobility or grasping impairments, but it also makes them easier to reach and use for everyone in the facility, including children and short people.

ADAAG References
205, Operable Parts
305, Clear Floor or Ground Space
308, Reach Ranges
309, Operable Parts

Where Applicable
All wall-mounted items, controls, and alarms operated or used by building occupants, except when only used by service or maintenance personnel. See 205.1 for exceptions.

Design Requirements
- Within reach range: 48" maximum to highest operable part. ADAAG 308 provides dimensions for reaching over obstructions and counters.
- All controls operable without tight grasping, pinching, or twisting of the wrist.
- 30" × 48" clear floor space in front of control.
- Operating force no greater than five pounds/foot.

Design Suggestions
Fire alarms can be covered with plastic shields to help prevent false alarms, but shields must comply with ADAAG 309 and be openable without tight grasping, pinching, or twisting of the wrist. Locate all items at least 18" from an inside corner, 24" preferred. For items in an alcove over 24" deep, the alcove must be 36"

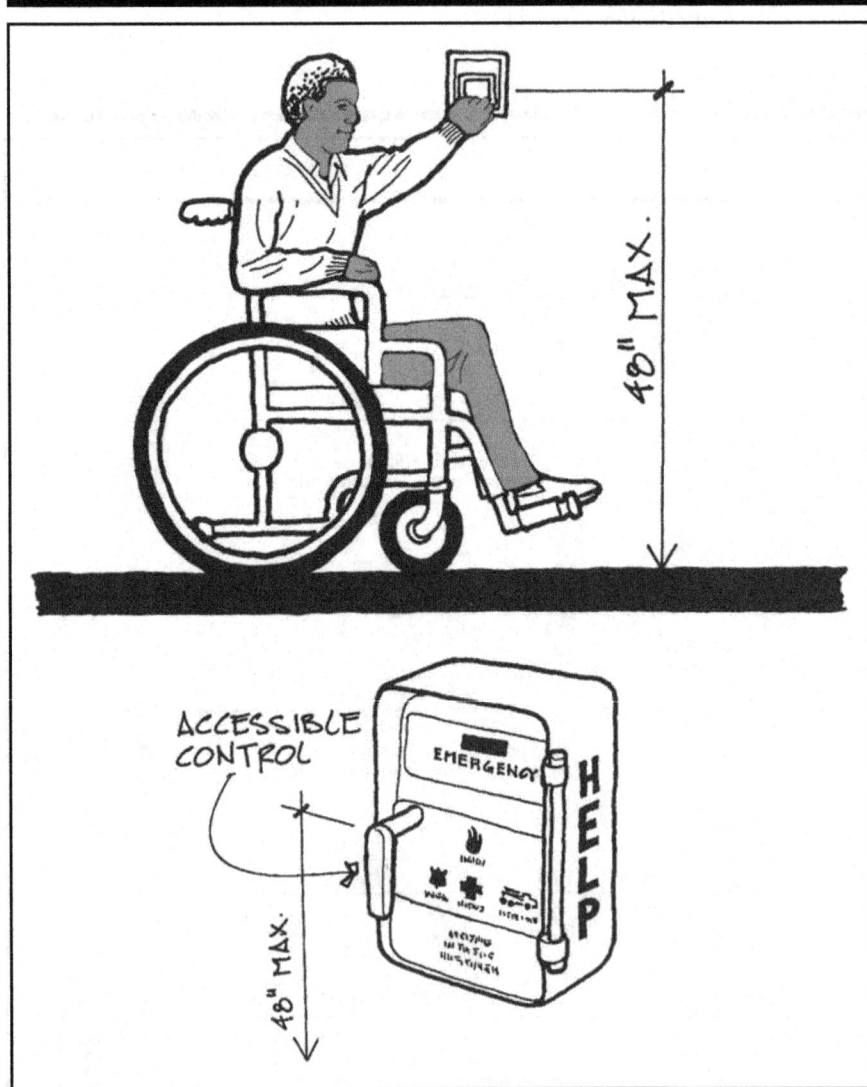

48" MAX.

ACCESSIBLE CONTROL

EMERGENCY

HELP

48" MAX.

wide for a front approach. For a parallel approach, the alcove must be 60" wide if it is over 15" deep. All items should be centered in the clear floor space or alcove.

Key Items

Controls, wiring, specialties; finishes to match existing.

Level of Difficulty

Moderate. Involves electrical and finish work.

Estimates

Lower 5 fire alarm boxes (surface-mounted conduit)

Description	Quantity	Unit	Labor-Hours	Material
No. 700 wiremold raceway	10.000	L.F.	0.800	8.20
#18 fire alarm conductor	0.100	C.L.F.	0.100	6.00
Labor minimum	1.000	Job	2.000	0.00
Totals			2.900	14.20

Total per each including general contractor's overhead and profit _____ **$53**

Total per 5 alarm boxes including general contractor's overhead and profit _____ **$266**

Copyright Reed Construction Data, 2004

41
Install or Modify Outlets

While placing outlets in accessible locations is essential for people who use wheelchairs, it also makes them easier for everyone to use. Outlets located close to the floor require stooping even for people who do not use wheelchairs. Installing or relocating outlets to accessible heights is a relatively simple modification that may only require minor electrical and finish work to complete.

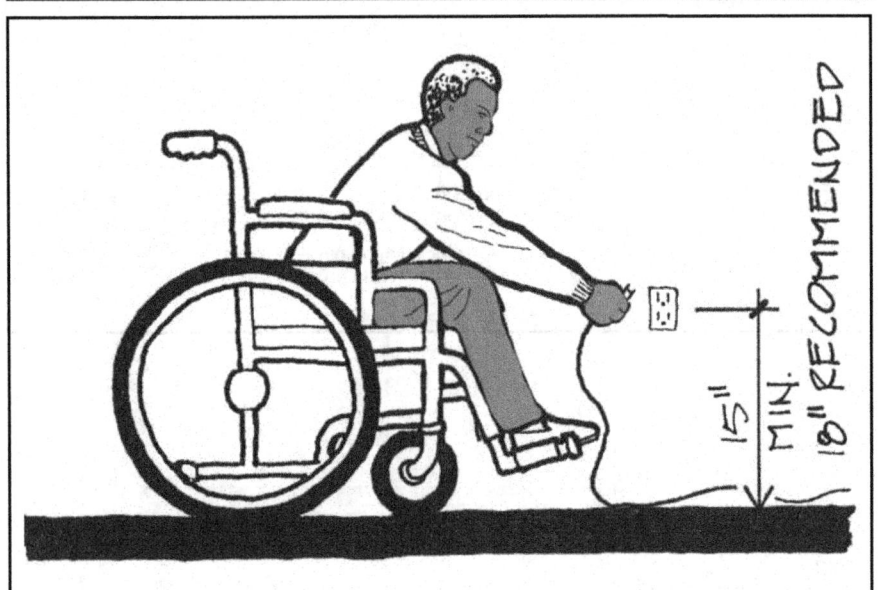

ADAAG References
205, Operable Parts
305, Clear Floor or Ground Space
308, Reach Ranges
309, Operable Parts

Where Applicable
All outlets operated by building occupants. See 205.1 for exceptions.

Design Requirements
- Within reach range: 48" maximum, 15" minimum a.f.f. for electrical or convenience outlets and telephone and data jacks.
- 30" × 48" clear floor space in front of control that allows either a parallel or a perpendicular approach. See ADAAG 308.2.2 and 308.3.2 for cases where there is an obstruction.

Design Suggestions
Try to locate all devices at least 18" to 24" from an inside corner, and mount them at 18" a.f.f whenever possible. In general, try to keep outlets below 44", particularly when over an obstruction.

Key Items
Outlets, wiring, finishes to match existing.

Level of Difficulty
Low to moderate. Involves electrical and finish work.

Estimates

Raise 5 outlets, gypsum/metal stud wall

Description	Quantity	Unit	Labor-Hours	Material
Remove plate and receptacle/switch/fixture	5.000	Ea.	0.156	0.00
Cutout demolition of partition	5.000	Ea.	1.667	0.00
Conductor	0.100	C.L.F.	0.296	1.23
Install junction box	5.000	Ea.	2.000	38.75
Re-install outlet	5.000	Job	1.000	0.00
Install plate	5.000	Ea.	0.500	8.80
Misc. materials for gypsum board painting and repair	1.000	Ea.	1.000	0.28
Repair gypsum board	1.000	Job	4.000	0.00
Paint gypsum board - minimum	1.000	Job	2.000	0.00
Totals			**12.619**	**49.06**

Total per each including general contractor's overhead and profit _____ **$200**

Total per 5 outlets including general contractor's overhead and profit _____ **$1,002**

Install 5 outlets, gypsum/metal stud wall

Description	Quantity	Unit	Labor-Hours	Material
Cutout demolition of partition	5.000	Ea.	1.667	0.00
Conductor	0.100	C.L.F.	0.296	1.23
Install junction box	5.000	Ea.	2.000	38.75
Install outlet	5.000	Ea.	1.482	35.25
Install plate	5.000	Ea.	0.500	8.80
Misc. materials for gypsum board painting and repair	1.000	Ea.	1.000	0.28
Repair gypsum board	1.000	Job	4.000	0.00
Paint gypsum board - minimum	1.000	Job	2.000	0.00
Totals			**12.945**	**84.31**

Total per each including general contractor's overhead and profit _____ **$218**

Total per 5 outlets including general contractor's overhead and profit _____ **$1,089**

42
Install or Modify Switches

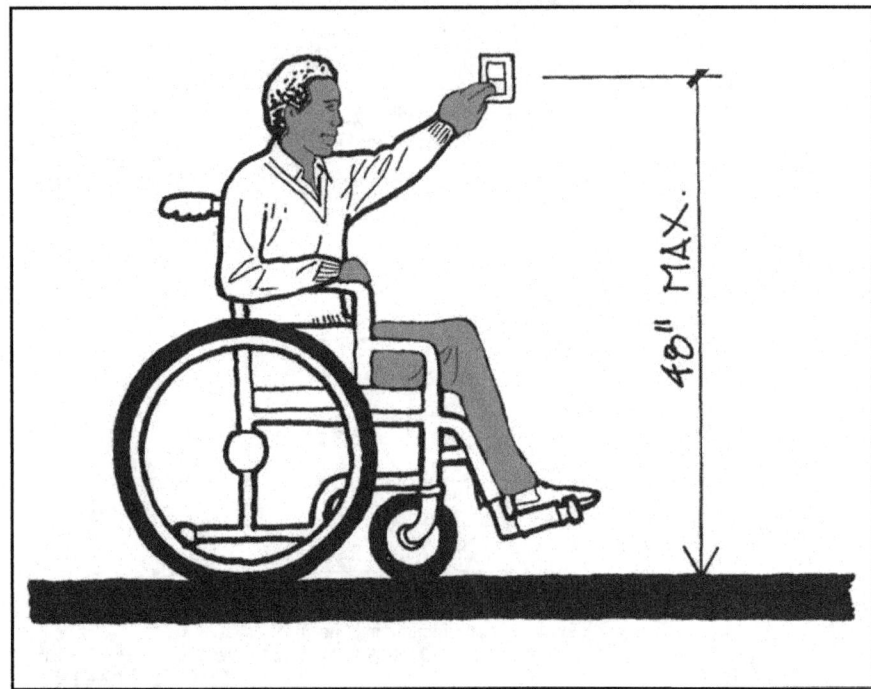

It is now common in new construction to locate light switches within accessible reach range, but in older facilities and in some new construction, switches are placed too high or too close to a corner, making them unusable by people in wheelchairs and others with limited range of arm movement. As with outlets, locating a switch in a location that meets ADAAG in new construction or renovations costs the same as placing it in an inaccessible location, and relocating a switch might involve only minor electrical and finish work.

ADAAG References
205, Operable Parts
305, Clear Floor or Ground Space
308, Reach Ranges
309, Operable Parts

Where Applicable
All switches operated by building occupants. See 205.1 for exceptions.

Design Requirements
- Within reach range: 48" maximum, 15" minimum a.f.f. See ADAAG 308.2.2 and 308.3.2 for situations where there are obstructions, such as counters.
- All controls operable without tight grasping, pinching, or twisting of the wrist.
- 30" × 48" clear floor space in front of control.

Design Suggestions
If you can, install controls at 44" a.f.f. even if side reach is possible and there is no obstruction or counter in the way. This may keep the switch within reach even if someone puts a chair or bookcase below it. Consider installing rocker switches (easier to use for people with limited fine motor control). Locate all controls at least 18" from an inside corner.

Key Items

Switches, wiring, finishes to match existing.

Level of Difficulty

Low to moderate. Involves electrical and finish work.

Estimates

Lower 5 light switches, gypsum/metal stud wall

Description	Quantity	Unit	Labor-Hours	Material
Remove plate and receptacle/switch/fixture	5.000	Ea.	0.156	0.00
Cutout demolition of partition	5.000	Ea.	1.667	0.00
Conductor	0.100	C.L.F.	0.296	1.23
Install junction box	5.000	Ea.	2.000	38.75
Re-install switch	5.000	Job	1.000	0.00
Install plate	5.000	Ea.	0.500	8.80
Misc. materials for gypsum board painting and repair	1.000	Ea.	1.000	0.28
Repair gypsum board - minimum	1.000	Job	4.000	0.00
Paint gypsum board - minimum	1.000	Job	2.000	0.00
Totals			12.619	49.06

Total per each including general contractor's overhead and profit **$200**

Total per 5 switches including general contractor's overhead and profit **$1,002**

Install 5 rocker switches, gypsum/metal stud wall

Description	Quantity	Unit	Labor-Hours	Material
Cutout demolition of partition	5.000	Ea.	1.667	0.00
Conductor	0.100	C.L.F.	0.296	1.23
Install junction box	5.000	Ea.	2.000	38.75
20 amp rocker switch	5.000	Ea.	1.482	58.50
Install plate	5.000	Ea.	0.500	9.00
Misc. materials for gypsum board painting and repair	1.000	Ea.	1.000	0.28
Repair gypsum board	1.000	Job	4.000	0.00
Paint gypsum board - minimum	1.000	Job	2.000	0.00
Totals			12.945	107.76

Total per each including general contractor's overhead and profit **$226**

Total per 5 switches including general contractor's overhead and profit **$1,129**

43
Install Audible and Visual Fire Alarms and Smoke Detectors

Maximum fire safety is based on assuming the worst case scenario and taking steps to prevent it. Accessible fire safety includes providing a means of warning people with visual and hearing impairments of an emergency wherever they may be in an accessible facility. This means that both visual and audible alarms are necessary. Installing alarms has the added benefit of increasing the level of fire safety for all users of a facility.

ADAAG References
215, Fire Alarm Systems
222.4, Guest Rooms
702.1 Fire Alarm Systems
806.3.1, Guest Room Alarms
809.5, Residential Building Fire Alarm Systems

Where Applicable
All public use areas and common use areas, such as file rooms, copy rooms, lounges/kitchenettes, restrooms, etc., as well as conference rooms, lobbies, and corridors, where fire alarms are provided (wiring only for visual alarms in employee work areas); transient lodging, and 2% of residential units in a project.

Design Requirements
- Follow NFPA 72, Chapter 4 for technical requirements.
- Alarms to be both audible and visual.
- Visual alarms no more than 50' (line of sight) from any point.
- Smoke detectors located between 80" and 96" a.f.f., or 6" below the ceiling, whichever is lower.
- 110 db maximum, contrary to NFPA 72.

Design Suggestions
Because of the 50' sight line maximum, more visual alarms may be required than audible alarms.

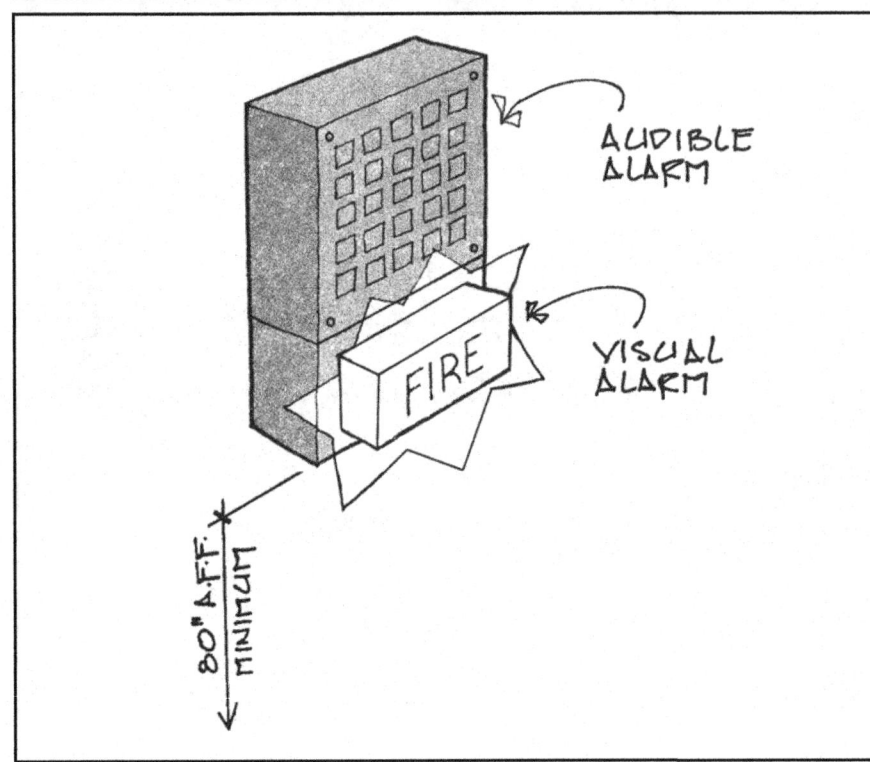

AUDIBLE ALARM

VISUAL ALARM

80" A.F.F. MINIMUM

FIRE

Key Items

Audible or visual components added to existing alarm, new audible and visual alarm, electrical conduit to main wiring system.

Level of Difficulty

High. Requires hard-wiring alarms to existing system, and may substantially increase power requirements. May require new wiring system.

Estimates

Install new visual alarm

Description	Quantity	Unit	Labor-Hours	Material
Fire alarm light	1.000	Ea.	1.509	95.00
#18 fire alarm conductor	0.100	C.L.F.	0.100	6.00
Totals			1.609	101.00

Total per each including general contractor's overhead and profit **$307**

Add visual alarm to existing audible alarm

Description	Quantity	Unit	Labor-Hours	Material
Fire alarm light	1.000	Ea.	1.509	95.00
#18 fire alarm conductor	0.100	C.L.F.	0.100	6.00
Totals			1.609	101.00

Total per each including general contractor's overhead and profit **$307**

Install new audible/visual alarm

Description	Quantity	Unit	Labor-Hours	Material
Fire alarm light and horn	1.000	Ea.	1.509	95.00
#18 fire alarm conductor	0.200	C.L.F.	0.200	12.00
Totals			1.709	107.00

Total per each including general contractor's overhead and profit **$326**

44
Install Signage

REQUIRED

RECOMMENDED

SIGNAGE INSTALLED
AT INACCESSIBLE ENTRANCE

Providing information is a vital component of any facility, and installing signs is often considered a readily achievable modification in businesses. Signage and wayfinding are often insufficient even for people with full vision, and such deficiencies are increased dramatically for those with visual or cognitive impairments (as well as for people not fully literate or fluent in English).

Each facility should be carefully studied in order to provide the clearest directional and informational signage.

ADAAG References
216, Signs
703, Signs
810. 4, Bus Signs
810.6, Rail Station Signs

Where Applicable
Building entrance signs and other signs that convey information about or directions to facilities on the site and interior spaces, including accessible elements. Follow visual specifications concerning contrast, character, height, font and proportions.

Temporary signs (seven days or fewer), building directories, signs with building addresses, corporate logos and names, and occupant names have no access requirements.

Signs designating permanent rooms and spaces (such as restrooms, exits, room numbers, etc.). Follow visual specifications and provide tactile and Braille elements.

Transportation facilities have other special specifications in ADAAG 810.4 and 810.6

Design Requirements
Visual characteristics:
- High contrast between characters/ pictures and background (dark on light, or light on dark).

- Matte or other non-glare background and characters.
- Simple font, conventional in form. Italic, script, oblique, or highly decorative fonts not allowed.
- Character height is a function of viewing distance (see ADAAG table 703.5.5), but as a rule of thumb, figure 1/8" per foot. 5/8" is the minimum.
- International symbol of accessibility at entrances if all are not accessible, with directions to the nearest accessible entrance; same for rest rooms.
- Mounted at least 40" a.f.f.

For tactile signs that are also visual:

- In addition to the visual characteristics in the previous section, characters must be raised at least 1/32", and have a 2" maximum height.
- Characters must be in sans serif font and upper case.
- Braille and characters must be at least 3/8" from raised borders.
- Braille text must be below the corresponding text, separated by 3/8".

- Braille must be Grade 2, rounded shape, and meet the spacing requirements of ADAAG Table 703.3.1.
- Signs to be located on wall adjacent to the door, on the latch side, and 60" a.f.f. to the centerline of sign. Users must be able to approach the sign, and the sign must be clear of the door swing when there are double doors with no wall beside the doors.

Design Suggestions

Regarding contrast, evidence suggests that light characters on a dark background are easier to read. Since there is no firm definition of high contrast, choose colors that leave as little doubt as possible, e.g., white characters on a dark gray or dark red background. Although room signs have been customarily placed on doors, this practice is not effective for people who must read the sign by touch or approach within inches of it to see the letters. If a sign is on a door, adding an accessible sign adjacent to the door can be effective for both

groups of users. Signs throughout the building should always be at the same height and location in relation to doors.

Informational signage also should always be the same height and in a consistent location. Raised characters and Braille are not required for informational or directional signs, but, if possible, a method of wayfinding should be provided for people with severe visual impairments.

The advisories in ADAAG provide useful information on making signs as helpful as possible.

Key Items

Signs, mounting or adhering mechanism.

Level of Difficulty

Low to moderate. While interior signs can be quite easy to install with self-adhesive backings, attaching exterior signs or signs on masonry may require difficult anchoring.

Estimates

Anchor plastic exterior signage to masonry wall

Description	Quantity	Unit	Labor-Hours	Material
Plastic signage	1.000	Ea.	0.267	26.50
Layout & drilling of anchor holes, per inch (4 holes, 2" deep)	8.000	Ea.	0.853	0.64
Wedge anchors	4.000	Ea.	0.213	1.44
Totals			1.333	28.58

Total per each including general contractor's overhead and profit **$146**

Estimates

Anchor metal exterior signage to masonry

Description	Quantity	Unit	Labor-Hours	Material
Metal signage	1.000	Ea.	0.457	31.50
Layout & drilling of anchor holes, per inch (4 holes, 2″ deep)	8.000	Ea.	0.853	0.64
Wedge anchors	4.000	Ea.	0.213	1.44
Totals			**1.523**	**33.58**

Total per each including general contractor's overhead and profit   **$175**

Anchor metal interior sign to gypsum board/stud wall

Description	Quantity	Unit	Labor-Hours	Material
Signage	1.000	Ea.	0.457	31.50
Layout & drilling of anchor holes	4.000	Ea.	0.213	0.04
Plastic shields and screws	4.000	Ea.	0.133	0.16
Totals			**0.803**	**31.70**

Total per each including general contractor's overhead and profit   **$119**

Anchor metal room sign to gypsum board/stud wall

Description	Quantity	Unit	Labor-Hours	Material
Signage	1.000	Ea.	0.457	31.50
Layout & drilling of anchor holes	4.000	Ea.	0.213	0.04
Plastic shields and screws	4.000	Ea.	0.133	0.16
Totals			**0.803**	**31.70**

Total per each including general contractor's overhead and profit   **$119**

Notes

45
Install Electronic Display Signage

In facilities where changing information is conveyed to the public on a regular basis, electronic display signage assists all users, especially those with hearing impairments. The ADA requires public entities and accommodations to ensure that communications with individuals with disabilities are as effective as communications with others. To this end, spoken information must be available in some form to users who can't hear. Electronic signage is an effective way of meeting that requirement. Signs can usually be retrofitted into an existing facility.

ADAAG References

218.5, Public Address Systems in Transportation Facilities
221.6, Public Address Systems in Assembly Areas
703.5, Visual Characters
810.7, Public Address Systems (Transportation Facilities)

Where Applicable

At transit facilities (bus, rail, and air) that broadcast information to the public. At places such as convention centers or hotels that have audible announcements. Public announcements in stadiums, arenas, and grandstands. Wherever there are audible pre-recorded messages or real-time audible public announcements.

Design Requirements

Audible information broadcast to the public must also be available to people with hearing impairments.

Design Suggestions

While ADAAG is not specific about design requirements for electronic signs, it is recommended that electronic signs comply with ADAAG 703.5, the standards for permanent signs.

- High contrast between characters/ pictures and background (dark on light, or light on dark).
- Simple serif or sans serif font conventional in form. Avoid italic, script, oblique, or highly decorative fonts.

- Character width-to-height ratio between 3:5 and 1:1 based on the upper case "O."
- Overhead signs with 80" minimum clearance below.
- Character height is a function of viewing distance (see ADAAG table 703.5.5), but as a rule of thumb, 1/8" per foot of horizontal distance.
- Avoid scrolling signs because they are often confusing and hard to read. Use blocks of print instead. Make sure the block message stays in view long enough for it to be read by slow readers.

Key Items

Signs, light-emitting diode (LED) boxes, mounting or adhering mechanism.

Level of Difficulty

Low to moderate. Involves electrical work and finish work.

Estimates

Install electric signage (2 lines of text, 3″ high characters, indoor appliances)

Description	Quantity	Unit	Labor-Hours	Material
Electric display signage (2 lines of text, 3″ high)	1.000	Ea.	1.000	73.00
Cutout demolition of ceiling	1.000	Ea.	0.333	0.00
Support for and installation of sign	1.000	Ea.	8.000	202.00
Conductor	0.100	C.L.F.	0.296	1.23
Install junction box	1.000	Ea.	0.400	7.75
Install outlet	1.000	Ea.	0.296	7.05
Install plate	1.000	Ea.	0.100	1.76
Misc. materials for gypsum board painting and repair	1.000	Job	0.800	0.14
Repair gypsum board	1.000	Job	4.000	0.00
Paint gypsum board - minimum	1.000	Job	2.000	0.00
Totals			17.225	292.93

Total per each including general contractor's overhead and profit **$1,970**

46
Install or Modify Grab Bars

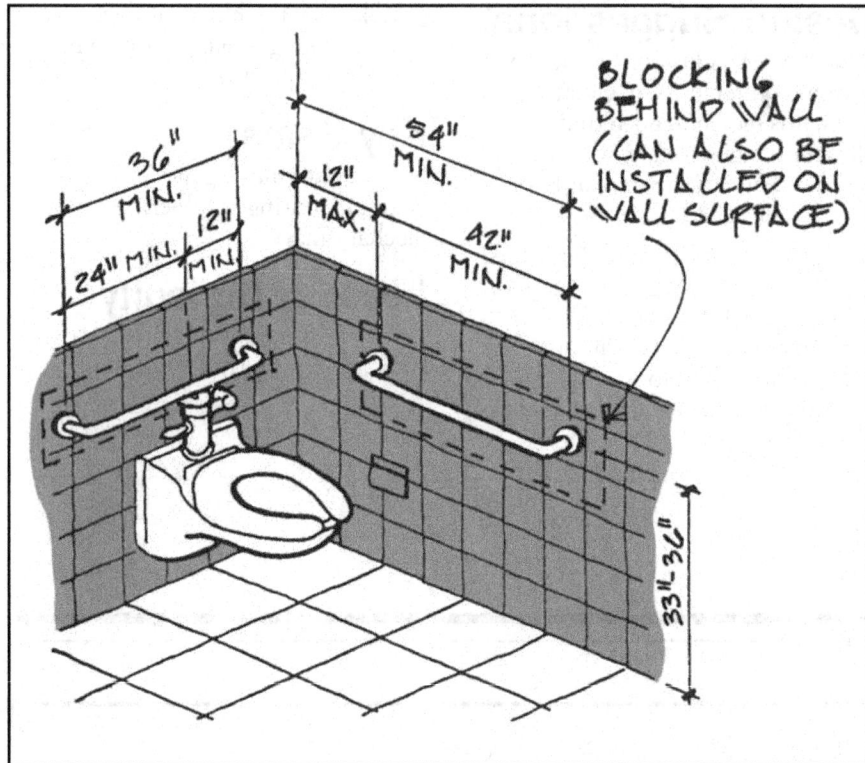

BLOCKING BEHIND WALL (CAN ALSO BE INSTALLED ON WALL SURFACE)

36" MIN.

54" MIN.

12" MAX.

42" MIN.

24" MIN.

12" MIN.

33"-36"

Grab bars are essential in enabling many people to use toilets, tubs, and showers. They are also one of the simplest access modifications to install. The types of available grab bars have increased, and colored, non-institutional models are available to fit into an existing decor. Different configurations are possible to assist people who might need bars in addition to those required by the ADA.

ADAAG References

604.5, 604.8.1.5, and 604.8.2.3, Grab Bars (at toilets)
607.4, Grab Bars (at tubs)
608.3, Grab Bars (at showers)
609, Grab Bars

Where Applicable

All accessible toilets, tubs, and showers.

Design Requirements

- 1-1/4" to 2" in diameter for circular cross section; 4" to 4.8" perimeter, 2" maximum diameter for non-circular cross section.
- 1-1/2" from wall, 12" minimum from protruding objects above except for shower controls, 1-1/2" minimum from protruding objects below and at ends.
- Smooth surface, free of sharp or abrasive elements behind and adjacent to grab bars.
- Capable of resisting 250 lbs. of vertical or horizontal force—tight in their fittings.
- 33" to 36" a.f.f. mounting height.
- For wheelchair-accessible water closet or toilet compartment: 42" long minimum at side wall starting 12" maximum from rear wall, 36" long minimum at back wall extending 12" from center of water closet or toilet compartment in one direction and 24" in the other.
- For ambulatory-accessible water closet or toilet compartment: 42"

long minimum at each side wall starting 12" maximum from rear wall.
- In transfer showers (typical, 36" deep), extend across control wall and across back wall to a point 18" from the maximum of control wall.
- In roll-in showers, on three walls to within 6" minimum of adjacent walls, but never behind seat.
- For tub with permanent seat: on foot end wall, 24" minimum grab bar at front edge of tub; on back wall, 12" maximum from foot end wall and 15" maximum from head end wall, plus an additional grab bar the same length mounted 9" above the tub rim.
- For tub without a permanent seat: on foot end wall, 24" minimum grab bar at front edge of tub; on head end wall, 12" minimum grab bar at front edge of tub; on back wall, 24" minimum grab bar 24" maximum from head end wall and 12" maximum from foot end wall, plus an additional grab bar the same length mounted 9" above the tub rim.

Design Suggestions
There are a variety of colors other than standard steel now available for grab bars. Textured grab bars provide a better gripping surface than smooth bars. If studs or blocking are insufficient or difficult to locate, it might be possible to attach to the studs a painted wood 1 × 6 on the face of the wall, and attach the grab bar to that. In tight spaces, a fold-down grab bar allows flexibility in use, but should be used only in addition to the required fixed grab bars. Grab bars are vital for safety, and installation methods must meet strength requirements.

Key Items
Grab bars, fasteners, either blocking in wall or anchors, and, possibly, finish materials.

Level of Difficulty
Moderate. May require finish work.

Estimates

Install grab bar, gypsum/metal stud wall

Description	Quantity	Unit	Labor-Hours	Material
Cutout demolition of partition	1.000	Ea.	0.333	0.00
2″ x 4″ blocking	0.005	M.B.F.	0.286	2.33
Miscellaneous materials for gypsum board repair	1.000	Job	0.800	0.14
Labor minimum to repair and paint gypsum board	1.000	Job	2.000	0.00
Grab bars	1.000	Ea.	0.400	51.50
Totals			**3.819**	**53.97**

Total per each including general contractor's overhead and profit **$345**

Install grab bar, gypsum/metal stud wall with ceramic tile

Description	Quantity	Unit	Labor-Hours	Material
Cutout demolition of partition	1.000	Ea.	0.333	0.00
2″ x 4″ blocking	0.005	M.B.F.	0.286	2.33
Misc. materials for gypsum board and ceramic tile repair	1.000	Job	0.800	0.14
Labor minimum to repair gypsum board and ceramic tile	1.000	Job	0.080	0.13
Grab bars	1.000	Ea.	0.400	51.50
Totals			**1.899**	**54.10**

Total per each including general contractor's overhead and profit **$225**

47
Install New Toilet

In some facilities that are being modified for access, an old toilet (WC) might not be accessible, or its location might not be accessible. In such situations, it might be easier (and less expensive) to install a new one meeting ADA rather than modify the existing one. At the same time, other upgrade features can be added, such as a low-flush water-saving model and other general upgrade features.

ADAAG Reference
604, Water Closets and Toilet Compartments

Where Applicable
Toilets in accessible stalls or in accessible single-use toilet rooms.

Design Requirements
- 56" from wall behind wall-hung WC to obstruction or wall in front of WC (59" if floor mounted WC); 60" from wall beside WC to obstruction on other side of WC.
- WC centerline 16" to 18" from the wall.
- Top of toilet seat 17" to 19" a.f.f. Seats that spring automatically to upright position are not acceptable.
- Flush valve on the approach side.
- Grab bars 33" to 36" a.f.f., 1-1/2" to 2" diameter, 1-1/2" from wall, 42" long minimum at side wall starting 12" maximum from rear wall, 36" long minimum at rear wall extending 12" from center of WC in one direction and 24" in other.
- 12" clearance above grab bars, 1-1/2" clearance below from any protruding object.
- Grab bar fitings tight so grab bars will not turn.
- Toilet paper dispenser 7" to 9" in front of WC, 14" to 19" a.f.f.

Design Suggestions
If the toilet is far from the side wall and is being replaced but not

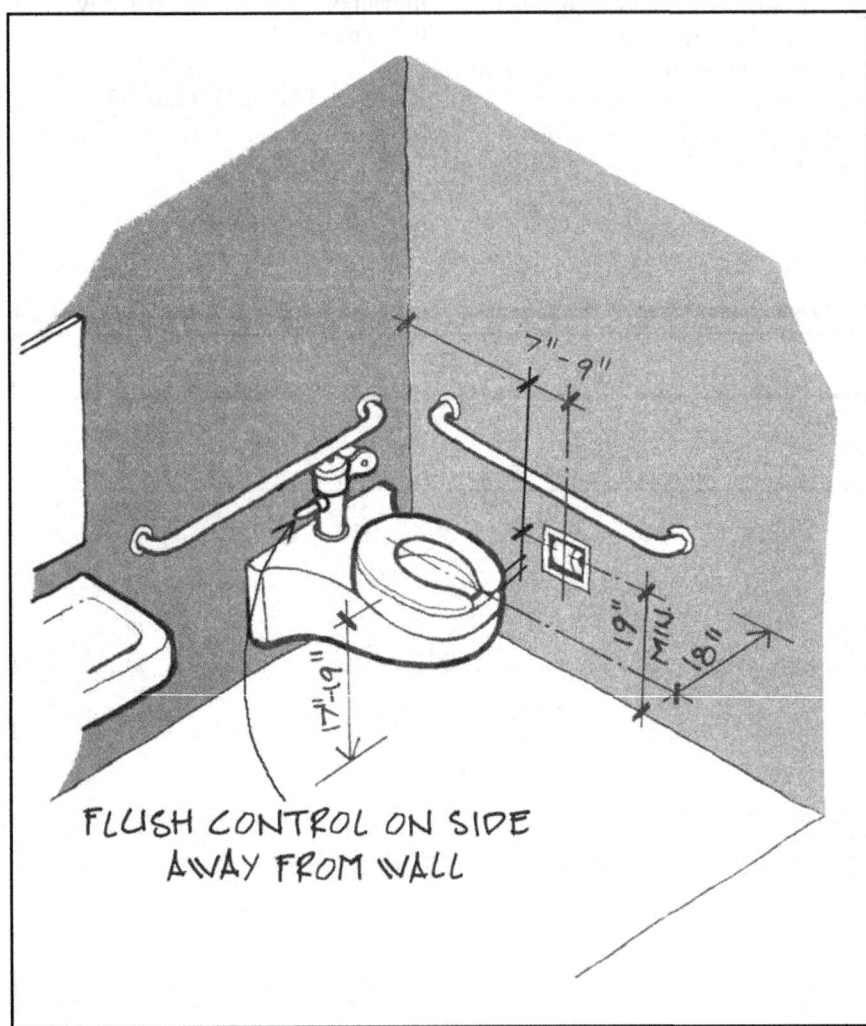

7"-9"

19" MIN.

18"

17"-19"

FLUSH CONTROL ON SIDE AWAY FROM WALL

relocated, blocking can be added to the wall surface to bring the grab bar closer to the toilet, or a new low wall can be added against the existing wall to provide a mounting surface for the grab bar. Wall-hung models provide more maneuvering space than floor-mounted ones.

Key Items

Toilet, piping, finish materials.

Level of Difficulty

Moderate to high. Requires plumber; can involve finish/tile work.

Estimates

Replace toilet in an existing location

Description	Quantity	Unit	Labor-Hours	Material
Remove wall-mounted water closet	1.000	Ea.	1.143	0.00
Labor minimum to disconnect and reconnect plumbing	1.000	Job	4.000	0.00
Water closet	1.000	Ea.	3.200	291.00
Totals			8.343	291.00

Total per each including general contractor's overhead and profit **$1,140**

Relocate toilet to new location where plumbing is available

Description	Quantity	Unit	Labor-Hours	Material
Labor minimum for toilet relocation	1.000	Job	4.000	0.00
Rough in supply, waste and vent	1.000	Ea.	5.634	139.00
Misc. materials for gypsum board and ceramic tile repair	1.000	Ea.	1.000	0.28
Labor minimum to repair gypsum board and ceramic tile	1.000	Job	0.080	0.13
Totals			10.714	139.41

Total per each including general contractor's overhead and profit **$1,045**

Relocate flush valve from wall side to approach side

Description	Quantity	Unit	Labor-Hours	Material
Labor minimum to relocate flush valve	1.000	Job	2.000	0.00
Totals			2.000	0.00

Total per each including general contractor's overhead and profit **$169**

Lower urinal

Description	Quantity	Unit	Labor-Hours	Material
Labor minimum for urinal relocation	1.000	Job	4.000	0.00
Rough in supply, waste, and vent	1.000	Ea.	5.654	92.00
Misc. materials for gypsum board and ceramic tile repair	1.000	Ea.	1.000	0.28
Labor minimum to repair gypsum board and ceramic tile	1.000	Job	0.080	0.13
Totals			10.734	92.41

Total per each including general contractor's overhead and profit **$974**

48
Modify Existing Toilet or Urinal

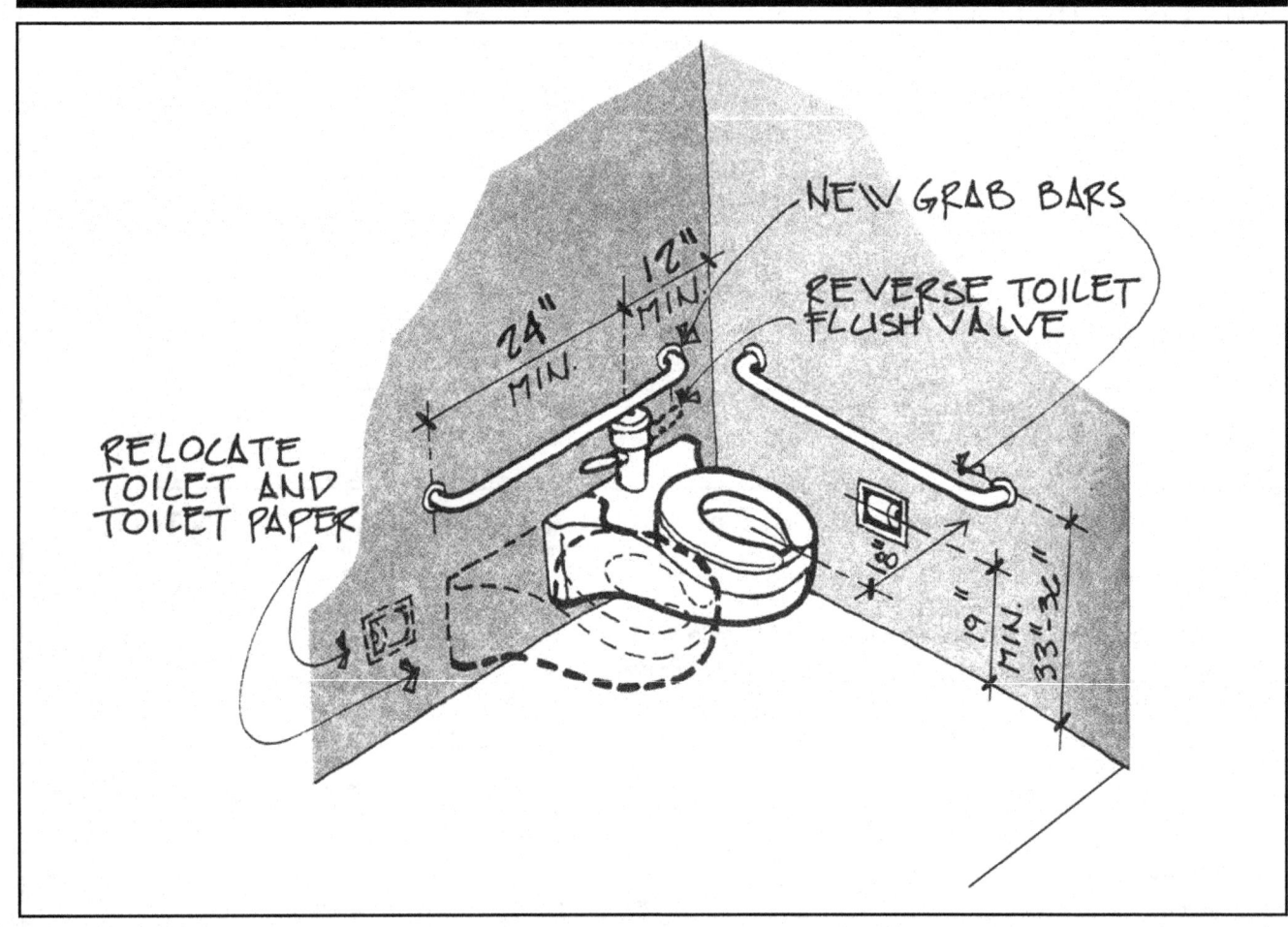

NEW GRAB BARS

REVERSE TOILET FLUSH VALVE

24" MIN.

12" MIN.

RELOCATE TOILET AND TOILET PAPER

18"

19" MIN.

33"-36"

If an existing toilet or urinal is in an accessible location but has inaccessible features, or would be too expensive or difficult to move, several elements can still be modified to increase accessibility. For a toilet, reversing the flush valve, removing obstacles in front of or beside the fixture, adding a seat height extender, or relocating the toilet paper holder and grab bars can all increase compliance and ease of use. For a urinal, relocating partitions or removing obstacles, lowering the urinal, or modifying the flush valve might be possible.

ADAAG Reference
See references for Project 47. 605, Urinals

Where Applicable

Toilets in accessible stalls or in accessible single-use toilet rooms. Toilet rooms with more than one urinal.

Design Requirements

See Requirements for Project 47.

Design Suggestions

Several features of an existing toilet can be modified to make it accessible, including the toilet location, seat height, flush valve location, and toilet paper location. Grab bars can also be added, using one of several approaches.

It might be possible to add a permanently installed seat height extender if the seat is low. If the toilet is far from the side wall, and it would be expensive to move it (typically the case), blocking can be added to the wall surface to bring the grab bar closer to the toilet. A new low wall can also be added against the existing wall to provide a mounting surface for the grab bar. (It might be possible to add floor-mounted grab bars 18" from the centerline of the toilet, but it may be very difficult to meet structural requirements. If installed, floor-mounted grab bars should be braced to the side wall to prevent wobbling.) Lowering a urinal is possible, but expensive, since it involves both plumbing and finish work. If a toilet is accessible, modifying the urinal is usually unnecessary. A floor-mounted urinal easily complies with the rim height requirement, but care must be taken to ensure that the flush valve is within reach range.

If the toilet room has a number of urinals, consider making the rim of the accessible one at 15" so it can be used by children. Also consider locating a grab bar over one of the regular height urinals.

Key Items

Toilet and/or urinal, piping and plumbing, grab bars, finish materials.

Level of Difficulty

Moderate to high. Requires plumber and may involve finish/tile work. Lowering the urinal requires structural modification as well.

Estimates

Modify toilet

Description	Quantity	Unit	Labor-Hours	Material
Labor minimum for toilet and flush valve relocation	1.000	Job	4.000	0.00
Rough in supply, waste and vent	1.000	Ea.	5.634	139.00
Misc. materials for gypsum board and ceramic tile repair	1.000	Ea.	1.000	0.28
Labor minimum to repair gypsum board and ceramic tile	1.000	Job	0.080	0.13
Grab bars	3.000	Ea.	1.200	154.50
Double-roll toilet tissue dispenser	6.000	Ea.	2.000	93.30
Totals			13.914	387.21

Total per each including general contractor's overhead and profit **$1,702**

49
Create Accessible Stall

Many toilet stalls are at the end of a path of travel between the row of stalls and the wall. It is sometimes possible to create an accessible stall by extending the partition across the path of travel to the opposite wall and entering straight on, but only if the stall has adequate maneuvering space. In a multiple-stall toilet room with typical inaccessible stalls, it is often possible to create an accessible stall by removing one toilet (WC) and combining the space of two stalls. If this is prohibited by local plumbing codes that require a certain number of fixtures for a given building population, try to seek a variance.

ADAAG Reference
See references for Project 47.

Where Applicable
All accessible toilet rooms with more than one stall.

Design Requirements
- See Requirement for Project 47.
- Stall door with 32" clear opening, accessible hardware, self-closing hinge if out-swinging.
- 60" wide by 56" minimum clear inside dimension for wall-mounted WC (60" × 59" for floor-mounted WC), with door swing not swinging into these dimensions.
- 9" minimum a.f.f. toe clearance extending 6" under the front and one side partitions (not required if stall is 66"+ wide and 65"+ deep).
- WC seat 17" to 19" a.f.f., centerline 16" to 18" from wall, flush valve on open side. Locate WC farthest corner from door.
- Grab bars 33" to 36" a.f.f., 1-1/4"-2" diameter, 1-1/2" from wall, 42" long minimum at side wall starting 12" maximum from rear wall, 36" long minimum at rear wall extending 12" from center of WC in one direction and 24" in other.
- Toilet paper dispenser 7" to 9" in front of WC, 14" to 19" a.f.f.

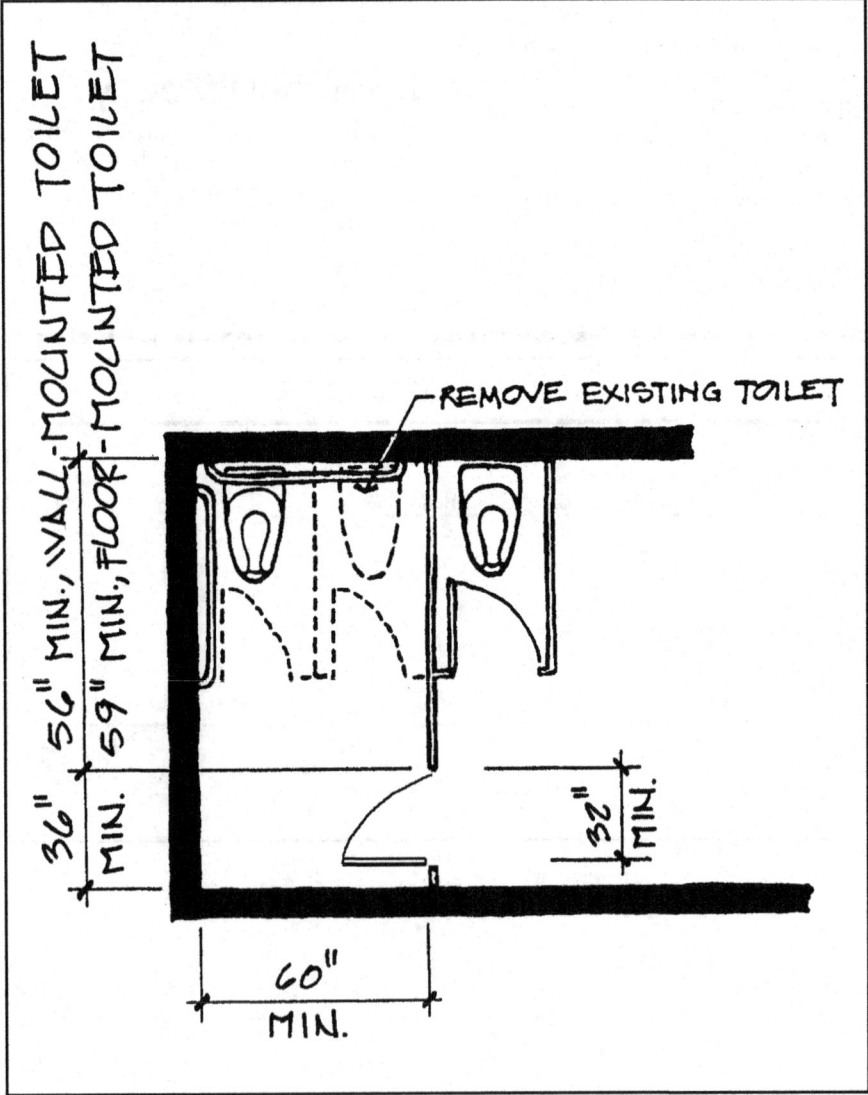

36" MIN., WALL-MOUNTED TOILET
56" MIN., WALL-MOUNTED TOILET
59" MIN., FLOOR-MOUNTED TOILET
MIN.

REMOVE EXISTING TOILET

32" MIN.

60" MIN.

- Coat hook within reach range 48" a.f.f.

Design Suggestions

It is usually recommended that the end toilet be used for the accessible stall, so the door can swing in and there is sufficient space on the latch side. Also, grab bars can be attached to the wall rather than to a partition.

Be sure that grab bars are anchored to either framing or blocking, so that they meet the minimum structural strength requirements (250 lbs. minimum). If the toilet can be reused, but is too low, a raised seat can be installed. If the flush valve is on the wrong side, you may have to replace the tank or valve stem.

Key Items

Minimum: grab bars, partitions, door hardware; possible a new toilet and/or raised seat.

Level of Difficulty

Low, moderate, or high. Requires plumbing and finish work.

Estimates

Install accessible stall

Description	Quantity	Unit	Labor-Hours	Material
Remove hollow metal toilet partitions	1.000	Ea.	1.000	0.00
Remove wall-mounted water closet	1.000	Ea.	1.143	0.00
Labor minimum to disconnect plumbing	1.000	Job	4.000	0.00
Water closet	1.000	Ea.	3.200	291.00
Rough in supply, waste and vent	1.000	Ea.	5.634	139.00
Misc. materials for gypsum board and ceramic tile repair	1.000	Ea.	1.000	0.28
Labor minimum to repair gypsum board and ceramic tile	1.000	Job	0.080	0.13
Painted metal toilet partitions	1.000	Ea.	2.667	390.00
Grab bars	3.000	Ea.	1.200	154.50
Totals			**19.924**	**974.91**

Total per each including general contractor's overhead and profit   **$3,149**

Install grab bar, gypsum/metal stud wall with ceramic tile

Description	Quantity	Unit	Labor-Hours	Material
Cutout demolition of partition	1.000	Ea.	0.333	0.00
2" x 4" blocking	0.005	M.B.F.	0.286	2.33
Misc. materials for gypsum board and ceramic tile repair	1.000	Ea.	1.000	0.28
Labor minimum to repair gypsum board and ceramic tile	1.000	Job	0.080	0.13
Grab bars	1.000	Ea.	0.400	51.50
Totals			**2.099**	**54.24**

Total per each including general contractor's overhead and profit   **$240**

Add new toilet partition with 36" door

Description	Quantity	Unit	Labor-Hours	Material
Ceiling-hung painted metal cubicle	1.000	Ea.	4.000	430.00
Add for accessible options	1.000	Ea.	0.000	320.00
Totals			**4.000**	**750.00**

Total per each including general contractor's overhead and profit   **$1,571**

50
Install New Sink

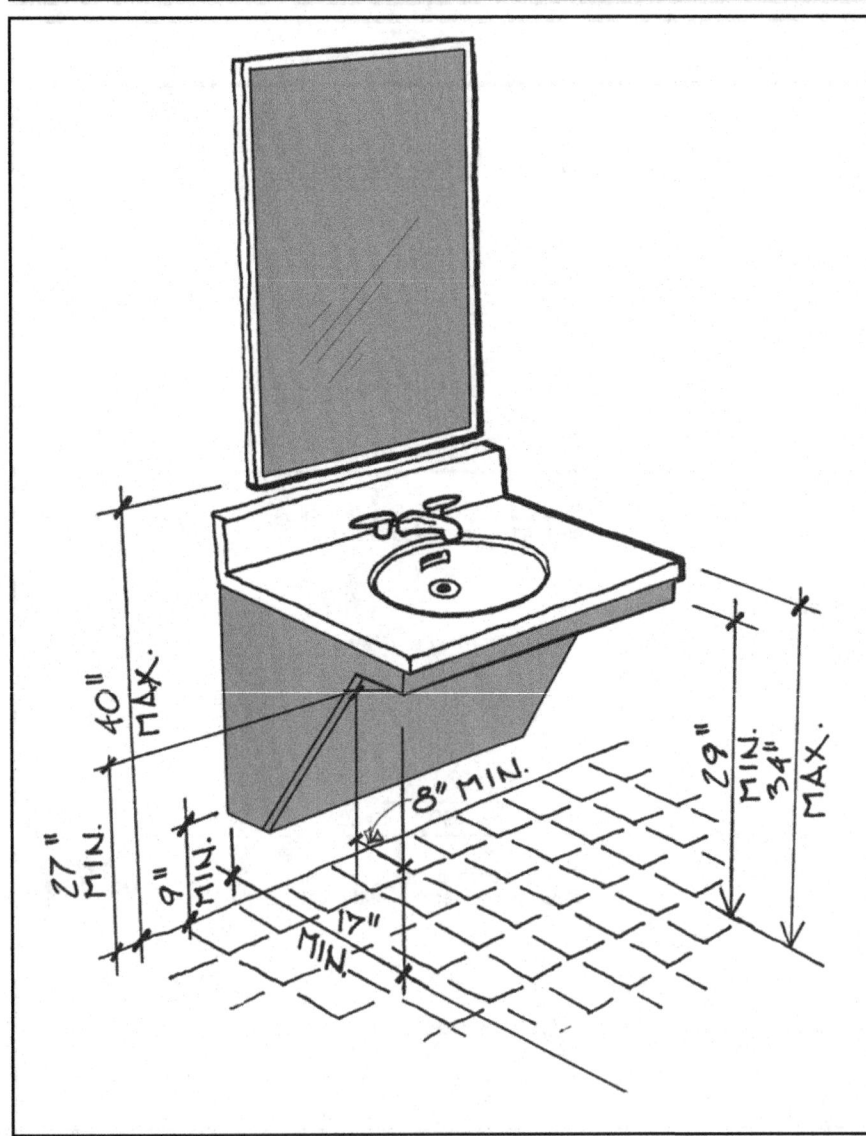

In situations where a sink is inaccessible but is too difficult to modify (such as an old pedestal sink, a sink with an exceptionally deep bowl, one resting on a vanity), replacing it with an accessible version might be the easiest way to achieve accessibility and compliance. In many situations, it can be replaced using the existing plumbing and perhaps structural bracing, but replacing the lavatory also allows the possibility of moving it to a more accessible location. Replacing the sink does not require using a designated "handicap" model. There are many attractive, standard models that comply with ADAAG that can be installed.

ADAAG References
213.3.4, Lavatories
305, Clear Floor or Ground Space
306, Knee and Toe Clearance
308, Reach Ranges
309, Operable Parts
603.2, Clear Floor Space
603.3, Mirrors
603.4, Coat Hooks and Shelves
606, Lavatories and Sinks

Where Applicable
At least one in all public toilet rooms.

Design Requirements
- 30" wide by 48" front approach to sink extending 19" maximum under sink.
- 34" maximum to rim, 27" clear knee space below bowl starting at 8" from front.
- Pipes wrapped with insulation or configured to protect against contact.
- Faucets operable with closed fist (electronic sensor faucets acceptable); self-closing faucets to remain on for at least ten seconds.
- Mirrors and shelves over sink 40" a.f.f. maximum.

- Dispensers (such as soap dispensers) operable with a closed fist, 48" a.f.f. if sink protrudes no more than 20" from wall. Otherwise 44".

Design Suggestions

If a sink is located in the only accessible bathroom or in a high-use bathroom, and is not accessible due to insufficient knee clearance below, it might be cheaper to replace it than to modify it or change its height. It is possible to use standard models to meet apron height and knee space requirements. They are not only less costly, but are also easier to install and minimize the stigma and visual disruption of having one so called "handicapped" sink in a row of standard sinks. (This is also true for tilted "HP" mirrors.) At least one mirror must be installed with its bottom edge no more than 40" a.f.f., if over a counter. The accessible height mirror does not have to be above the accessible sink. If it is not above a sink, it must start at 35" a.f.f., maximum.

Since soap dispensers are almost always over the sink, they might have to be lowered.

Key Items

Sink, piping, faucets, insulation, and under-sink or wall bracing.

Level of Difficulty

Moderate to high. Involves plumbing, finish, possibly structural work.

Estimates

Install new sink, faucet, handles and shelf w/ tilted mirror (plumbing available)

Description	Quantity	Unit	Labor-Hours	Material
Remove lavatory	1.000	Ea.	0.800	0.00
Rough in supply, waste and vent	1.000	Ea.	9.639	220.00
Wall-hung porcelain enamel lavatory (22" x 19")	1.000	Ea.	2.000	405.00
Faucet, handles and drain	1.000	Ea.	0.800	107.00
Misc. materials for gypsum board and ceramic tile repair	1.000	Ea.	1.000	0.28
Labor minimum to repair gypsum board and ceramic tile	1.000	Job	0.080	0.13
Mirror with stainless steel shelf	1.000	Ea.	0.400	90.50
Totals			14.719	822.91

Total per each including general contractor's overhead and profit **$2,535**

51
Modify Existing Sink

A sink on an accessible route can almost always be modified to increase accessibility. Removing a vanity or apron, replacing the existing faucet with a single-lever or paddle handles, or even adjusting the bowl height can help make a sink not only comply with the ADA, but also more convenient for all users. (Ask anyone who has ever tried to use a ball faucet with soapy hands.)

ADAAG Reference
See references for Project 50.

Where Applicable
At least one sink in all public toilet rooms.

Design Requirements
See Project 50 for design requirements.

Design Suggestions
If a sink is required to be accessible, there may be many minor modifications necessary to bring it into compliance, such as raising the sink, replacing faucets with lever faucets, modifying the apron below to create sufficient knee space, and so forth. At least one mirror must be installed with its bottom edge no more than 40" a.f.f., but it is not required to be above the accessible sink. If not over a sink, the bottom of the mirror at 35" a.f.f. or lower.

Key Items
Sink, piping, faucets, insulation, mirror, and under-sink or wall bracing.

Level of Difficulty
Varies. May involve insulation, plumbing, faucets, and possibly finish and structural work.

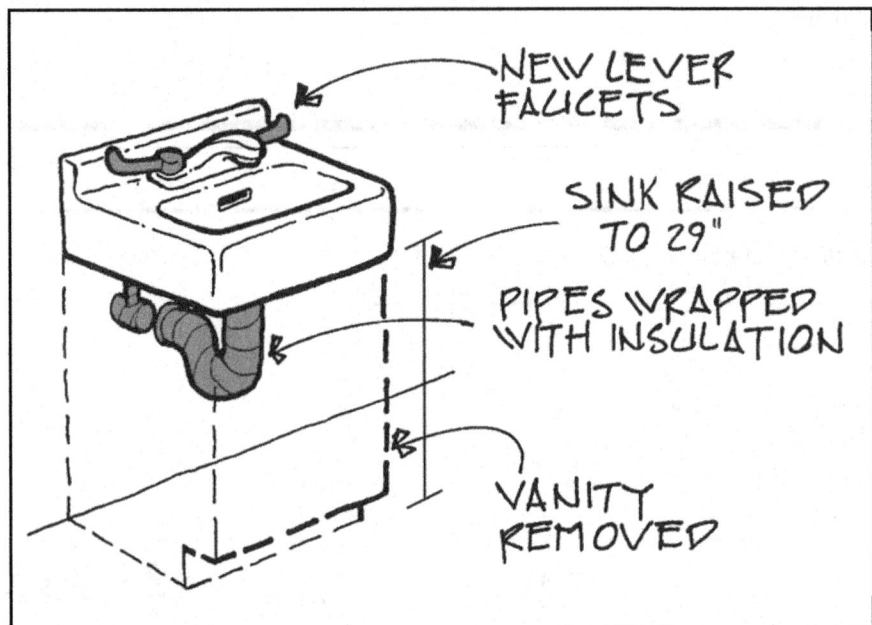

NEW LEVER FAUCETS

SINK RAISED TO 29"

PIPES WRAPPED WITH INSULATION

VANITY REMOVED

Estimates

Lower existing sink

Description	Quantity	Unit	Labor-Hours	Material
Labor minimum for lowering sink	1.000	Ea.	2.000	0.00
Rough in supply, waste and vent	1.000	Ea.	9.639	220.00
Misc. materials for gypsum board and ceramic tile repair	1.000	Ea.	1.000	0.28
Labor minimum to repair gypsum board and ceramic tile	1.000	Job	0.080	0.13
Totals			12.719	220.41

Total per each including general contractor's overhead and profit **$1,358**

Replace knob faucets with paddle faucets

Description	Quantity	Unit	Labor-Hours	Material
Labor minimum to remove faucets	1.000	Job	2.000	0.00
Paddle faucets with spout	1.000	Ea.	0.800	107.00
Totals			2.800	107.00

Total per set including general contractor's overhead and profit **$419**

Wrap pipe with insulation

Description	Quantity	Unit	Labor-Hours	Material
1" wall rubber tubing, flexible closed cell foam insulation	4.000	L.F.	0.381	7.60
Totals			0.381	7.60

Total per each including general contractor's overhead and profit **$44**

Remove base cabinets, add additional bracing

Description	Quantity	Unit	Labor-Hours	Material
Labor minimum to remove base cabinets	1.000	Job	2.000	0.00
Blocking	0.010	M.B.F.	0.471	4.65
Bracing cover (plywood)	32.000	S.F.	1.138	54.08
Additional labor required for cutting and fitting plywood bracing cover	1.000	Job	4.000	0.00
Totals			7.609	58.73

Total per each including general contractor's overhead and profit **$626**

Remove apron below plastic laminate counter

Description	Quantity	Unit	Labor-Hours	Material
Labor minimum to remove apron	1.000	Job	2.000	0.00
Totals			2.000	0.00

Total per each including general contractor's overhead and profit **$115**

51. Modify Existing Sink *(continued)*

Remove apron below synthetic stone counter

Description	Quantity	Unit	Labor-Hours	Material
Labor minimum to remove apron	1.000	Job	2.000	0.00
Totals			2.000	0.00

Total per each including general contractor's overhead and profit **$115**

Remove apron below granite counter

Description	Quantity	Unit	Labor-Hours	Material
Labor minimum to remove apron	1.000	Job	2.000	0.00
Totals			2.000	0.00

Total per each including general contractor's overhead and profit **$115**

Copyright Reed Construction Data, 2004

Notes

52
Install or Modify Toilet Room Dispensers

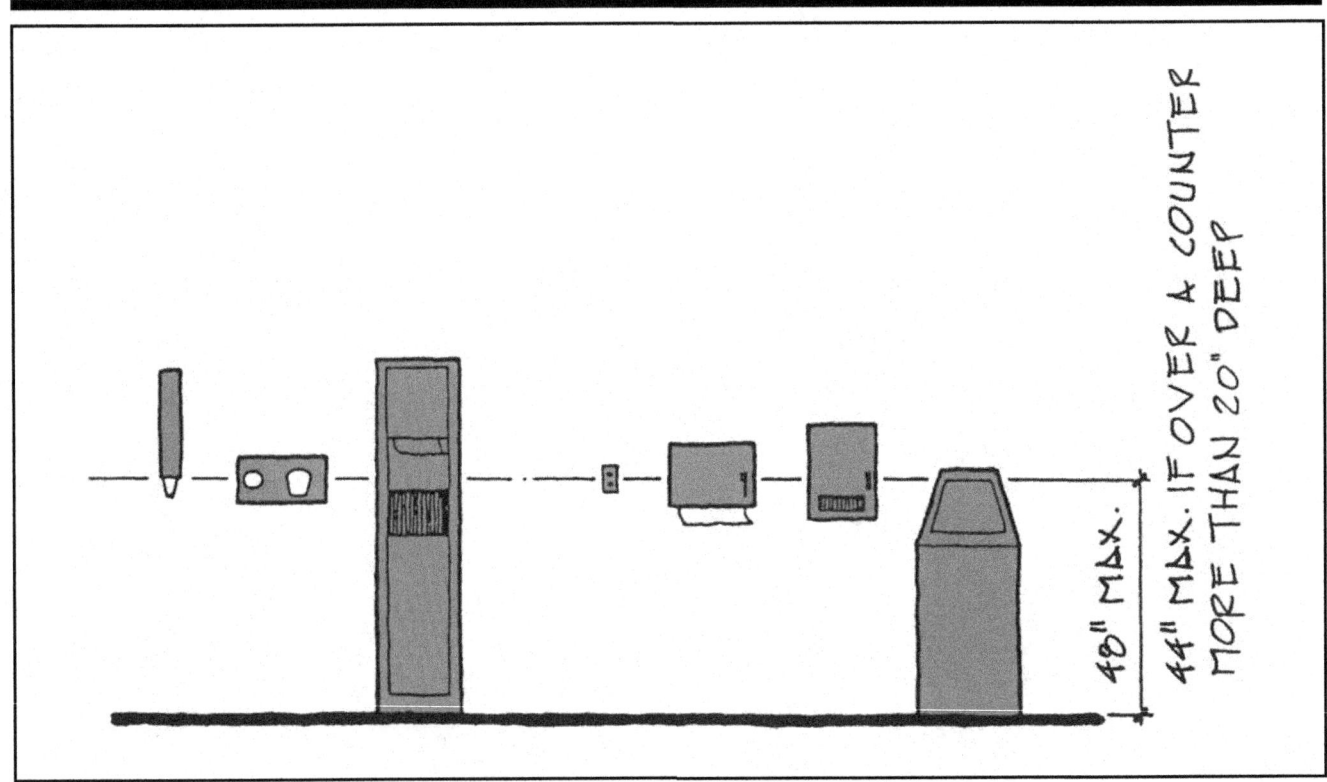

The most common problem with toilet room dispensers is that they are simply out of reach for people with disabilities, or the waste basket is in the way. Relocating dispensers so that they are within reach ranges is an access modification that could make a restroom more accessible and usable. Some dispensers have controls that are difficult to operate for people with limited fine motor control. Replacing these is also a helpful modification.

ADAAG References
305, Clear Floor or Ground Space
308, Reach Ranges
309, Operable Parts

Where Applicable
At least one of each type of dispenser in accessible public toilet rooms.

Design Requirements
- Dispensers on an accessible route, 30" × 48" clear floor space in front.
- Dispensers 15" to 48" a.f.f., with 2" clearance minimum below grab bars and 12" clearance minimum above grab bars.
- Controls operable without tight grasping, pinching, or twisting.

Design Suggestions

Specify dispensers with easy-to-operate controls. Locate all dispensers as close as possible to accessible fixtures (but still within easy reach) and at least 18" from an inside corner in a location where no one will ever put a waste basket.

Consider making the highest operating part of dispensers at 42" so they are within the reach ranges of children. Consider locating the soap dispenser lower than 48".

Carefully check all controls to make sure they do not require pinching or excessive force. One should be able to operate them with one's fist.

Key Items

Dispensers, fasteners, anchors or blocking, and finish materials if necessary.

Level of Difficulty

Low to moderate. May involve finish or tile work.

Estimate

Lower dispenser, screwed to wall (gypsum board on metal studs)

Description	Quantity	Unit	Labor-Hours	Material
Cutout demolition of partition	1.000	Ea.	0.333	0.00
2" x 4" blocking	0.005	M.B.F.	0.286	2.33
Miscellaneous materials for gypsum board repair	1.000	Job	0.800	0.14
Labor minimum to repair and paint gypsum board	1.000	Job	2.000	0.00
Labor minimum to remove and re-install dispenser	1.000	Job	1.600	0.00
Totals			**5.019**	**2.47**

Total per each including general contractor's overhead and profit

$345

53
Modify Multiple-Stall Toilet Rooms

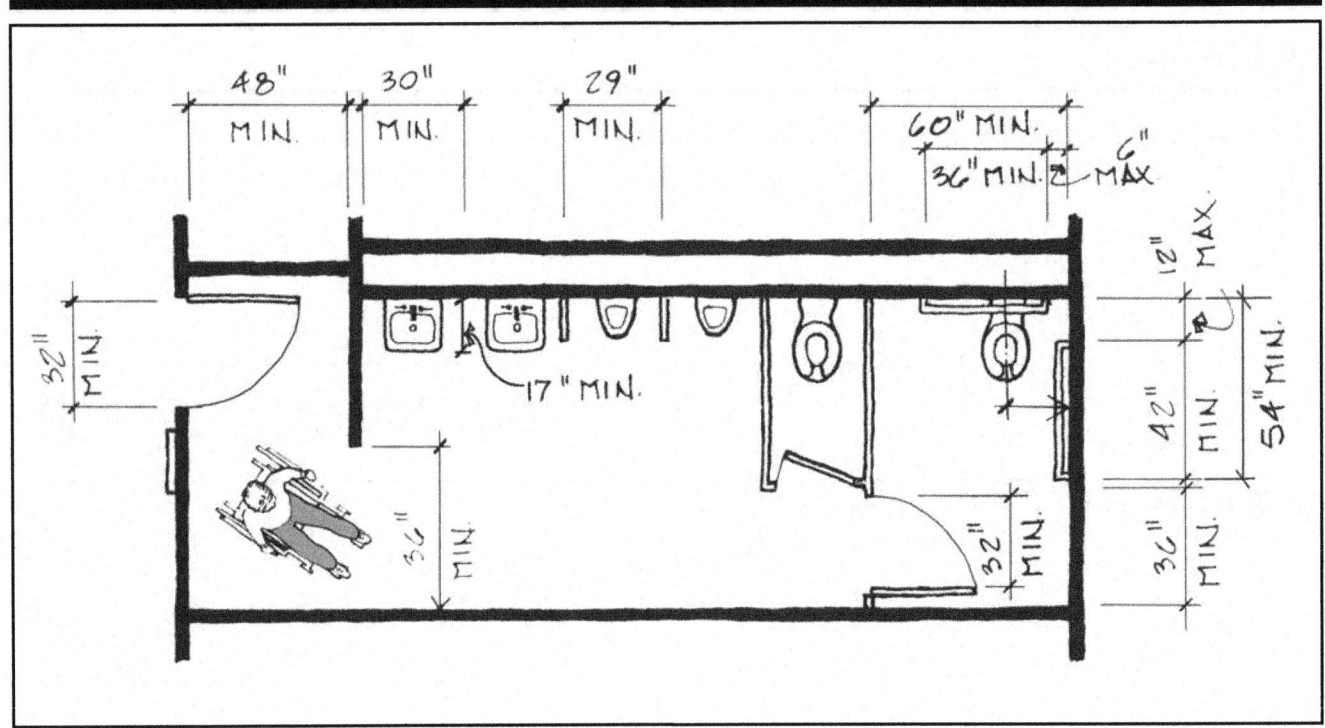

Many multi-stall toilet rooms can be made accessible by rearranging a few elements. Problems are often with entrances, dimensions of stalls, and lavatories without proper clearances. Faucets and accessories — mirrors, hooks, dispensers and grab bars — are usually easy to adjust or replace. Often it is more efficacious to build a family (unisex) toilet room nearby.

ADAAG References
See Project 46 for references.
605, Urinals

Where Applicable
All public and common-use, multiple-use toilet rooms.

Design Requirements
- Requirements of Project 46 apply.
- One accessible urinal and one fully accessible stall as described in

Project 46. If the sum of water closets and urinals is 6 or more an ambulatory stall is required.
- Ambulatory stall: 36" wide by 60" deep with self-closing, out-swinging door and grab bars on both sides of stall. WC centered in space.
- 9" minimum a.f.f. toe clearance extending 6" under the front and one side partitions (not required if stall is 66"+ wide and 65"+ deep).
- Accessible urinal: rim 17" a.f.f. maximum, 13-1/2" minimum

depth; accessible flush valve 44" a.f.f. maximum if hand operated; 30" wide by 48" deep space in front of urinal (29" minimum between privacy screens if they do not project beyond urinal).

Design Suggestions

Privacy partitions at entrances of toilet rooms can make it difficult to enter/exit a multi-stall toilet room, even when the clear widths comply. If possible, try to locate the entrance door so there is no need for a privacy partition, or use an entry vestibule with doors arranged so that tight turns are not necessary. But be certain to provide required maneuverability space at each door and between doors.

Often an accessible stall can be created by combining two stalls.

Similarly, extra space can be created at a tight entry by removing one lavatory and shifting the privacy partition.

It is possible (and preferred) to use standard sinks that are not designated "HP" if they are installed so that they meet the apron height and knee space requirements; they will be less costly, easier to install and use and all the sinks can be the same. The same is true of tilted "HP" mirrors; if a mirror above the sink cannot be lowered, add another mirror starting at 35" a.f.f. maximum in a different location. Tilting does not always create an accessible alternative. *Never* use an adjustable tilting mirror. If you do not install an alternate mirror, at least one mirror over the sink must be accessible with its bottom edge no

more than 40" above the floor.

Be sure that grab bars are anchored to either framing or blocking, so that they meet the minimum structural strength requirements.

Stall door hardware must also be accessible. Most knobs are not; slide or flip latches are usually easier to use.

Key Items

Plumbing fixtures/hookups, toilets, sinks with paddle faucets; toilets stall partitions; grab bars, mirror, dispensers; door, framing, partitions, finishes; ventilation.

Level of Difficulty

High. Requires plumbing, framing, electrical, and finish (usually tile) work.

Estimate

Remodel multi-use toilet room

Description	Quantity	Unit	Labor-Hours	Material
Remove metal stud/gypsum board partition for door opening	21.000	S.F.	0.969	0.00
Paddle faucets with spout	1.000	Ea.	0.800	107.00
1" wall rubber tubing, flexible closed cell foam insulation	10.000	L.F.	0.952	19.00
Remove hollow metal toilet partitions	1.000	Ea.	1.000	0.00
Remove wall-mounted water closet	1.000	Ea.	1.143	0.00
Labor minimum to disconnect plumbing	1.000	Job	4.000	0.00
Miscellaneous materials for gypsum board and ceramic tile repair	1.000	Ea.	1.000	0.28
Labor minimum to repair gypsum board and ceramic tile	1.000	Job	0.080	0.13
Painted metal toilet partitions	1.000	Ea.	2.667	390.00
Grab bars	3.000	Ea.	1.200	154.50
Combined soap/towel dispenser/mirror/shelf	1.000	Ea.	0.800	330.00
Water closet	1.000	Ea.	3.200	291.00
Rough in supply, waste and vent	1.000	Ea.	5.634	139.00
Interior door frame	1.000	Ea.	1.000	75.00
Hollow metal flush door, 3'-0" x 6'-8"	1.000	Ea.	0.941	166.00
Totals			25.386	1,671.91

Total per each including general contractor's overhead and profit ___ **$4,740**

54
Construct Family/ Single-User Accessible Toilet Rooms

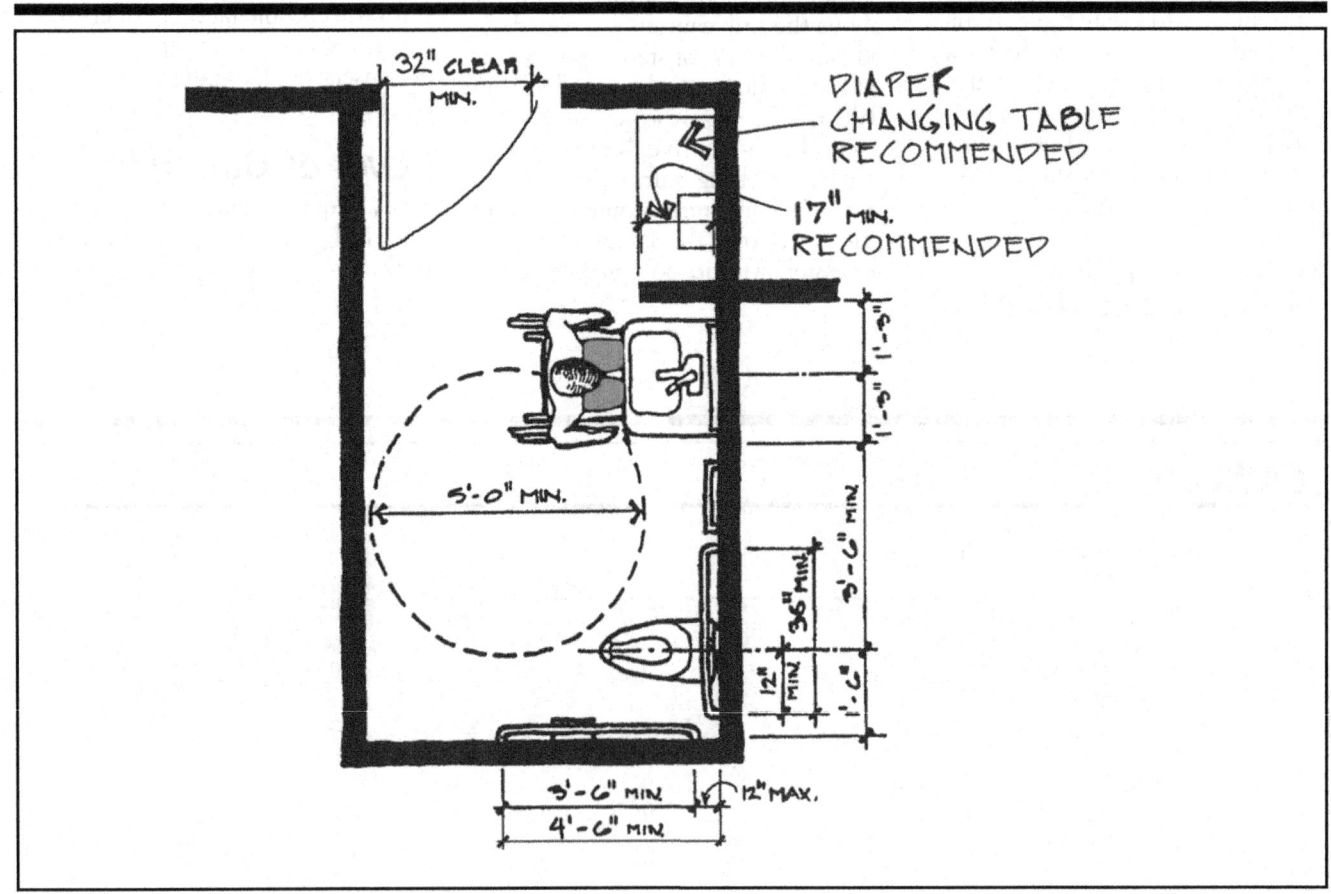

As an access modification, single-use restrooms should be considered even where existing, multiple-fixture toilet rooms can be made to be, or are, accessible. If the toilet room is to be unisex, verify that local plumbing codes allow them. Another advantage of the family toilet room is that it can be used by parents who need to accompany small children.

ADAAG References
205, Operable Parts
213, Toilet Facilities and Bathing Rooms
216.8, Signs

304, Turning Space
305, Clear Floor or Ground Space
306, Knee and Toe Clearance
308, Reach Ranges
309, Operable Parts
404, Doors, Doorways, and Gates
603, Toilet and Bathing Rooms
604, Water Closets and Toilet Compartments

606, Lavatories and Sinks
609, Grab Bars
703, Signs

Where Applicable
Where there are public and common-use, single-user or multi-fixture toilet rooms, including toilet rooms used by employees.

Design Requirements
- Located on an accessible route.
- Door must comply with width, hardware, pull weight, and maneuvering space requirements.
- Latch that meets section 309, Operable Parts, on door.
- Raised character, Braille, signage adjacent to door (latch side).
- If other toilet rooms are not in facility, use international symbol of accessibility on sign at accessible toilet rooms.
- 60" by 60" diameter clear turning space inside.
- 30" × 48" clear space in front of lavatory, must extend 17" under lavatory.
- Sink: paddle, lever, or other accessible faucets; 34" maximum to top, 27" clear knee space below bowl; exposed pipes wrapped with insulation or protected from contact.
- Mirror and shelf over sink, 40" maximum a.f.f.
- 56" from wall behind toilet to obstruction or wall in front of toilet (59" if wall-hung); 60" from wall beside WC to obstruction on other side of WC.
- Water Closet: top of seat 17" to 19" a.f.f., flush valve on open side.

- Grab bars 33" to 36" a.f.f.; 1-1/4" x 2" diameter, 1-1/2" from wall; 42" long minimum at side wall starting 12" maximum from rear wall, 36" long minimum at back wall extending 12" from center of toilet in one direction and 24" in other.
- 12" clearance above grab bars and 1-1/2" clearance below from any projecting object.
- Grab bars shall not rotate in their fittings and shall support 250 lbs at any place or angle.
- Toilet paper dispenser 7" to 9" in front of toilet, 15" to 48" a.f.f., with 2" clearance minimum from grab bars.
- Accessories: mirror 40" a.f.f. maximum to bottom; hook 48" a.f.f max.; operable parts of all dispensers 48" a.f.f. maximum and accessible in operation; waste container not in clearance spaces, door maneuverable spaces, or turning radius.
- Slip-resistant flooring.

Design Suggestions
Adding a family toilet room is one of the most useful modifications one can make, particularly when there are families involved or the other restrooms are on another floor. But it should not be limited to disabled folk only; many people can benefit from the size and easy access. The dimensions given are minimum; make the toilet room large enough for two people—a person in a wheelchair and an attendant, or a father and young daughter. Consider installing a diaper changing station, and even a small bench. Try to use a design where the

door swings into the room so that it can be closed more easily by a person in a wheelchair. Use standard lavatories that are not designated "HP" units, mounted to meet the apron height and knee space requirements. They will be less costly, more attractive, and easier to install and use.

The same is true of tilted "HP" mirrors; having tilted mirrors does not meet accessibility requirements if they are too high. Consider mounting dispensers and devices at 42" a.f.f. to highest operating part so they can be used by children. Consider installing a full length mirror, top at 74", bottom at 18". It can be used by everyone. If a mirror is installed that is not over a sink, it must start at 35" a.f.f. maximum.

If you can, avoid flush valves and related plumbing behind the toilet to prevent imbalance or injury when a person leans back. If plumbing must be exposed, check if the local authorities will allow a seat back.

Be sure that grab bars are anchored to either framing or blocking, so that they meet the minimum weight resistance requirements.

Key Items
Plumbing fixtures/hook-ups, toilet, sink with paddle faucets; door with privacy lock, framing, finishes; grab bars, mirror, dispensers; ventilation.

Level of Difficulty
High. Requires plumbing, framing, electrical, and finish (usually tile) work.

Estimate

Install single-user toilet room

Description	Quantity	Unit	Labor-Hours	Material
Remove metal stud/gypsum board partition for door opening	21.000	S.F.	0.969	0.00
Remove gypsum board from stud face at room interior	55.000	S.F.	0.440	0.00
Remove vinyl flooring	60.000	S.F.	0.960	0.00
Ceiling demolition (suspended A.C.T.)	60.000	S.F.	0.640	0.00
Metal stud partition (3-5/8″ wide, 16″ O.C.)	230.000	S.F.	3.834	48.30
1/2″ gypsum board, taped and finished	230.000	S.F.	3.813	52.90
2″ x 4″ blocking	0.030	M.B.F.	1.714	13.95
Water-resistant 1/2″ gypsum board	300.000	S.F.	2.400	72.00
Ceramic tile cove base	30.000	L.F.	5.275	92.10
Ceramic tile floor, thin-set 4″ x 4″	60.000	S.F.	5.053	125.40
Ceramic tile walls, thin-set 4″ x 4″	300.000	S.F.	25.263	627.00
Mineral fiber suspended ceiling	60.000	S.F.	1.391	100.80
Vinyl tile flooring	23.000	S.F.	0.368	38.64
Vinyl base, 6″ high	23.000	L.F.	0.584	18.40
Painting	230.000	S.F.	5.842	50.60
Interior door frame	1.000	Ea.	1.000	75.00
Hollow metal flush door, 3′-0″ x 6′-8″	1.000	Ea.	0.941	166.00
Hinges	1.500	Pr.	0.000	66.75
Lever-handled lockset	1.000	Ea.	0.800	121.00
Signage	1.000	Ea.	0.457	31.50
Layout & drilling of anchor holes	4.000	Ea.	0.213	0.04
Plastic shields and screws	4.000	Ea.	0.133	0.16
Grab bars	3.000	Ea.	1.200	154.50
Combined soap/towel dispenser/mirror/shelf	1.000	Ea.	0.800	330.00
Double-roll toilet tissue dispenser	1.000	Ea.	0.333	15.55
13-gallon waste receptacle	1.000	Ea.	0.800	161.00
Accessible lavatory	1.000	Ea.	2.286	320.00
Rough in supply, waste and vent	1.000	Ea.	9.639	220.00
Water closet	1.000	Ea.	3.200	291.00
Rough in supply, waste and vent	1.000	Ea.	5.634	139.00
Totals			**85.982**	**3,331.59**

Total per square foot including general contractor's overhead and profit	**$407**
Total per each including general contractor's overhead and profit	**$9,358**

Copyright Reed Construction Data, 2004

Notes

55
Children's Accessible Bathroom Fixtures

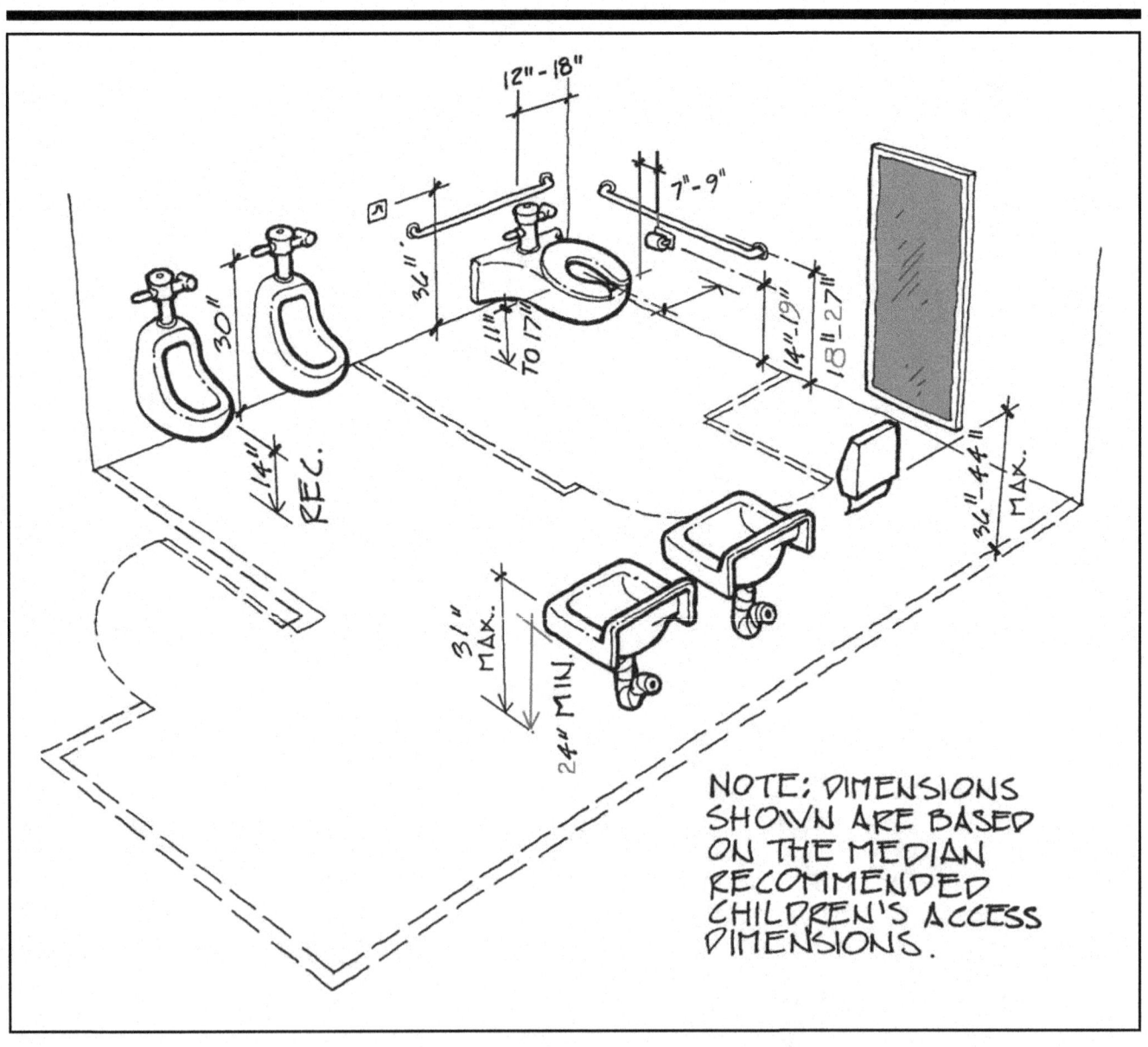

NOTE: DIMENSIONS SHOWN ARE BASED ON THE MEDIAN RECOMMENDED CHILDREN'S ACCESS DIMENSIONS.

The specifications given in the preceding sections are based on adult dimensions and anthropometrics. Designing toilet rooms for children with disabilities requires different dimensions, as discussed below.

ADAAG References

305, Clear Floor or Ground Space
306, Knee and Toe Clearance
308.1, Reach Ranges for Children
309, Operable Parts
604.3, Clear Floor Space at Water Closet
604.8.1, Wheelchair Accessible Compartments
604.9, Water Closets and Toilet Compartments for Children's Use
606, Lavatories and Sinks
609, Grab Bars

Where Applicable

There are no scoping requirements for children's toilet rooms. Common sense must be used to decide where and when to provide fixtures accessible to a child. Children's accessibility design guidelines, cited below, will be most useful in settings that have regular use by younger children, such as in daycare centers and elementary schools. The sinks and toilets in a kindergarten room should be appropriately sized, as should be the toilet fixtures in an elementary school.

Design Requirements

- Located on an accessible route, with accessible entrance and maneuvering space inside.
- Accessible compartments: 60" wide by 59" deep, with 12" toe clearance for 6" under front and on side partition (no required side toe clearance if width is greater than 66", no required front toe clearance if depth is greater than 65").
- Door hardware, hooks, grab bars, and dispensers should meet age appropriate requirements. Flush no higher than 36" a.f.f. and toilet paper dispenser no higher than 19".
- Grab bars to withstand 250 lbs force applied anywhere and in any direction, and be tight in their fittings.
- Sinks, drains, and faucets comply with adult standards, except 24" knee clearance and 31" a.f.f. rim or counter surface allowed for ages 6 through 12.

Design Suggestions

When designing bathrooms for children's access, it is important to allow greater maneuvering space than usual, since many children who use wheelchairs are not yet able to control them as well as adults, and frequently an adult will assist the child. Also, some children who use wheelchairs may need assistance in the bathroom after the age that other children are able to use the bathroom unassisted.

Privacy is an important consideration, and visual privacy should be provided with full-height partitions where possible. If a school is being modified for an individual child, it is vital to get input on the student's needs, although the restroom should also allow for future use by other children with various needs. In situations where restrooms are not divided by age, the middle dimensions (such as 15" a.f.f. for toilet seat height) provide a usable compromise.

Key Items

Standard plumbing installations: fixtures, water supply, drain pipes, faucets, insulation, counters, mirrors, dispensers.

Level of Difficulty

Moderate to high.

	Children's Fixtures			
	Toilet Seat Height	Grab Bar Height	Dispenser Height	W.C. Centerline to Wall
Ages 3-4	11"-12"	18"-20"	14"	12"
Ages 5-8	12"-15"	20"-25"	14"-17"	12"-15"
Ages 9-12	15"-17"	25"-27"	17"-19"	15"-18"

Estimate

Install children's accessible bathroom fixtures (plumbing available)

Description	Quantity	Unit	Labor-Hours	Material
Two-piece floor-mounted water closet	1.000	Ea.	3.019	160.00
Junior-size toilet seat	1.000	Ea.	0.333	31.00
Rough in supply, waste and vent	1.000	Ea.	5.634	139.00
16" x 14" porcelain enamel on cast iron lavatory	1.000	Ea.	2.000	315.00
Paddle faucets with spout	1.000	Ea.	0.800	107.00
Rough in supply, waste and vent	1.000	Ea.	9.639	220.00
Grab bars	1.000	Ea.	0.348	23.50
Totals			**21.773**	**995.50**

Total for 2 fixtures as described including general contractor's overhead and profit **$3,361**

Notes

56
Modify Existing Tub

Many people who use wheelchairs transfer to a seat in a tub. However, a bathtub does not necessarily have to be part of a fully accessible bathroom to be modified for increased accessibility. Accessible grab bars, controls, and an in-tub seat can be added to assist people with mobility, balance, and grasping impairments.

ADAAG References
309, Operable Parts
607, Bathtubs
609, Grab Bars
610, Seats

Where Applicable
Accessible bathing facilities: at least one accessible tub or shower.

Design Requirements
- On accessible route.
- Clear floor space at least 60" wide and at least 30" deep (a lavatory protruding into this space at the foot end is allowed). If end seat, clear space 12" beyond end of seat.
- Either an in-tub seat or end seat; end seat 15" wide minimum, full depth of tub and 17"-19" a.f.f..
- Grab bars 1-1/4" to 2" in diameter, 1-1/2" from wall.
- Grab bars for tub without permanent seat: two grab bars on rear wall, top bar 33" to 36" a.f.f., bottom 8"-10" above rim of tub; bars 12" max. from control end of tub and 24" max. from head end. One 24" long minimum grab bar on wall at foot at the front edge of the tub; one 12" minimum grab bar on wall at head (opposite controls) at the front edge of the tub.
- Grab bars for tub with end seat: two grab bars on rear wall, 15" maximum from head end wall and 12" from foot end wall. Top bar 33" to 36" a.f.f., bottom 8"-10" above rim of tub. Both grab bars 48" minimum. One 24" minimum long grab bar on foot wall (wall with controls) at the front edge of the tub.

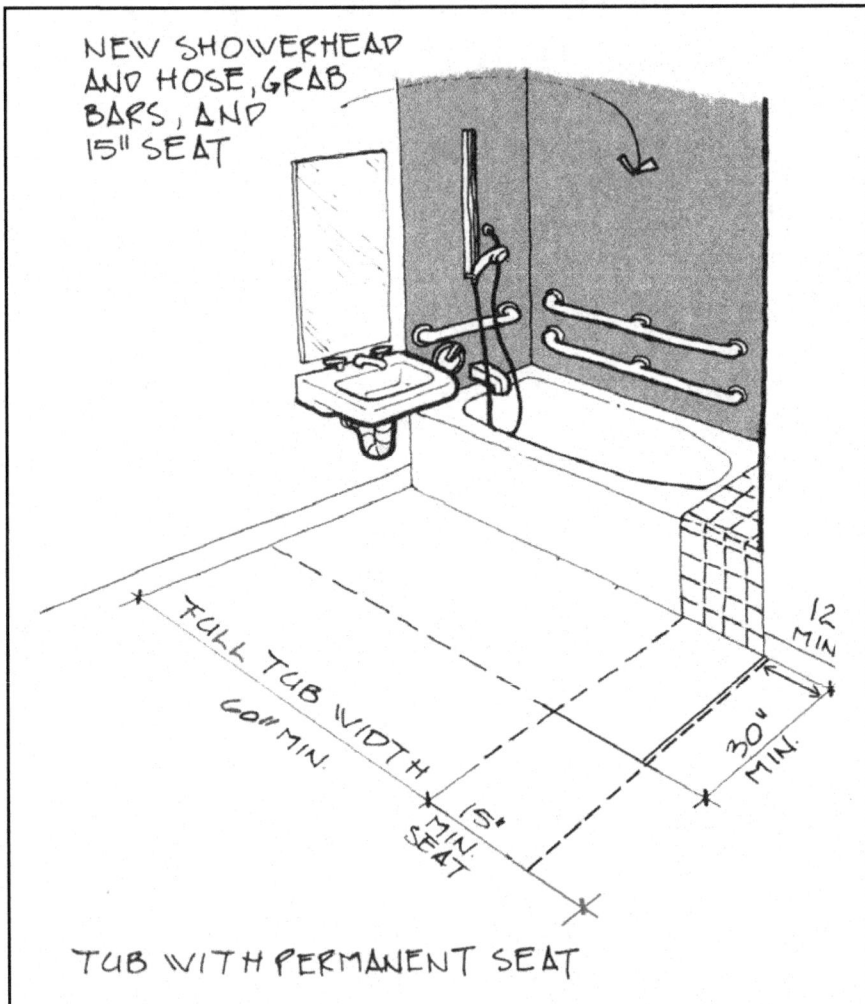

NEW SHOWERHEAD AND HOSE, GRAB BARS, AND 15" SEAT

FULL TUB WIDTH
60" MIN.
15" MIN. SEAT
30" MIN.
12 MIN

TUB WITH PERMANENT SEAT

- Grab bars to withstand 250 lbs. applied anywhere in any direction and be tight in their fittings.
- Controls operable with a closed fist, on foot wall, below grab bar, and between center-line of tub and front edge of tub.
- Shower operable both as fixed and hand-held shower with a flexible hose at least 59" long. Spray unit must have an on/off control with a non-positive valve.
- Water thermal shock protected with maximum temperature of 120°.

- No tub enclosure obstructing controls or transfer to a seat, or mounted on a rim. No enclosure tracks on tub rim.

Design Suggestions

In-tub seats prevent a shower curtain from closing fully; for an in-tub seat, two curtains would minimize spillage. Some sliding doors with tracks overhead permit transfer to a seat.

Key Items

Shower head on a hose, tub seat, grab bars.

Level of Difficulty

Low to moderate. Replacing shower head and faucets might require a plumber; grab bar installation may require finish (tile) work. Extending tub enclosure to include end seat requires additional framing.

Estimates

Extend tub enclosure

Description	Quantity	Unit	Labor-Hours	Material
Remove wood stud/gypsum board partition	36.000	S.F.	1.661	0.00
Remove partition finishes	12.000	S.F.	0.107	0.00
Wood stud partition	3.000	L.F.	0.480	9.75
Water-resistant 5/8" gypsum board	36.000	S.F.	0.288	9.72
Seat framing	0.030	M.B.F.	1.412	13.95
Seat sheathing	18.000	S.F.	0.192	11.70
Ceramic tile covering for seat	9.000	S.F.	0.758	18.81
Ceramic tile walls, thin-set 4-1/4" x 4-1/4"	45.000	S.F.	3.789	94.05
Totals			8.687	157.98

Total per each including general contractor's overhead and profit **$508**

Add in-tub seat

Description	Quantity	Unit	Labor-Hours	Material
Portable in-tub or in-shower seating	1.000	Ea.	0.267	129.00
Totals			0.267	129.00

Total per each including general contractor's overhead and profit **$240**

56. Modify Existing Tub (continued)

Add grab bars

Description	Quantity	Unit	Labor-Hours	Material
Cutout demolition of partition	1.000	Ea.	0.333	0.00
2" x 4" blocking	0.005	M.B.F.	0.286	2.33
Miscellaneous materials for gypsum board and ceramic tile repair	1.000	Ea.	1.000	0.28
Labor minimum to repair gypsum board and ceramic tile	1.000	Job	0.080	0.13
Grab bars	1.000	Ea.	0.400	51.50
Totals			**2.099**	**54.24**

Total per each including general contractor's overhead and profit   **$240**

Replace fixed shower head with shower head on hose

Description	Quantity	Unit	Labor-Hours	Material
Labor minimum to remove and install shower head	1.000	Job	2.000	0.00
Bar-mounted hand-held shower head	1.000	Ea.	0.364	63.50
Totals			**2.364**	**63.50**

Total per each including general contractor's overhead and profit   **$309**

Relocate controls, tiled wall

Description	Quantity	Unit	Labor-Hours	Material
Cutout demolition of partition	1.000	Ea.	0.333	0.00
Labor minimum to remove & replace mixing valve	1.000	Job	2.000	0.00
Miscellaneous materials for gypsum board and ceramic tile repair	1.000	Ea.	1.000	0.28
Labor minimum to repair gypsum board and ceramic tile	1.000	Job	0.080	0.13
Totals			**3.413**	**0.41**

Total per each including general contractor's overhead and profit   **$267**

Notes

57
Replace Tub with Roll-in Shower

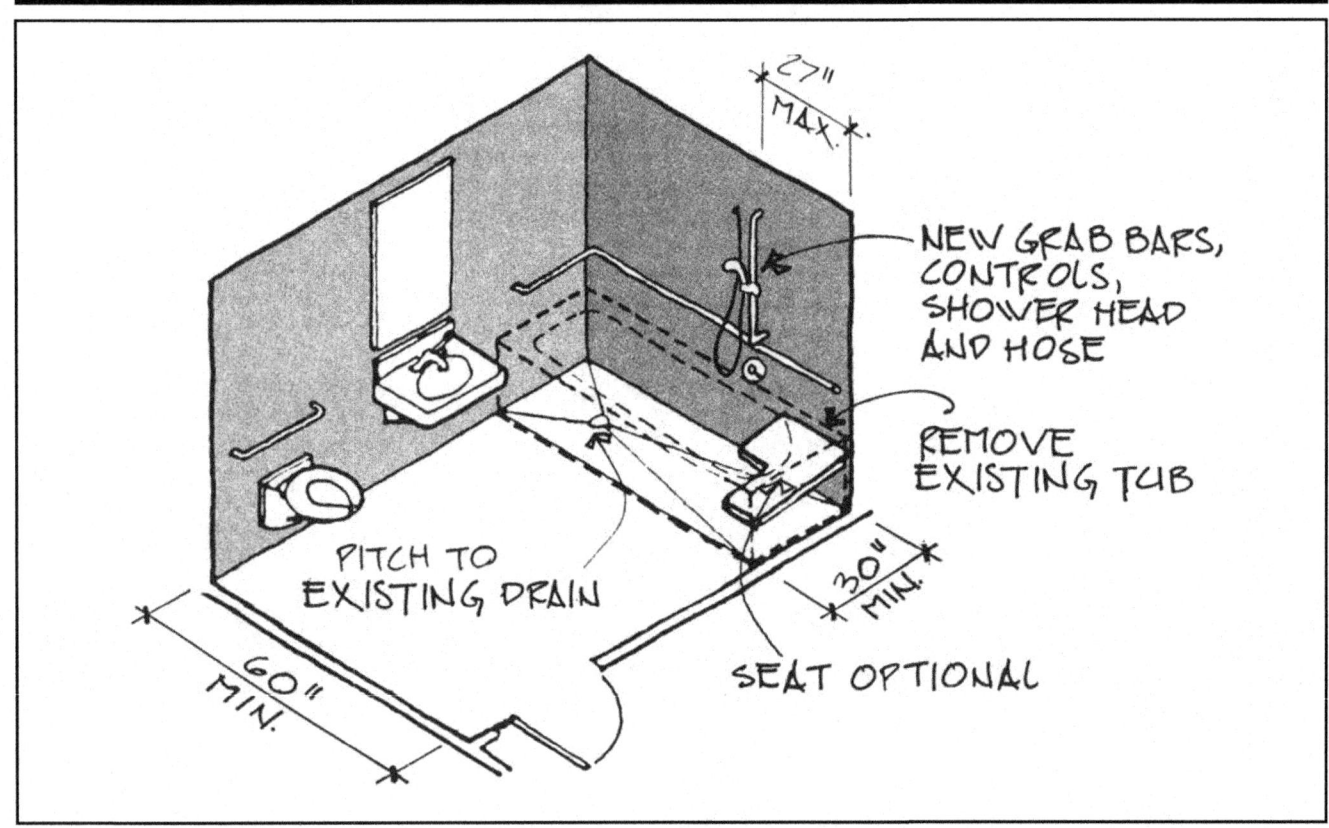

27" MAX.

NEW GRAB BARS, CONTROLS, SHOWER HEAD AND HOSE

REMOVE EXISTING TUB

PITCH TO EXISTING DRAIN

SEAT OPTIONAL

30" MIN.

60" MIN.

The floor space of a standard 30" × 60" tub is the same as what is required for a roll-in shower. In an existing structure with plumbing and drain locations in place, it is often possible to create an accessible roll-in shower by replacing an existing tub. In an accessible bathroom, this modification can create a shower usable by a wide range of people, including those who use wheelchairs.

ADAAG References
303, Changes in Level
309, Operable Parts
608, Shower Compartments
609, Grab Bars
610, Seats

Where Applicable
Where there already is at least one tub, and accessible bathing facilities are required.

Design Requirements
- On accessible route.
- Curbs, 1/2" maximum and beveled.
- Roll-in 30" × 60" shower with 36" × 60" clear space in front.

- Seat (optional) 15" to 16" deep, mounted 17" to 19" above floor of stall and 1-1/2" maximum from adjacent wall. Seat must withstand a 250 lb. force applied anywhere.
- Grab bars 33" to 36" a.f.f., 1-1/4" to 2" in diameter, 1-1/2" from wall, on both sides and rear walls (omit behind seat if seat is installed).
- Grab bars to withstand 250 lbs. applied anywhere in any direction and be tight in their fittings.
- Shower (fixed or hand-held) with flexible hose at least 59" long.
- Water thermal shock protected with maximum temperature of 120°.

- Controls operable with a closed fist, 3" above the grab bar to 48" a.f.f. maximum. When there is a seat, controls mounted on wall adjacent to, and within 27" from the seat wall.

Design Suggestions

Although the 30" x 60" dimensions preclude the shower being a transfer-type shower, consider installing a fold-down seat anyway to allow for people who transfer to a seat and those who shower in a wheelchair. Recessing the floor pan below floor level allows for smooth roll-in. This is relatively easy to do in new construction, but can be difficult in existing buildings, since it requires strengthening the existing floor structure. The shower curtain should reach all the way to the floor to prevent spillage.

Key Items

Either prefab shower stall or custom-framed stall and floor pan, water supply and drainage, and wall/ceiling finish materials.

Level of Difficulty

High. Requires plumbing, finish, and possibly framing work.

Estimates

Replace tub with accessible fiberglass shower

Description	Quantity	Unit	Labor-Hours	Material
Remove bathtub	1.000	Ea.	2.000	0.00
Labor to disconnect plumbing and remove faucets/handles, etc.	1.000	Job	4.000	0.00
Remove partition finishes	66.000	S.F.	0.587	0.00
Fiberglass shower, corner seat, grab bars and nonskid floor	1.000	Ea.	8.000	640.00
Bar-mounted hand-held shower head	1.000	Ea.	0.364	63.50
Rough in supply, waste and vent	1.000	Ea.	13.445	141.00
Totals			**28.396**	**844.50**

Total per each including general contractor's overhead and profit **$3,652**

Replace tub with custom roll-in shower

Description	Quantity	Unit	Labor-Hours	Material
Remove bathtub	1.000	Ea.	2.000	0.00
Labor to disconnect plumbing and remove faucets/handles, etc.	1.000	Job	4.000	0.00
Remove partition finishes	66.000	S.F.	0.587	0.00
Water-resistant 5/8" gypsum board	110.000	S.F.	0.880	29.70
Copper shower pan	18.000	S.F.	1.440	55.98
Ceramic tile floor (pitched to drain)	15.000	S.F.	1.311	134.25
Ceramic tile walls, thin-set 4-1/4" x 4-1/4"	110.000	S.F.	9.263	229.90
Ceramic bath accessories	2.000	Ea.	0.390	19.00
Tub grab bar	1.000	Ea.	0.571	78.50
Grab bar verticle arms	2.000	Ea.	1.333	173.00
Bar-mounted hand-held shower head	1.000	Ea.	0.364	63.50
Rough in supply, waste and vent	1.000	Ea.	13.445	141.00
Totals			**35.584**	**924.83**

Total per each including general contractor's overhead and profit **$4,116**

58
Install Roll-in Shower

A shower that allows a person using a wheelchair to roll in is one of the most versatile access modifications, since it permits use both by people who use wheelchairs and those who do not. For people who shower while in their wheelchairs, it is vital. Prefab fiberglass roll-in showers are now standard and make installation relatively simple.

ADAAG References
309, Operable Parts
608, Shower Compartments
609, Grab Bars

Where Applicable
Accessible public bathing facilities.

Design Requirements
- On accessible route.
- No vertical lip at entry, but 1/2" threshold with a 1:2 slope is permitted.
- Standard compartment 30" minimum deep by 60" wide, with a 60" minimum wide entry and an adjacent 36" × 60" clear floor space in front. A sink may protrude into the clear floor space.
- Alternate compartment 36" by 60", with 36" wide entrance either at end or at one end of 60" side; controls mounted on end wall farthest from opening.
- Grab bars 33" to 36" a.f.f. (all grab bars the same height), 1-1/4" to 2" in diameter, 1-1/2" from wall, on three walls (omit behind seat), and extending to 6" maximum from adjacent walls.
- See seat requirements in Project 57 if optional seat is installed.
- Grab bars to withstand 250 lbs. force applied anywhere in any direction and be tight in their fittings.
- Controls located above grab bar, 38"-48" a.f.f. If a roll-in shower has a seat, controls to be located on wall adjacent to seat and within 27" of the seat wall.

36" X 60" MIN.
OPEN SPACE

36" MIN.

MINIMUM

36" RECOMMENDED

30"
MINIMUM

6" MAX

60"
MINIMUM

6" MAX

PLAN VIEW,
ROLL-IN SHOWER

- Shower spray unit that can be either fixed or hand-held with an on/off, non-positive shut-off valve and a flexible hose at least 59" long.
- Water temperature 120° maximum and thermal shock protected.

Design Suggestions

Recessing the floor pan below floor level allows for smooth roll-in. This can be difficult as a retrofit, since it requires strengthening the existing floor structure, but it is easier to do in new construction. Shower curtains should reach to the floor to prevent spillage. An adjustable shower head on a 24" vertical bar can be used by both tall and short people, but if it is left in the high position, it will be out of the reach of a person in a wheelchair. If space permits, consider installing a permanent seat for people who can transfer or need sitting as a bathing option. A permanent seat (or fold-down seat) has an advantage in that it is more stable than a portable seat that might shift under a person making a transfer.

Key Items

Either prefab shower stall or custom-framed stall and floor pan, water supply and drainage, ventilation, wall/ceiling finish materials.

Level of Difficulty

High. Requires plumbing, framing, finish work.

Estimates

Install prefabricated roll-in shower (plumbing available)

Description	Quantity	Unit	Labor-Hours	Material
Remove metal stud/gypsum board partition for shower opening	35.000	S.F.	1.615	0.00
Remove vinyl flooring	20.000	S.F.	0.320	0.00
Metal stud partition	77.000	S.F.	1.284	16.17
1/2" gypsum board, taped and finished	110.000	S.F.	1.824	25.30
Insulation	110.000	S.F.	0.652	23.10
Vinyl tile flooring	11.000	S.F.	0.176	18.48
Vinyl base, 6" high	12.000	L.F.	0.305	9.60
Painting	110.000	S.F.	2.794	24.20
Bar-mounted hand-held shower head	1.000	Ea.	0.364	63.50
Fiberglass shower, corner seat, grab bars and nonskid floor	1.000	Ea.	8.000	640.00
Rough in supply, waste and vent	1.000	Ea.	13.445	141.00
Totals			**30.779**	**961.35**

Total per each including general contractor's overhead and profit **$3,905**

Install custom roll-in shower

Description	Quantity	Unit	Labor-Hours	Material
Remove metal stud/gypsum board partition for shower opening	35.000	S.F.	1.615	0.00
Remove vinyl flooring	20.000	S.F.	0.320	0.00
Wood stud partition	11.000	L.F.	1.760	35.75
Insulation	110.000	S.F.	0.652	23.10
Water-resistant 5/8" gypsum board	110.000	S.F.	0.880	29.70
5/8" gypsum board, taped and finished	110.000	S.F.	1.824	28.60
Vinyl tile flooring	11.000	S.F.	0.176	18.48
Vinyl base, 6" high	12.000	L.F.	0.305	9.60
Painting	110.000	S.F.	2.794	24.20
Copper shower pan	18.000	S.F.	1.440	55.98
Gypsum board ceiling	15.000	S.F.	0.314	3.90
Ceramic tile floor (pitched to drain)	15.000	S.F.	1.311	134.25
Ceramic tile walls, thin-set 4-1/4" x 4-1/4"	110.000	S.F.	9.263	229.90
Ceramic bath accessories	2.000	Ea.	0.390	19.00
Tub grab bar	1.000	Ea.	0.571	78.50
Grab bar verticle arms	2.000	Ea.	1.333	173.00
Bar-mounted hand-held shower head	1.000	Ea.	0.364	63.50
Rough in supply, waste and vent	1.000	Ea.	13.445	141.00
Totals			**38.757**	**1,068.46**

Total per each including general contractor's overhead and profit **$4,473**

Notes

59
Modify Existing Shower to a Transfer-Type

Many existing showers have the approach and space to be accessible, but need modifications to comply with ADAAG. Even if small, some showers have a non-plumbing wall that can be relocated to increase the shower's size, and usually a 36" x 36" prefab shower can be installed. Creating a shower with grab bars, accessible controls, and a seat can make an existing shower into a transfer-type shower that can be accessible to a person in a wheelchair, and more useful to anyone with a mobility or balance impairment.

ADAAG References
309, Operable Parts
608, Shower Compartments
609, Grab Bars
610, Seats

Where Applicable
Public and residential bathing facilities.

Design Requirements
- On accessible route.
- Curbs, 1/2" maximum and beveled (2" curbs allowed in transfer-type, existing situations where 1/2" curbs cannot be achieved because of structural reasons).
- 36" × 36" clear shower stall with seat and 30" × 48" minimum clear floor space in front (long dimension parallel to shower entrance and protruding 12" minimum beyond seat end).
- Seat 15" to 16" deep, mounted 17" to 19" above floor of stall, 1-1/2" maximum from adjacent wall. Seats must withstand a 250 lb. force applied anywhere in any direction.
- Grab bars 33" to 36" a.f.f., 1-1/4" to 2" in diameter, 1-1/2" from wall; in 36" × 36" stalls, grab bars on side wall and wall opposite seat.
- Grab bars to withstand 250 lbs. applied anywhere in any direction and be tight in their fittings.

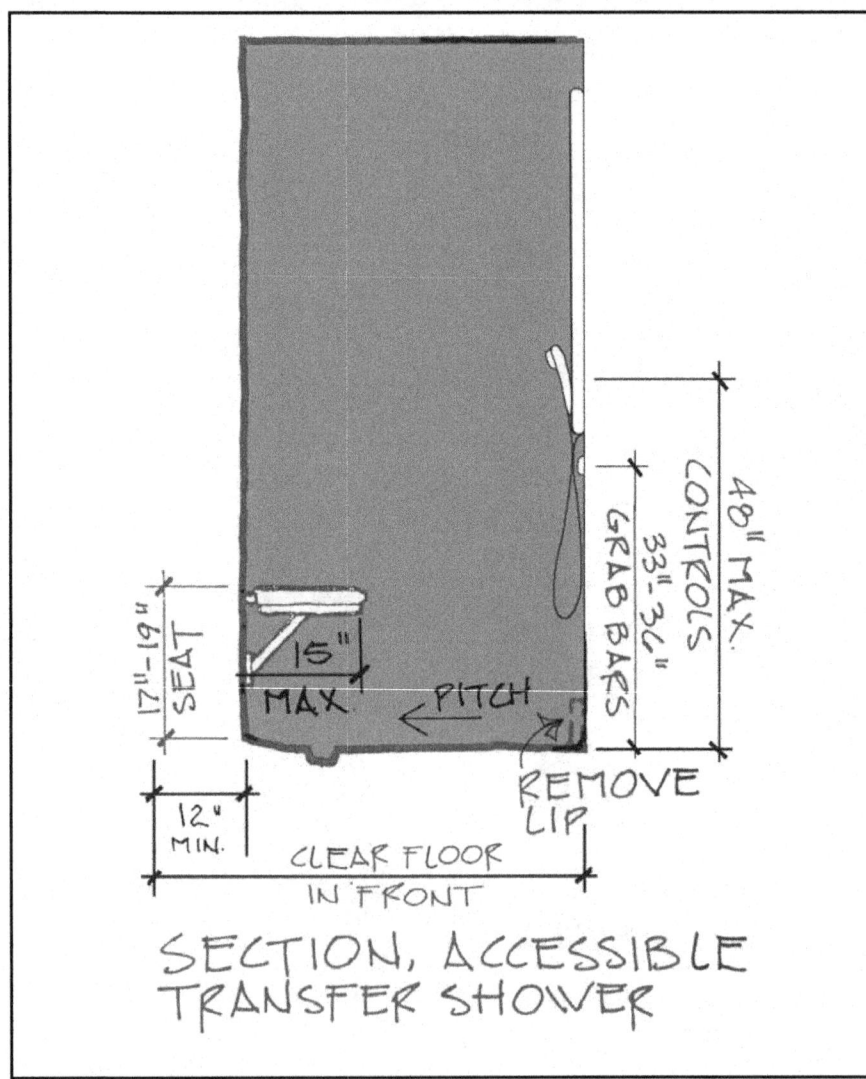

SECTION, ACCESSIBLE TRANSFER SHOWER

- Shower (fixed or hand-held) with flexible hose at least 59" long.
- Water thermal shock protected with maximum temperature of 120° F.
- Controls operable with a closed fist, 3" above the grab bar to 48" a.f.f. maximum. Controls mounted on wall opposite seat.

Design Suggestions

All 36" × 36" showers require a seat, since they are not roll-in showers. A transfer-type shower is often preferable because the low threshold of a roll-in type shower can result in wet floors. However, recessing the floor pan below floor level allows for a smooth roll-in, which increases options. This is relatively easy to do in new construction, but can be difficult in existing buildings, since it requires strengthening the existing floor structure.

Key Items

Grab bars, shower head on hose, fold-down seat, bevel at entrance, possibly new floor pan, new tile work.

Level of Difficulty

Moderate to high. Requires plumbing, seat installation, finish work. Removing the lip requires extensive modification to an existing shower.

Estimates

Remove concrete lip at entry and patch flooring

Description	Quantity	Unit	Labor-Hours	Material
Labor minimum for demolition and tile installation	1.000	Job	4.923	0.00
Ceramic tile floor	5.000	S.F.	0.437	44.75
Totals			5.360	44.75

Total per linear foot including general contractor's overhead and profit	**$78**
Total per each 5 linear feet including general contractor's overhead and profit	**$390**

Replace existing floor pan with pan recessed below floor

Description	Quantity	Unit	Labor-Hours	Material
Selective demolition of floor (concrete)	4.000	C.F.	3.556	0.00
Remove thin-set ceramic tiles	18.000	S.F.	0.427	0.00
Remove shower pan	1.000	Job	2.000	0.00
Install new copper shower pan	18.000	S.F.	1.440	55.98
Ceramic tile floor	15.000	S.F.	1.311	134.25
Ceramic tile walls, thin-set 4-1/4" x 4-1/4"	9.000	S.F.	0.758	18.81
Shower drain	1.000	Ea.	2.000	163.00
Totals			11.492	372.04

Total per each including general contractor's overhead and profit	**$1,379**

Copyright Reed Construction Data, 2004

59. Modify Existing Shower to a Transfer-Type *(continued)*

Replace fixed shower head with shower head on hose

Description	Quantity	Unit	Labor-Hours	Material
Labor minimum to remove and reset	1.000	Job	2.000	0.00
Bar-mounted hand-held shower head	1.000	Ea.	0.364	63.50
Totals			**2.364**	**63.50**

Total per each including general contractor's overhead and profit — **$309**

Add fold-down seat and grab bars to tiled shower wall

Description	Quantity	Unit	Labor-Hours	Material
Demolition of ceramic tile wall and gypsum board	3.000	Ea.	1.000	0.00
2" x 4" blocking	0.005	M.B.F.	0.286	2.33
Misc. materials for gypsum board and ceramic tile repair	1.000	Ea.	1.000	0.28
Labor minimum to repair gypsum board and ceramic tile	1.000	Job	0.080	0.13
Ceramic tile walls, thin-set 4-1/4" x 4-1/4" (material)	3.000	S.F.	0.253	6.27
Grab bars	1.000	Ea.	0.400	51.50
Fold-down seat	1.000	Ea.	0.267	129.00
Totals			**3.286**	**189.51**

Total per each including general contractor's overhead and profit — **$545**

Notes

60
Create Accessible Gang Showers

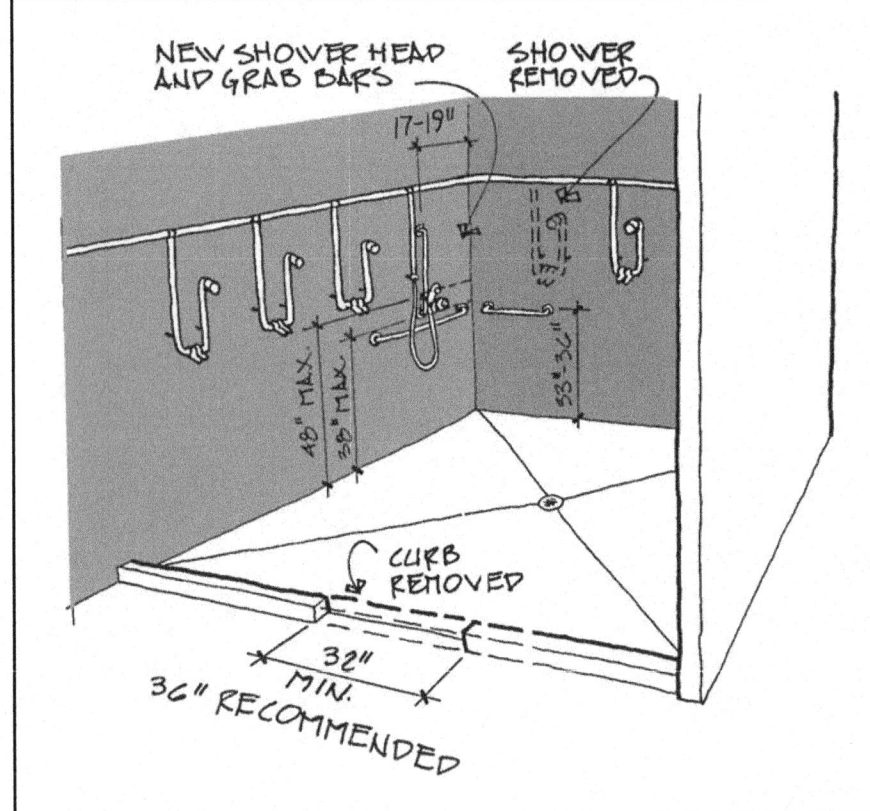

It is impossible to predict the precise access needs of all people who use a shower. Some people shower in a wheelchair, while some transfer to a seat. Others with mobility impairments who do not use a wheelchair can walk to a shower, but still need a seat and/or grab bars for assistance.

ADAAG addresses single-user roll-in and transfer stalls, but does not specifically address gang showers. Since gang showers are often located in facilities open to the public, they are important to modify for access in order to make them usable by as wide a range of people as possible. Such showers often have adequate maneuvering space along the approach and in the shower. A gang shower can be made accessible by creating a wide entrance, adding grab bars and possibly a seat, and modifying the controls.

Where there are multiple stalls, it is often possible to modify the front lip and add the other required elements. Even where existing materials are difficult to modify (such as marble stall partitions or concrete shower pans), it is often possible to add accessible features.

ADAAG References
303, Changes in Level
309, Operable Parts
608, Shower Compartments
609, Grab Bars
610, Seats

Where Applicable
At least one shower or shower stall in public shower rooms.

Design Requirements
- On an accessible route. Opening into the shower at least 32" wide and clear.
- Slip-resistant flooring.

- Curbs, 1/2" maximum and beveled. (Higher curbs allowed if entered/ exited by 1:12 sloped threshold.)
- Seat (optional), fixed or fold down, 15" to 16" deep, mounted 17" to 19" above floor and 1-1/2" maximum from adjacent wall. Seat must withstand a 250 lb. force applied anywhere in any direction.
- Grab bars, 36" minimum length, 33" to 36" a.f.f., 1-1/4" to 2" in diameter, 1-1/2" from wall; center on shower. If shower is adjacent to aside wall, a similar bar on that wall. (Omit behind seat if seat is installed on the adjacent wall.)
- Grab bars to withstand 250 lbs. applied anywhere in any direction and be tight in their fittings.
- Controls on an accessible route, 3" above the grab bar but no higher than 48", operable with a closed fist, with clear floor space 30" × 48" in front. If a seat is installed, controls mounted on wall adjacent to and within 27" from the seat wall.

- Shower spray unit with a hose at least 59" long, which can be used as both a shower head and hand held; in unmonitored facilities, a fixed shower head mounted at 48" a.f.f. and 18" from adjacent wall.
- Water thermal shock protected with maximum temperature of 120° F.

Design Suggestions

It is preferable to locate the accessible shower in a corner to allow for grab bars on two walls and offer the possibility of adding a seat. Often there is a continuous lip a certain distance from the shower wall. A section will have to be removed to create an accessible route to the shower. An alternative is a ramp up to the lip, and installing a grating flush with the top of the lip. This involves less demolition and will maintain existing drainage, but can be expensive, since the grating will have to be strong enough to support an

adult using a wheelchair. (A lip or some sort of protection will be needed on one side for safety.)

A row of individual stalls is often more difficult to make accessible, since stalls are rarely large enough to allow a roll-in situation. If the stall is at least 36" × 36", a transfer shower can be installed if there is an accessible route to the stall. (See Project 57.) This can involve the same solutions as creating access at the gang shower described above: removing a lip or ramping up and installing a grating across the shower pan. Another option is removing one wall between two stalls and creating one extra-wide stall.

Key Items

Shower hose, seat, grab bar(s), accessible shower controls, and finish floor materials.

Level of Difficulty

Moderate to high.

Estimate

Modify gang showers

Description	Quantity	Unit	Labor-Hours	Material
Labor minimum for demolition and tile installation	1.000	Job	4.923	0.00
Ceramic tile floor	5.000	S.F.	0.437	44.75
Cutout demolition of partition	3.000	Ea.	1.000	0.00
2" x 4" blocking	0.005	M.B.F.	0.286	2.33
Remove and reset mixing valve, remove shower head	1.000	Ea.	2.000	0.00
Miscellaneous materials for gypsum board and ceramic tile repair	1.000	Ea.	1.000	0.28
Labor minimum to repair gypsum board and ceramic tile	1.000	Job	0.080	0.13
Fold-down seat	1.000	Ea.	0.267	129.00
Grab bars	1.000	Ea.	0.400	51.50
Bar-mounted hand-held shower head	1.000	Ea.	0.364	63.50
Totals			**10.757**	**291.49**

Total per each including general contractor's overhead and profit **$1,216**

61
Modify Check-out and Food Service Counters

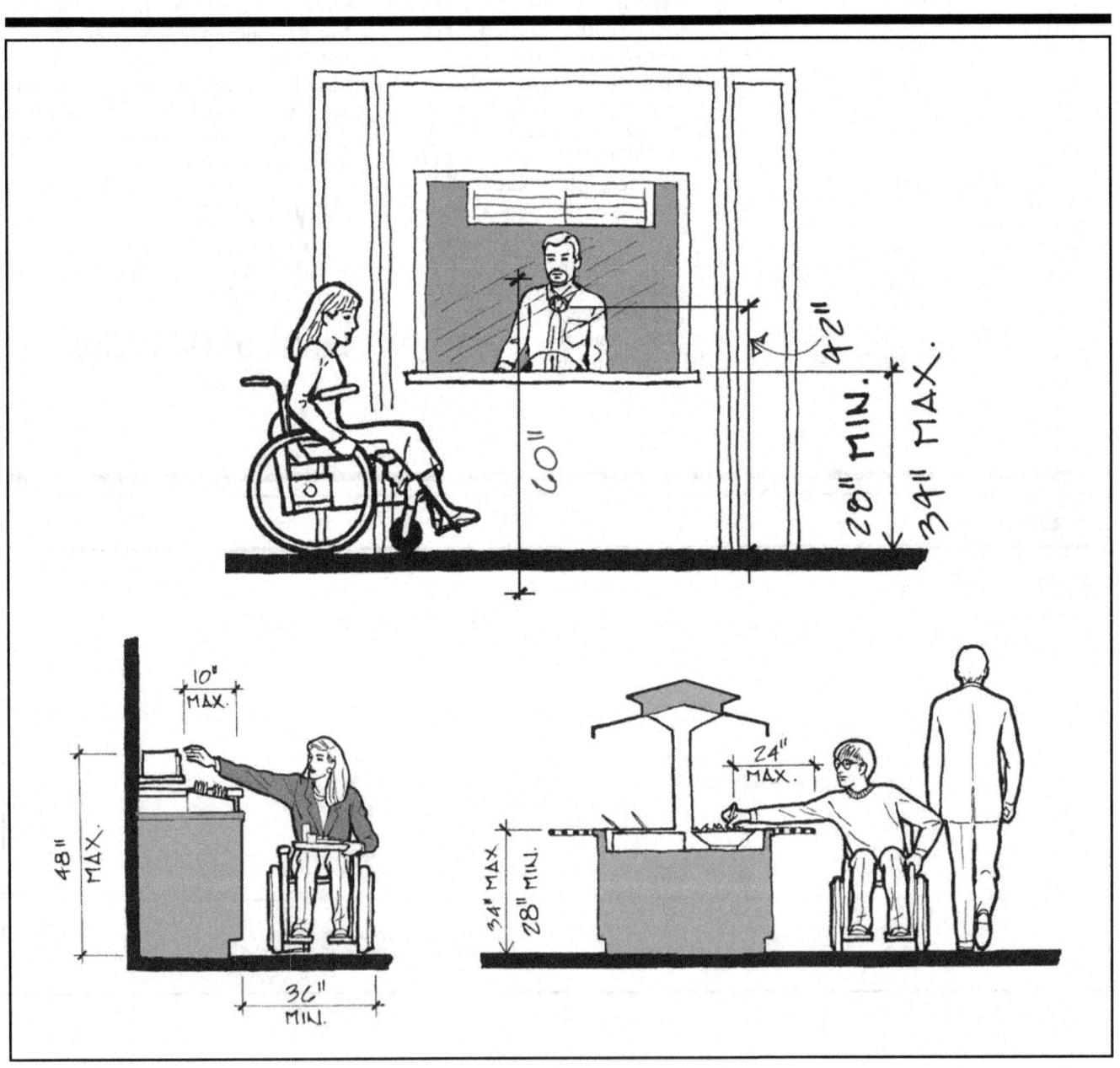

Just about every public facility has a service counter. Counters that are too high, even in accessible spaces, are difficult to use for people who are in wheelchairs or who have difficulty reaching over a certain height. Since the height given in ADAAG is usable by most adults (and can actually be more convenient for transferring goods across a surface), creating an additional counter that is accessible, or adding an accessible shelf, will greatly increase the usability of a public space.

Self-service food not only depends on counter heights, but also the ability of people to slide their tray, select food, get silverware and napkins, and pay a cashier.

ADAAG References

216.11, Signs at Check-out Aisles
226, Dining Surfaces and Work Surfaces
227, Sales and Services
Table 227.2, Minimum Number of Accessible Check-Out Aisles
308, Reach Ranges
403.5, Clearances
902, Dining Surfaces and Work Surfaces
904, Sales and Service Counters

Where Applicable

Service windows, check-out counters, counters where sales or distribution of goods, foods, or services take place including counters for ordering, pick-up, express, and return.

Design Requirements

- On accessible route. Aisle 36" wide minimum. See Project 62 for other requirements.

- Sign with international access symbol identifying accessible aisles if there are non-accessible aisles.
- Height: At least one 36" long portion of the check-out counter between 28" and 34" a.f.f. where there is check writing, and up to 36" a.f.f. where there is not check writing. A 2" maximum edge protection is allowed.
- For a forward approach, width of openings under work surfaces at least 36" wide. (In alterations, 24" centered on the 30" by 48" clear space is permitted.)
- Auxiliary counters of the same dimensions are allowed if counters cannot be modified, provided that the service is the same as at other counters.
- Clear floor space in front of the accessible counter at least 30" × 48" (positioned either for a parallel approach or forward approach).
- 50% of food service shelving and at least one of each of dispensing device for tableware, dishware, etc., within reach ranges. (Ranges vary if reaching over a counter; suggest using 44" a.f.f and 20" maximum depth.)
- Tray slides: 28" minimum, 34" maximum a.f.f.

Design Suggestions

If it is not possible to lower a section of counter due to historic preservation requirements or because of existing conditions (if the counter is made of marble or granite, for instance), it might be possible to install a fold-down shelf on the face of the existing counter or a pass-through window. Some areas behind the counter are entered through a door on a public route, and it might be possible to replace this with a split door that can be used on an as-needed basis.

Sometimes a free-standing table added adjacent to the counter can meet the requirements. This can be particularly advantageous to customers if there is also a chair off to a side. Although there is no single height that will accommodate everyone, it is recommended that the accessible counter be 34" to facilitate use by the greatest number of people and meet all counter requirements.

While the number of check-out aisles required is based on specific ratios (see Table 227.2), if at all possible, make all the aisles accessible.

For places where there is a glass shield or wall above a service counter, make sure there is one speaking hole at 42" a.f.f. Alternatively, a speaker system or telephone receiver can be used if the operable parts meet ADAAG and there is a volume control for the handset. (See 704.3.)

Provide a continuous shelf the same height as the tray slide from the tray and utensil area to the check-out to avoid having to carry the tray from one function to another.

Key Items

Low counter, or fold-down counter if a low counter is technically infeasible.

Level of Difficulty

Varies. Depends on material and extent of renovation.

Estimates

Lower 36″ wide section of counter

Description	Quantity	Unit	Labor-Hours	Material
Remove 36″ wide section of counter	3.000	L.F.	0.400	0.00
Remove counter support	1.000	Job	2.000	0.00
End caps at counter cuts (pine trim)	4.000	L.F.	0.133	2.52
Counter supports	0.030	M.B.F.	1.412	13.95
Additional labor for end caps and counter support	1.000	Job	2.000	0.00
Totals			**5.945**	**16.47**

Total per each including general contractor's overhead and profit **$427**

Install fold-down counter to face of existing counter

Description	Quantity	Unit	Labor-Hours	Material
Plastic laminate counter	3.000	S.F.	0.800	24.90
Continuous steel hinge	3.000	L.F.	0.750	32.40
Totals			**1.550**	**57.30**

Total per linear foot including general contractor's overhead and profit **$71**

Total per each 3 linear feet including general contractor's overhead and profit **$212**

Notes

62
Create Accessible Aisles

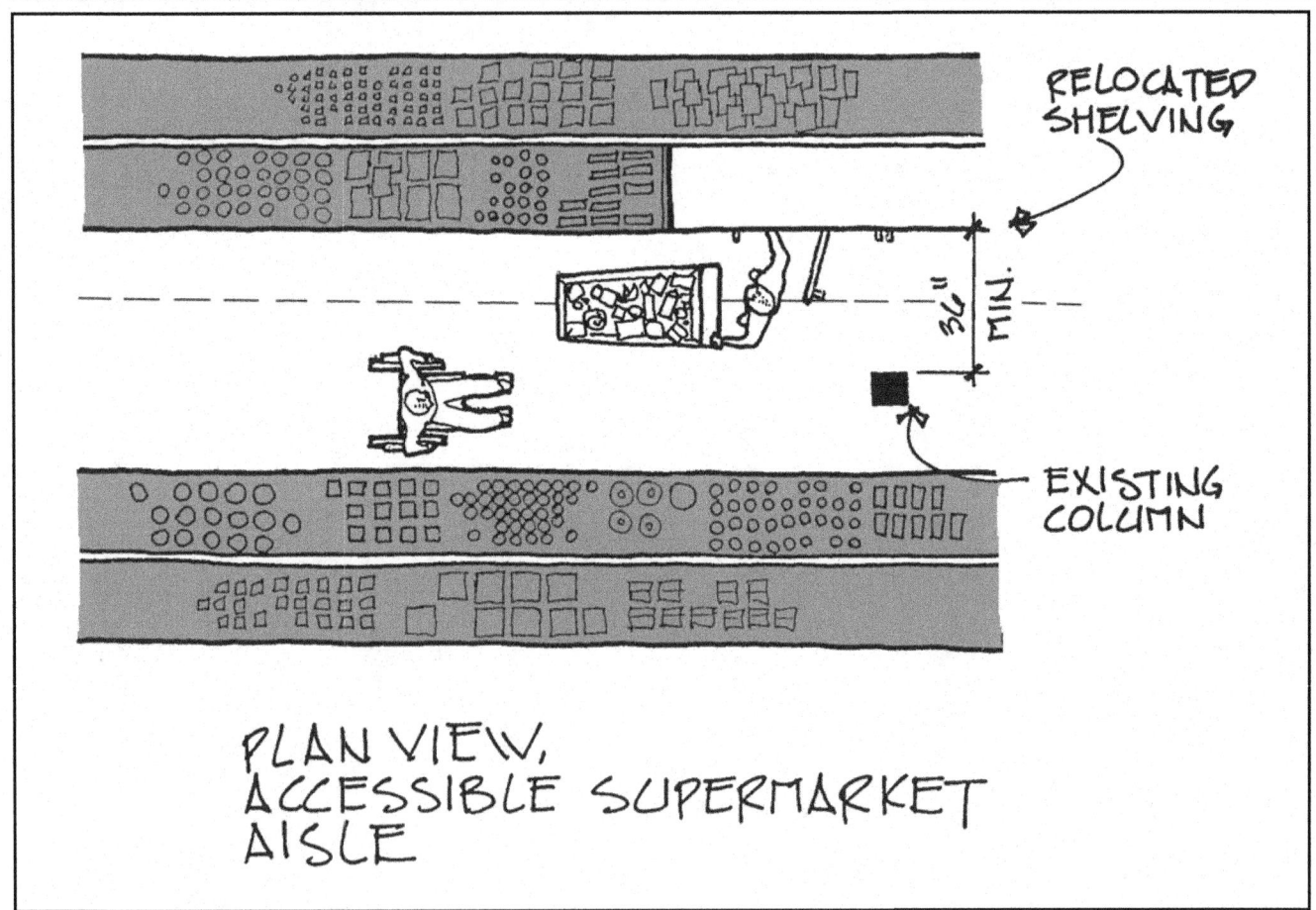

RELOCATED SHELVING

36" MIN.

EXISTING COLUMN

PLAN VIEW,
ACCESSIBLE SUPERMARKET
AISLE

An accessible entrance is the first step in creating an accessible commercial space; the next is ensuring that people can get around in the facility. Providing a 36" minimum width route of travel with adequate space to turn at the end of aisles makes it possible for people using wheelchairs to get to goods or services. It also makes it easier for others, since any point narrower than 36" is difficult for many people to use (including those carrying packages). All that is often required is moving or modifying shelving or display cases, with some possible replacement of the flooring surface.

ADAAG References
225.2.2, Self-Service Shelving
227, Sales and Services

305, Clear Floor Space
307, Protruding Objects
308, Reach Ranges
403.5, Clearances
811, Storage
904, Sales and Service Counters

Where Applicable

All aisles on accessible routes where goods or services are offered to the public: libraries, mail box banks, food service lines, queues and waiting lines, check-out counters.

Design Requirements

- Aisles, including check-out aisles, 36" wide minimum route of travel to *all* goods and services (may have constrictions down to 32" for a 24" distance). The preferred food service line width is 42" or more.
- The width of the aisle at its end must be as follows:
  (1) 60" minimum wide.
  (2) If the display shelves between two 36" wide parallel aisles are at least 48" wide, the end aisle must be at least 36" wide.
  (3) If the parallel aisles are at least 42" wide but the display shelves between them are less than 48", the width of the end aisle must be 48" wide minimum. (See ADAAG 403.5.2, Clear Width at Turn.)
- Protruding objects, 4" maximum unless higher than 80" or lower than 27" (12" overhang is allowed for signs on posts, but it is not recommended).
- Route must comply with slope and slip-resistant surface requirements.
- Shelving has no fixed percentage that must be within reach requirements.

Design Suggestions

The most frequent complaint about aisles is that someone has placed a display that narrows the width to less than 32", or there is two-way traffic in a 36" aisle. If at all possible, aisles should be at least 60" wide. Not only does this allow passing, but small constrictions, displays, etc., can be added here and there without compromising the width. The second most frequent complaint is not being able to exit the aisle. Either it is a dead end, or the cornering dimensions are too small because of added displays. To mitigate these types of problems, try to avoid dead ends and have extra wide end aisles.

There is no height restriction on shelving, but some method must be provided for obtaining goods out of reach range (below 15" or above 48", although people using wheelchairs that have full functioning of their upper bodies may be able reach farther than these ranges). One solution is to display goods, such as food, vertically instead of horizontally, making the same goods available at different heights. Where informational material is offered for distribution, such as maps, schedules, or forms, it should be within reach range (between 15" and 48" a.f.f.).

Key Items

Shelving, information displays, rearrangement of furnishings.

Level of Difficulty

Low. Relocating information displays can involve finish work.

Estimates

Make aisles accessible (patching of flooring after moving of F.F.E. type items by owner)

Description	Quantity	Unit	Labor-Hours	Material
Heavy duty vinyl sheet goods	1.000	S.F.	0.035	2.39
Totals			0.035	2.39

Total per square foot including general contractor's overhead and profit **$6**

62. Create Accessible Aisles *(continued)*

Make aisles accessible (moving of unfixed F.F.E. type items)

Description	Quantity	Unit	Labor-Hours	Material
Labor minimum	1.000	Job	2.000	0.00
Totals			**2.000**	**0.00**

Total per each including general contractor's overhead and profit   **$115**

Lower wall-mounted information rack

Description	Quantity	Unit	Labor-Hours	Material
Labor minimum for lowering signage	1.000	Job	2.000	0.00
Miscellaneous materials for gypsum board repair	1.000	Ea.	1.000	0.28
Labor minimum to repair and paint gypsum board	1.000	Job	2.000	0.00
Totals			**5.000**	**0.28**

Total per each including general contractor's overhead and profit   **$347**

Copyright Reed Construction Data, 2004

214

Notes

63
Modify Dining Area

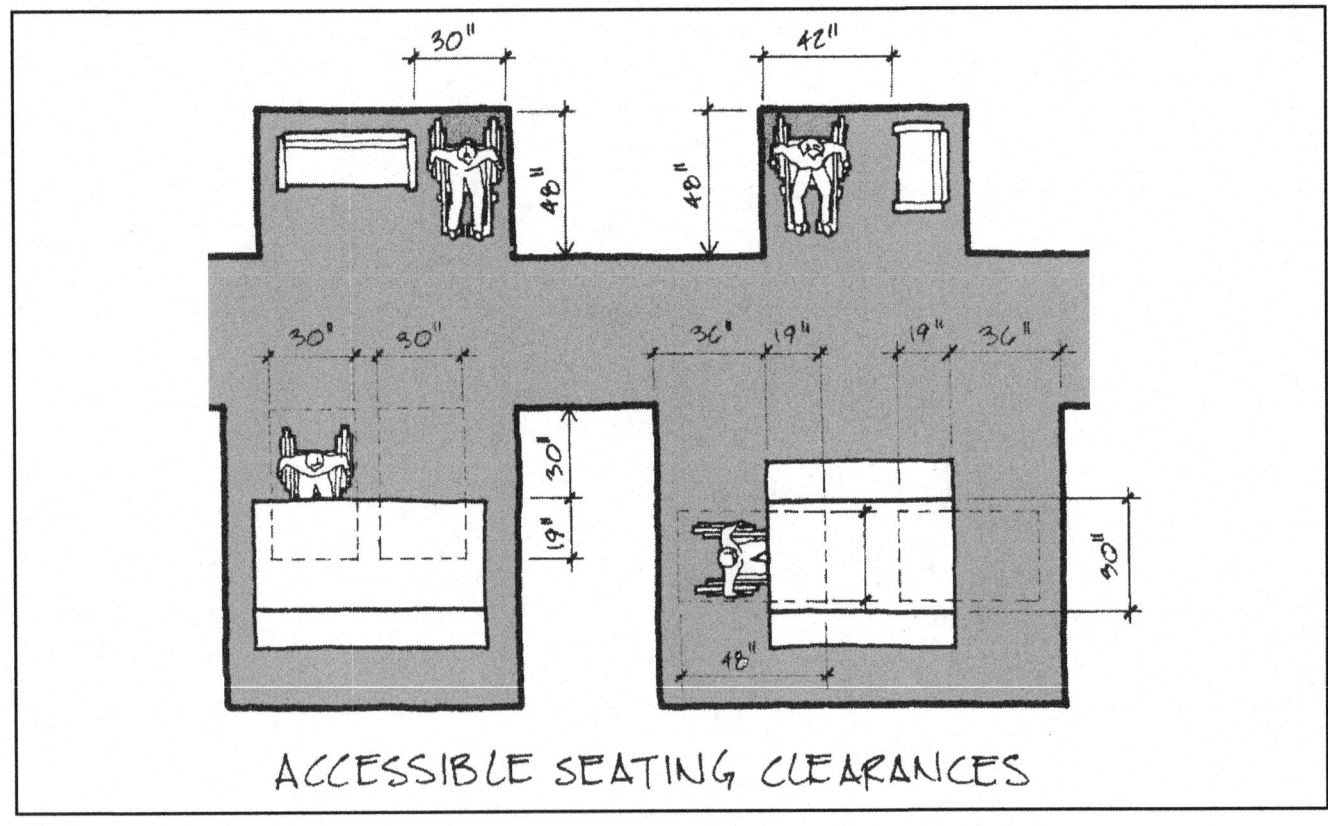

ACCESSIBLE SEATING CLEARANCES

Many dining areas are designed to maximize seating capacity, which can minimize access. In some dining areas or restaurants, it is often possible to create accessible seating and services without major modifications to the space by relocating or removing some seating and widening aisles. This can make the space easier to use for all customers, as well as the cleaning and wait staff.

ADAAG References
206.2.5, Restaurants and Cafeterias
226, Dining Surfaces and Work Surfaces
305, Clear Floor or Ground Space
306, Knee and Toe Clearance
902, Dining Surfaces and Work Surfaces

Where Applicable
Public and common-use dining areas and bars.

Design Requirements
- 36" clear interior route of travel; floor surfaces slip-resistant and stable.
- Minimum of 5% of the standing and seating area accessible (in all

sections: smoking, non-smoking, and range of quality) and dispersed throughout the facility.

- 30" × 48" clear space for parallel approach to tables and counters.
- 30" × 48" deep clear space for front approach to table or counter. Up to 11" of that may be toe space at least 9" high; a total of 17" may be under the counter or table for up to 11" knee space that is at least 27" high.
- Table and counter heights between 28" and 34".

Design Suggestions

Try to have all movable tables with correct height and adequate space below.

In reaching the 5% criterion for the total number of spaces, try to have 5% of window seats accessible, 5% of spaces at the bar accessible, 5% smoking spaces accessible, etc.

For bar accessible spaces, try to have a 6' section of the bar at 34" a.f.f. It may be open underneath and without a rail.

If the dining areas are used mostly by children, see section 902.4 for height recommendations.

Key Items

Existing features to be relocated and counters lowered; new movable seating.

Level of Difficulty

Low to moderate. Removing fixed seating may require finish work to existing floors and/or walls. Could require schematic planning and design drawings.

Estimates

Lower tray slide (assume slide is mounted with brackets every 2 feet)

Description	Quantity	Unit	Labor-Hours	Material
Remove existing brackets and slide	1.000	Job	0.500	0.00
Layout and drill new mounting holes (4 per bracket)	2.000	Job	0.500	0.00
Re-install brackets and slide	1.000	Job	1.000	0.00
Totals			**2.000**	**0.00**

Total per linear foot including general contractor's overhead and profit   **$122**

Widen food service line (assume pipe rail line delineator, uprights 4′ O.C.)

Description	Quantity	Unit	Labor-Hours	Material
Remove existing pipe rail (bolted to floor through flanges)	1.000	Job	0.252	0.00
Layout and drilling of new bolt holes in concrete floor (per inch of depth)	2.000	Ea.	0.320	0.14
Expansion anchors and shields (1/2″ diameter)	1.000	Ea.	0.107	3.74
Re-install pipe rail	1.000	Job	0.252	0.00
Vinyl tiles 12″ x 12″ (1 tile at each former upright location)	0.250	S.F.	0.004	0.19
Totals			**0.935**	**4.07**

Total per linear foot including general contractor's overhead and profit   **$85**

63. Modify Dining Area *(continued)*

Replace fixed seating with movable tables and chairs

Description	Quantity	Unit	Labor-Hours	Material
Remove four-seat, 24" x 48" table	1.000	Job	2.000	0.00
Patching concrete floor	3.000	S.F.	0.240	12.15
Vinyl tile	3.000	S.F.	0.048	9.87
Tile layer minimum	1.000	Job	2.000	0.00
Totals			4.288	22.02

Total per each including general contractor's overhead and profit **$334**

Lower section of bar (wood)

Description	Quantity	Unit	Labor-Hours	Material
Remove 60" section of bar/counter	5.000	L.F.	0.667	0.00
Remove supporting framework for bar/counter	1.000	Job	2.000	0.00
Miscellaneous millwork demolition	5.000	L.F.	1.000	0.00
Install new framework for bar/counter	0.030	M.B.F.	1.412	13.95
Install new wood bar/counter	5.000	L.F.	1.429	262.50
Additional labor for framework/millwork	1.000	Job	2.000	0.00
Totals			8.508	276.45

Total per linear foot including general contractor's overhead and profit **$208**

Total per each 5 foot section including general contractor's overhead and profit **$1,039**

Install ramp up to raised (6") area, 4' wide (carpet on wood, no rails)

Description	Quantity	Unit	Labor-Hours	Material
Ramp framing	0.040	M.B.F.	0.438	24.00
Ramp floor deck	24.000	S.F.	0.307	22.80
Additional labor required for ramp framing	1.000	Job	2.000	0.00
Additional carpet labor, minimum	1.000	Job	2.667	0.00
Carpet (minimum material quantity)	8.000	S.Y.	0.853	252.00
Totals			6.265	298.80

Total per square foot including general contractor's overhead and profit **$39**

Total per each including general contractor's overhead and profit **$940**

Notes

64
Create Accessible Dressing/Fitting Rooms

The standard dressing or fitting room is inaccessible to someone who uses a wheelchair. Increasing the size of the door and room, and adding a bench and hooks, makes it possible for people with mobility impairments to try on clothes, as well as easier for large, frail, or older people, and for parents with children to use the space.

ADAAG References
222, Dressing, Fitting, and Locker Rooms
306, Knee and Toe Clearance
404, Doors, Doorways, and Gates
703, Signs
803, Dressing, Fitting, and Locker Rooms
903, Benches

Where Applicable
Wherever dressing/fitting rooms are provided, at least 5%, but not less than one of each type in each cluster. If an alteration is "technically infeasible" to meet the 5% criterion, then one for each sex for each floor is acceptable.

Design Requirements
- On an accessible route.
- Door: 32" clear opening, hardware operable with a closed fist, meeting maneuvering clearances in Table 404.2.4.1.
- Curtain or café door high and low enough to provide privacy for children and tall people.
- 60" diameter clear floor space allowing a person in a wheelchair to make a 180° turn. Turning space may include toe space under a bench.
- Bench between 20" and 24" wide by 42" long, with back at least 18" high (wall is o.k.), and a clear floor space beside the bench to allow transfer from a wheelchair. Bench seat between 17" and 19" a.f.f. Door can swing into turning space, but not 30" x 48" clear floor space.

TRANSFER SPACE

TURNING SPACE

42" MIN.

BENCH

48"

36"

20" TO 24"

60" MIN.

32" MIN.

PLAN VIEW, ACCESSIBLE FITTING ROOM

- Coat hook(s) mounted within reach range (48" a.f.f. and not over bench).
- Shelves, if provided, between 40" and 48" a.f.f.

Design Suggestions

An accessible fitting room should allow as much maneuvering space as possible. If the floor plan permits, the turning space inside should be a clear 60" circle that does not overlap with the space underneath the bench. The disadvantage of having a large changing room is that it may also be used for storage. Care should be taken to ensure that this does not happen. Many people without physical limitations, such as parents with children, will find the larger changing room useful.

A double-acting door with gravity or spring hinges and pulls on each side is recommended. Many changing areas have partial doors (saloon type) that are open at the top and bottom. Care should be taken that they are both low and high enough to give privacy to children and to both short and tall people.

Mirrors, if provided, should be full-length, at least 18" wide by 54" high, located to allow a view for a person sitting on a bench or standing.

Put a tactile/Braille sign with the international symbol of accessibility on the corridor wall, adjacent to door latch.

Key Items

Wallboard and stud wall, wood bench, door with accessible hardware, full-length mirror, coat hooks.

Level of Difficulty

Moderate. Basic carpentry and finish work required.

Estimate

Install dressing rooms

Description	Quantity	Unit	Labor-Hours	Material
Metal stud partition (3-5/8" wide, 16" O.C.)	168.000	S.F.	2.801	35.28
1/2" gypsum board, taped and finished	336.000	S.F.	5.571	77.28
Painting	336.000	S.F.	8.534	73.92
Interior door frame	18.000	L.F.	0.768	71.82
Interior wood door, birch face, 3'-0" x 6'-8"	1.000	Ea.	1.143	87.50
Hinges	1.500	Pr.	0.000	66.75
Lever-handled lockset	1.000	Ea.	0.800	121.00
Bench, 24" x 4'	4.000	L.F.	1.143	210.00
Mirror	12.000	S.F.	1.200	80.40
Coat hook	1.000	Ea.	0.222	10.80
Totals			22.182	834.75

Total per square foot including general contractor's overhead and profit **$592**

Total per each including general contractor's overhead and profit **$2,961**

65
Create Assembly Seating Accommodating Wheelchairs

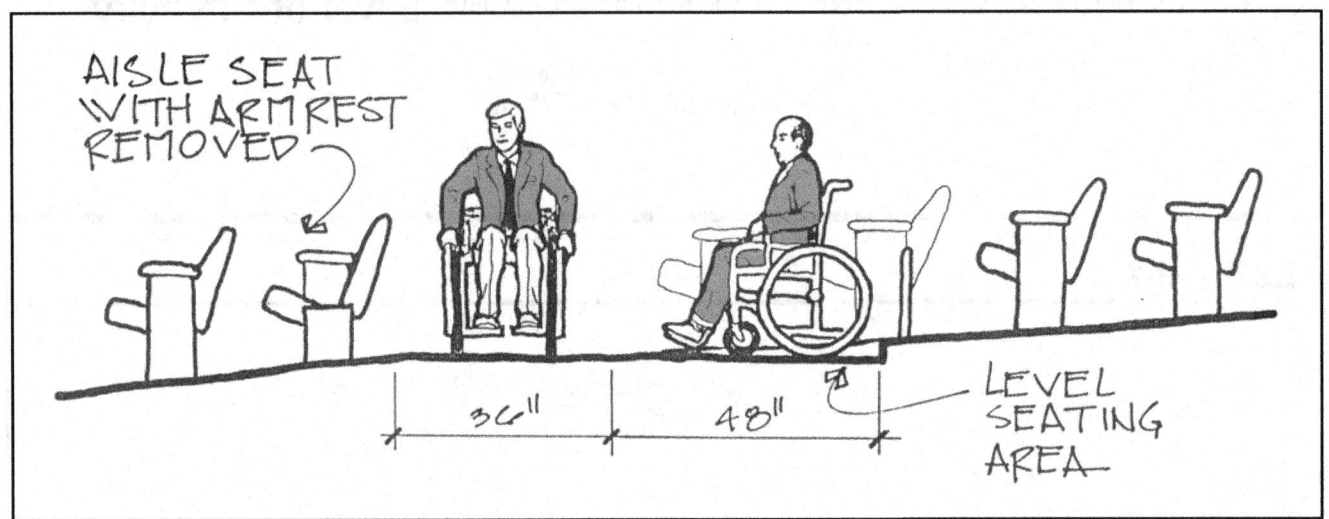

The design of seating for large gatherings is complicated, as are the regulations covering seat access. While many factors are taken into account when designing auditoriums — sight lines, acoustics, maximizing seating, and fire egress — accessibility has not traditionally been one of these. Since existing theaters (and stadiums) are complex facilities often built in structures that are difficult to modify with sloped floors and many stairs, integrating accessible seating into them is often a challenge. Nevertheless, it is critical to make these public facilities accessible. If a place of assembly has an accessible route of travel, it is often possible to create accessible seating and, depending on the availability of accessible fire egress, to distribute the accessible seating throughout the space.

ADAAG References
206.4, Accessible Routes, Assembly Areas
221, Assembly Seating
403, Walking Surfaces
802, Wheelchair Spaces, Companion Seats, and Designated Aisle Seats

Where Applicable
All public and common-use assembly areas with *fixed* seating, such as concert halls, theaters, and sporting facilities.

Design Requirements
- Number of wheelchair spaces varies from one accessible space for four seats, to six accessible spaces for 500 seats. Above 500 seats, there is a formula. (See Table 221.2.1.1.)
- Choice of admission prices for accessible seats comparable to the rest of the seating.
- Wheelchair spaces on an accessible route and not accessed through another accessible space.
- Wheelchair spaces with sight lines comparable to that of the rest of the seating. Where crowds are expected to stand, sight lines for people in wheelchairs comparable to standing viewers.

- Wheelchair spaces integrated horizontally into the general seating, and integrated vertically when the total number of seats exceeds 300 and when the viewing angles will be equivalent or better than the average seat. (Exceptions: wheelchair seating can be grouped for bleachers and balconies and other areas where sight lines require slopes greater than 5%.)
- If the other seats have armrests, designated seats to house folding armrests on the aisle side. 5% of total aisle seats provided must have a sign identifying them as a "designated aisle seat." These seats must be aisle seats closest to an accessible route.
- At least one fixed companion seat next to each accessible space, equivalent in size, quality, comfort and amenities. Companion seats may be movable.
- Width: single wheelchair space, 36" wide; two spaces side by side, 66" wide.
- Depth: 48" minimum when entered from front or rear; 60" deep when entered only from the side.
- Wheelchair spaces not to overlap required circulation space.
- Level area (2% maximum slope) for wheelchair seating.

Design Suggestions

Locate accessible seats adjoining an accessible route of fire egress when possible. Even when the total number of seats is less than 300, disperse the accessible seating as much as possible. If a theater has different price ranges for seating areas, accessible seats hould be located in each area, or policies be modified to offer the same price range for the accessible seats as for general seating.

Key Items

Varies depending on location. May require alteration(s) to accessible route: lifts, doors, or flooring; leveling of sloped areas to create accessible seating spaces.

Level of Difficulty

Varies. Low for seating removal, if accessible seating area(s) are level and on an accessible route. High for creating an accessible route, and for leveling seating areas on existing sloping surface. Could require schematic or design drawings to show compliance with local safety and building codes.

Estimate

Create accessible theater seating

Description	Quantity	Unit	Labor-Hours	Material
Remove existing seating	4.000	Job	2.000	0.00
Concrete formwork	20.000	SFCA	3.556	24.80
Cast in place concrete, finishing included	25.000	S.F.	0.659	72.25
Layout & drilling of anchor holes	4.000	Ea.	0.213	0.04
Plastic shields and screws	4.000	Ea.	0.133	0.16
Signage	1.000	Ea.	0.457	31.50
Totals			7.018	128.75

Total for 2 accessible locations including general contractor's overhead and profit **$719**

66
Install Assistive Listening Systems

ASSISTIVE LISTENING DEVICE, RECEIVER

ASSISTIVE LISTENING SYSTEM, SENDER

Assistive listening systems are amplification systems designed to reduce interference from background noise and room reverberations, and to pick up sound at the source, amplify it, and direct it to the ear of the listener. Certain assistive listening devices can be used with hearing aids. (Note that earphone jacks with variable volume controls benefit only people who have slight hearing loss and do not help people who use hearing aids.)

ADAAG References

216.10, Signs
219, Assistive Listening Systems
703.4, Installation Height and Location of Signs
703.7.2.4 Assistive Listening Systems
706, Assistive Listening Systems

Where Applicable

All assembly areas where audible information is integral to the use of the space.

Design Requirements

- Up to 500 seats: at least 4% of seats (but never less than two) must have receivers; 500 to 1,000 seats: 3% of seats plus 5 must have receivers; 1,000 to 2,000 seats: 2% plus 15 seats must have receivers; over 2,000 seats: 1% plus 35 seats must have receivers.
- At least one out of every four receivers (but never less than two) must be hearing aid compatible.
- Signage indicating availability of services located at ticket offices or windows. The signs must comply with Section 703.4 and must include the International Symbol of Access for Hearing Loss.

- Receivers must have 1/8-inch standard mono jacks or adapters.
- Receivers required to be hearing-aid compatible must interface with T-coils in hearing aids through the provision of neck loops.
- Between 110 dB and 118 dB with a 50 dB volume control.
- Signal to noise ratio of 18 dB or more.
- 18 dB maximum peak clipping.

Design Suggestions

Assistive listening systems are categorized by their mode of transmission. There are hard-wired systems and three types of wireless systems — induction loop, infrared, and FM radio transmission. Each system has advantages and disadvantages. For example, an *FM system* (which is broadcast to an entire area) may be better than an infrared system in some open-air assemblies, since infrared signals can be overpowered by the sun or movement of people.

On the other hand, an *infrared system* may be a better choice for confidential transmission. It is directional and uses infrared light beams to carry information from a transmitter connected to a PA system or microphone directly to a receiver worn by the listener. Infrared systems may benefit people with mild to moderate hearing loss. They can be useful in multi-theater complexes, where sound distraction from other theaters can be a problem.

An *audio loop system* consists of a microphone, amplifier, and wire loop that is placed around the listening area of the room. Sound is transmitted by a magnetic field created within the loop, and the

listener picks up the amplified sound either with a "T" switch on the hearing aid or with a special receiver. Audio loops can be used indoors and outdoors, in large or small rooms, but people must sit within the "loop." This system is helpful to people with mild to profound hearing loss.

The FM system is a wireless amplification system consisting of a microphone, transmitter, and receiver resembling a small pocket radio worn around the neck. The transmitter sends the sound to the listener's receiver using FM radio signals. FM systems can be used in any sized space and benefit people with mild to profound hearing loss. In multi-plex theaters, it is necessary to have different FM systems set at different frequencies, since FM systems transmit through walls into adjacent rooms.

There are several factors involved in the selection of audio systems—size, shape, and acoustics of the space; whether or not it is indoors; the type of sound being projected; and the level of background noise. It is strongly recommended that a manufacturer's representative and local agencies for people who are deaf or hard of hearing be consulted prior to the installation of any system.

Key Items

Wiring, outlets, assistive listening equipment.

Level of Difficulty

Varies, depending on the system. Infrared or audio loop systems can require electricians and skilled technicians for installation; FM system requires equipment and outlets.

Estimates

Provide magnetic induction loop assistive listening system

Description	Quantity	Unit	Labor-Hours	Material
Emitter, base, 20 receivers, 200' loop	1.000	Ea.	4.000	1,570.00
Totals			4.000	1,570.00

Total per each including general contractor's overhead and profit _____ **$6,007**

Provide infrared assistive listening system (existing P.A. system is necessary)

Description	Quantity	Unit	Labor-Hours	Material
Basic system (transmitter, emitter, receiver)	1.000	Ea.	4.000	3,833.33
Totals			4.000	3,833.33

Total per each including general contractor's overhead and profit _____ **$8,388**

Provide radio frequency assistive listening system (existing P.A. system is necessary)

Description	Quantity	Unit	Labor-Hours	Material
Basic system (transmitter, 3 receivers)	1.000	Ea.	1.000	1,761.19
Totals			1.000	1,761.19

Total per each including general contractor's overhead and profit _____ **$4,230**

Install electrical outlet for assistive listening system

Description	Quantity	Unit	Labor-Hours	Material
Cutout demolition of partition	1.000	Ea.	0.333	0.00
Conductor	0.100	C.L.F.	0.296	1.23
Install junction box	1.000	Ea.	0.400	7.75
Install outlet	1.000	Ea.	0.296	7.05
Install cover plate	1.000	Ea.	0.100	1.76
Miscellaneous materials for gypsum board repair	1.000	Ea.	1.000	0.28
Labor minimum to repair and paint gypsum board	1.000	Job	2.000	0.00
Totals			4.425	18.07

Total per each including general contractor's overhead and profit _____ **$341**

Install signage for assistive listening system

Description	Quantity	Unit	Labor-Hours	Material
Signage	1.000	Ea.	0.457	31.50
Layout & drilling of anchor holes	4.000	Ea.	0.213	0.04
Plastic shields and screws	4.000	Ea.	0.133	0.16
Totals			0.803	31.70

Total per each including general contractor's overhead and profit **$119**

67
Install Emergency Communication Device

Enabling all people to summon assistance in an emergency situation is one of the most critical elements of making a facility accessible. People with mobility impairments may not be able to leave the scene of a fire, and those with visual or hearing difficulties may not know where to go to call for assistance. Emergency communication devices must be usable not only by people with mobility impairments, but also by those who may not be able to communicate by voice or have limited vision.

ADAAG References
230, Two-Way Communication Systems
305, Clear Floor or Ground Space
308, Reach Ranges
309, Operable Parts
703, Signs

Where Applicable
Wherever emergency communication devices are installed for use by facility occupants, such as areas of refuge, elevators, fire call boxes, emergency call boxes in public spaces, and so forth.

Design Requirements
- Reachable (within 18") from an accessible route.
- Within reach range: 48" a.f.f. maximum, 18" minimum from corner, not over counter or other protruding object.
- Handset cord 29" or more, if handset provided.
- Accessible controls: operable with one hand without tight grasping, pinching, or twisting of the wrist.
- Audible and visual acknowledgement of receipt of request for aid.
- Non-voice communication, such as push-button emergency assistance call/LED.
- Identifying sign and instructions in raised characters and Braille meeting 703.

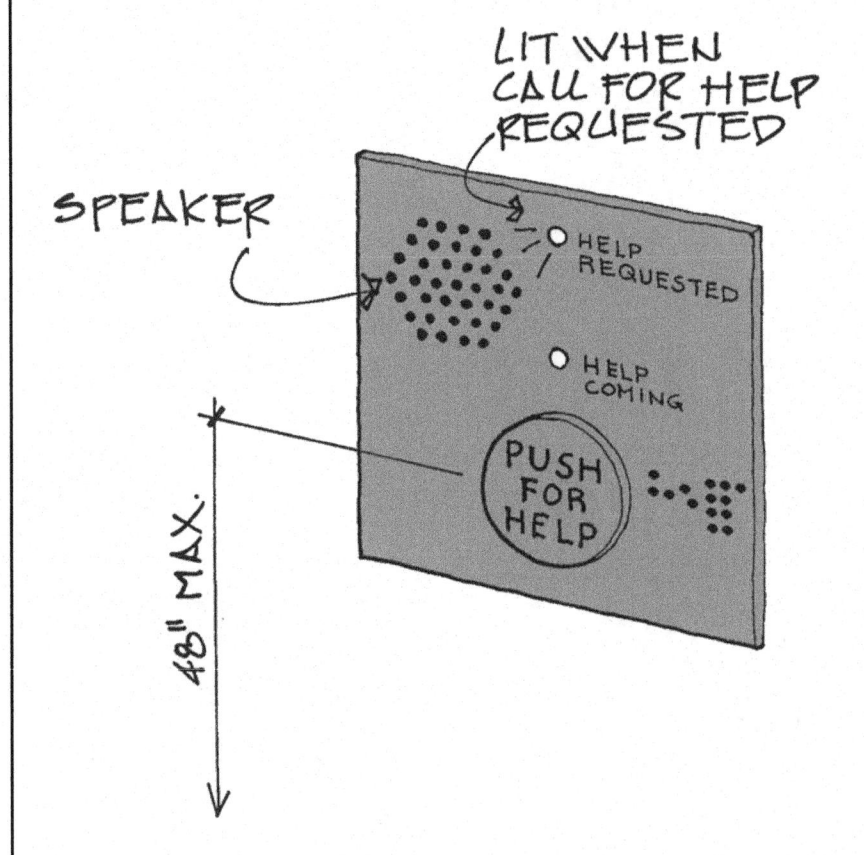

SPEAKER

LIT WHEN CALL FOR HELP REQUESTED

○ HELP REQUESTED

○ HELP COMING

PUSH FOR HELP

48" MAX.

Design Suggestions

An emergency communication box should be located on an accessible route and in a clearly visible area, with raised-letter signage, including instructions in Braille, identifying it as emergency communications and indicating how it operates. Particularly negligent are street-side fire call boxes, which often are too high, frequently off the accessible route, carry no signs in Braille or raised letters, and have no feedback mechanisms.

Voice communication is often preferred as a deterrent against false alarms, but it is essential that emergency communication not be dependent on voice communication alone. Someone who cannot communicate verbally must be able to obtain assistance in an emergency situation. Further, non-voice communication also enables people not fluent in English to summon help. At a minimum, the system must provide both an audio and visual indication that the message was received.

It is possible to install an emergency communication box that summons help just by pushing a button or pulling a lever (like a fire alarm), or that uses a TTY. When the message is answered, the visible signal requirement could be satisfied with something as simple as a beep and a light indicating that help is on the way. This all must be explained in raised letters and Braille.

A device that requires no handset is easier to use by people who have difficulty reaching. Further, small handles on handset compartment doors are not usable by people who have difficulty grasping. Any controls should be large and easily operable. Some call buttons are small and meant to be pushed with one finger, but this can be difficult for individuals with limited fine motor control. If the voice box is separate from the call buttons, it should be located 48" a.f.f. to allow access to people using a wheelchair. The color of the box should contrast with the wall finish to make it easier for people with reduced vision to see.

Key Items

Communications device, electrical wiring to emergency power source.

Level of Difficulty

Moderate.

Estimate

Install emergency communications device

Description	Quantity	Unit	Labor-Hours	Material
Master panel, four area stations, amplifier, battery back-up	1.000	Ea.	2.000	0.00
Conductor	5.000	C.L.F.	17.391	107.50
Cutout demolition of partition	5.000	Ea.	1.667	0.00
Labor minimum to repair and paint gypsum board	2.000	Job	4.000	0.00
Totals			25.058	107.50

Total per each including general contractor's overhead and profit **$2,100**

68
Modify Kitchenette

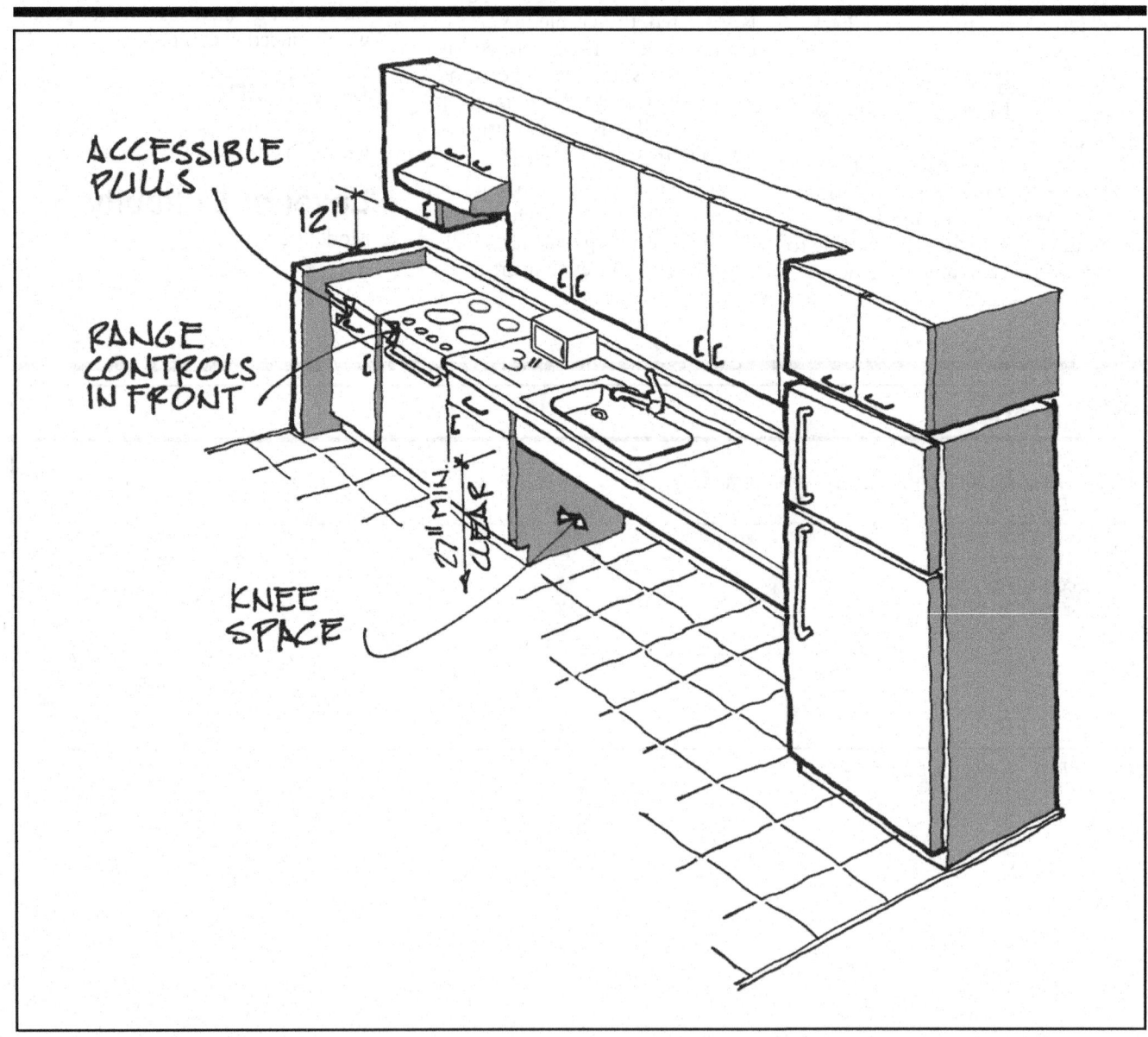

ACCESSIBLE PULLS

12"

RANGE CONTROLS IN FRONT

3"

27" MIN. CLEAR

KNEE SPACE

Students with disabilities in home economics classes, employees with disabilities in office kitchenettes, and residents of group homes or shelters, all may make use of accessible kitchens or kitchenettes. As with any other space, a kitchen located on an accessible route can have an entrance modified to create better access. Kitchen spaces not located on an accessible route can be altered to include an accessible sink, faucets, work counters, drawers and cabinet pulls, and outlets with relatively little modification to the space.

ADAAG References

212, Sinks, Kitchens, and Kitchenettes
305, Clear Floor or Ground Space
306, Knee and Toe Clearance
308, Reach Ranges
309, Operable Parts
606, Sinks
804, Kitchens and Kitchenettes

Where Applicable

All common-use kitchens, such as kitchenettes in employee common areas, hotel/motel common kitchens, common kitchens in transient facilities, and so forth.

Design Requirements

- On an accessible route, with an accessible entrance to the room.
- Clear floor space (30" x 48") for frontal or parallel approach to all features.
- U-shaped kitchens: 60" clearance between opposing base cabinets, appliances or walls.
- Pass-through, galley, kitchens (cabinets on opposing sides or parallel walls): open at each end with 40" clearance between base cabinets, appliances, or walls.
- 30" wide open area under sink.
- Minimum 50% shelf and refrigerator/freezer space within reach range.
- Accessible controls and handles (acceptable if operable with a closed fist).
- Slip-resistant surface on accessible route.
- Over counter outlets and switches 18" from corners and 44" a.f.f.

Design Suggestions

L-shaped or kitchen facilities along one wall work best, since they allow objects to slide without having to be picked up.

One possible renovation is to join two separate existing counter sections to create a single U-shaped counter. If the cabinet floor under the sink is not installed and the finish floor continues for the full depth, removing base cabinet doors can provide knee space. If the whole area is being modified, consider using a sink with an off-set drain, and leave the space under it and the adjacent 30" of counter surface beside the sink open (on the side opposite the off-set drain).

It is important to remember that for most users of kitchen areas lowered counters are not appropriate. People using crutches, those who have back problems, or those with fused spines often even have difficulty with counters at 36" a.f.f.

Usually, kitchenettes in common rooms have a microwave oven, which should sit on the counter and be accessible by a parallel approach.

A hose at the sink is recommended, even at sinks with a faucet located near the front. Wire pull or loop-type handles on drawers and cabinets allow for ease of use (cabinets with routed holds and tiny knobs are difficult to use for people with low fine motor control. Pull-out shelves and lazy Susans prevent the need for reaching to the back of storage spaces.

Some appliances are easier to use by a wide range of people. Side-by-side refrigerator/freezers allow a range of storage space on both sides. Stove and range controls should be located in front. Staggered burners prevent having to reach over a hot surface from a seated position.

It is important to make the kitchen usable for people with low vision. Lighting levels should be high at work stations, and placed so as not to cast shadows on the work space. Also, light-colored finishes and matte surfaces help people with low vision.

Key Items

Varies: drawer hardware, sink hardware, cabinets, slide drawers, and new appliances, possibly relocation of appliances and cabinets to create maneuvering space.

Level of Difficulty

Varies. Low for cabinet and storage modifications; moderate for sink and stove alterations; high for kitchen reconfiguration or total rehabs.

Estimate

Modify kitchens

Description	Quantity	Unit	Labor-Hours	Material
Remove cabinet hardware (labor minimum)	1.000	Job	2.000	0.00
Remove drawer hardware (labor minimum)	1.000	Job	2.000	0.00
Remove & replace knob handle faucet w/ lever handle faucet (labor only)	1.000	Job	4.000	0.00
Kitchen faucet, gooseneck spout, paddle handles	1.000	Ea.	0.800	107.00
Add for spray hose	1.000	Ea.	0.333	9.50
Install D pull drawer/cabinet handles	13.000	Ea.	0.650	45.24
Install new drawer guides	2.000	Pr.	0.333	11.80
Under-cabinet task lighting	4.000	Ea.	2.000	162.00
Conductor	0.300	C.L.F.	1.043	6.45
Outlet boxes	4.000	Ea.	1.391	12.00
Switch	1.000	Ea.	0.200	4.69
Labor minimum to repair and paint gypsum board	1.000	Job	2.000	0.00
Totals			**16.750**	**358.68**

Total per each including general
contractor's overhead and profit _____ **$1,891**

Notes

69
Modify Closet

Most closets can be modified with some basic carpentry work, and in many cases, without altering the doorway. Simple modifications, such as lowering a shelf and pole or installing them on adjustable brackets, are nevertheless vital in making common storage spaces accessible.

ADJUSTABLE POLE AND SHELF

48" MAX.

ADAAG Reference
225, Storage
305, Clear Floor or Ground Space
308, Reach Ranges
309, Operable Parts
811, Storage

Where Applicable
If storage is provided in an accessible space, at least one accessible closet or storage area.

Design Requirements
- On an accessible route.
- Clear floor space in front: 30" by 48" minimum.
- If closet is more than 24" deep, 32" minimum door opening width required.
- Shelves, poles, and hooks within reach range (48" a.f.f.).
- Accessible door hardware.

Design Suggestions
Adjustable shelves and poles, set at 12" heights, allow for flexibility in use. If the closet is deep, consider adding hooks to side walls or inside of doors for easy reach. The best design, if possible, is a "roll-in" closet with hooks or shelves on one side and hangers on the other. 32" minimum door clear opening width recommended for all closets.

Key Items
Shelves, brackets, poles, coat hooks, closet door hardware.

Level of Difficulty
Low.

Estimates

Install accessible wood shelf and pole

Description	Quantity	Unit	Labor-Hours	Material
Shelf supports	7.000	L.F.	0.233	4.41
Closet pole	3.000	L.F.	0.120	2.40
Shelving	6.000	L.F.	0.686	11.82
Additional labor required for work within closet area	1.000	Job	2.000	0.00
Totals			**3.039**	**18.63**

Total per each including general contractor's overhead and profit **$254**

Install adjustable wood shelf and pole

Description	Quantity	Unit	Labor-Hours	Material
Adjustable closet rod and shelf, 12" wide, 3' long	1.000	Ea.	0.400	8.50
Additional labor required for work within closet area	1.000	Job	2.000	0.00
Totals			**2.400**	**8.50**

Total per each including general contractor's overhead and profit **$191**

Lower 2 coat hooks

Description	Quantity	Unit	Labor-Hours	Material
Labor minimum to lower 2 coat hooks	1.000	Job	2.000	0.00
Totals			**2.000**	**0.00**

Total per each including general contractor's overhead and profit **$73**

Total per 2 coat hooks including general contractor's overhead and profit **$146**

70
Create Accessible Transient Lodging Guest Rooms

When inaccessible sleeping rooms are located on an accessible route, it is almost always possible to make at least some modifications to increase accessibility. Widened doorways, lever handles, visual and audible alarms, drawer pulls, closet poles, grab bars, and other modifications are changes that can make guest rooms in the hospitality industry even more accessible.

ADAAG References

206.5.3, Doors in Transient Facilities
215.4, Fire Alarms in Transient Lodging
224, Transient Lodging Guest Rooms
305, Clear Floor and Ground Space
306, Knee & Toe Clearance
308, Reach Ranges
404.2.3, Doorway Clear Width
Chapter 6, Plumbing Elements and Facilities
608.4, Shower Seats in Transient Guest Rooms
702.3, Guest Room Visual Alarms
804, Sinks, Kitchens, & Kitchenettes
806, Transient Lodging Guest Rooms

Where Applicable

All transient lodging buildings for rent or hire, including time-shares and dormitories. Exception for five rooms or less if the building is the owner's residence. Existing lodging facilities, as public accommodations, must remove barriers where it is readily achievable to do so, but are not required to make existing rooms in compliance unless they are altering them.

Calculate the minimum number of accessible rooms at 5% up to 100 rooms, and a sliding scale after that. Within a guest facility, having more than 25 beds, the requirement is that 5% of the beds be accessible.

Approximately 10% of the guest rooms must accommodate persons

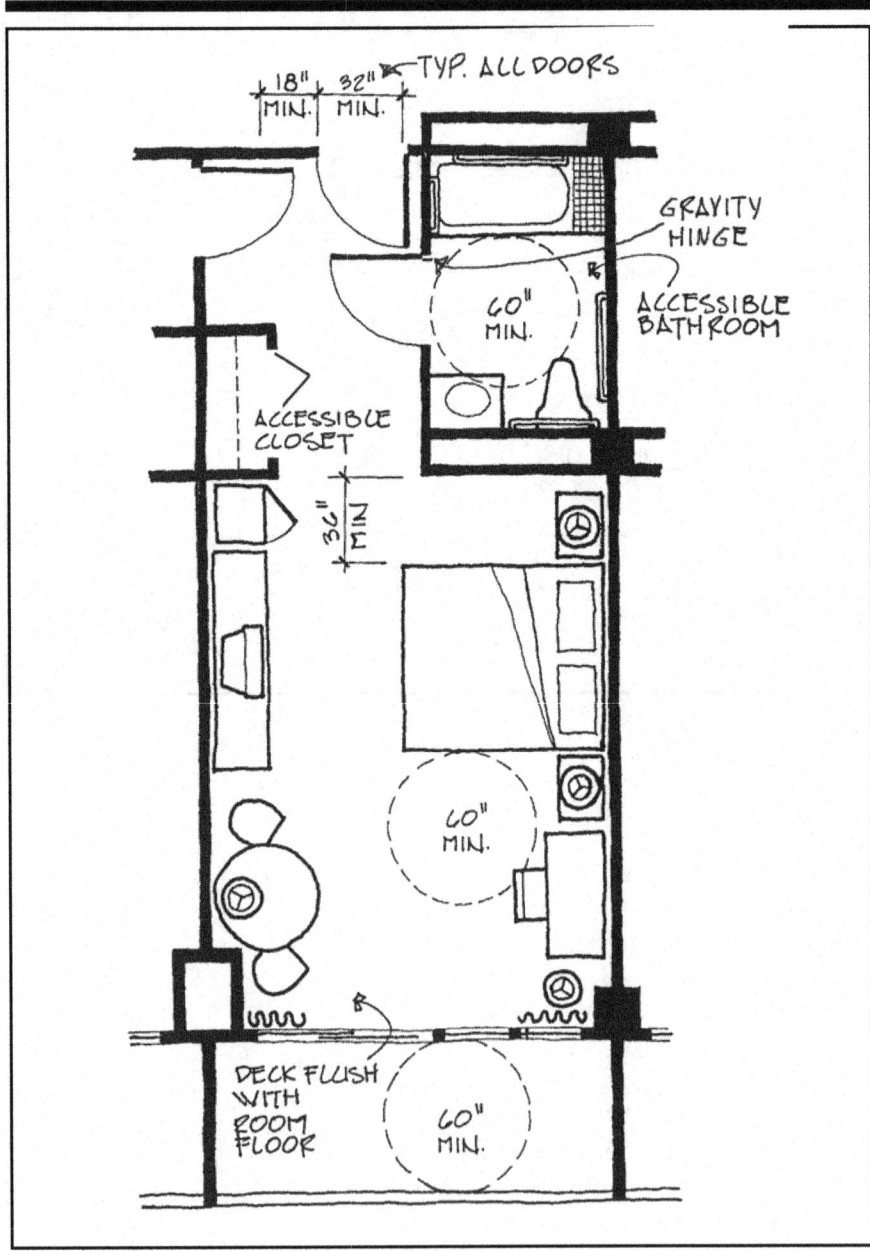

with hearing impairments according to 806.3. (Consult table 224.4 for the exact number.) The rooms providing mobility and communication features should be dispensed among different types of units.

Design Requirements

- Audible (110 dB, maximum) and visual alarms meeting NFPA 72.

Design Requirements, Guest Rooms, Mobility Features

- Accessible route (36" clear minimum, with no protruding objects) connecting all accessible spaces and elements (including phones) within the guest room, and a 60" minimum turning area.
- If balconies or other outdoor spaces are provided for the accessible guest room, they must be accessible.
- Accessible bed(s) must be on an accessible route (36" clear minimum, with no protruding objects) and must have a 30" by 60" clear floor area on both sides.
- At least one of each type of cabinet, dresser, or storage area on an accessible route and within reach range, hardware and controls 36" to 48" a.f.f. and operable with a closed fist.
- An accessible full bathroom (toilet, sink, tub, or roll-in shower); roll-in shower must have folding transfer seat. If only half baths are provided, an accessible half bath.
- If vanity tops are provided in any guest room, then the accessible room must have a comparable vanity top adjacent to a 30" by 60" clear area.
- If provided in other guest rooms, must include an accessible kitchen,

kitchenette, and wet bar, with sufficient clear floor space for a front or side approach at all cabinets, sinks, and appliances, with counters and sinks mounted at 36" high maximum. 50% of shelf space in cabinets and refrigerator/ freezers within reach range. Sufficient floor space to allow all doors to be accessible and usable, and controls and pulls operable with a closed fist.

- 32" minimum, clear openings/ doorways to and within all units, whether designated accessible or not, except for shower and closet doors in the rooms without mobility features.
- If carpeting is installed, it should be un-backed, with 1/2" pile maximum.

Design Requirements for Guest Rooms for Hearing-Impaired Users

- Visual smoke/fire alarms visible in all parts of the room either directly or by reflection mounted within 16' (measured horizontally) from the head of the bed.
- Visual telephone call alert devices and visitor alert devices, not connected to visual alarms.
- Volume controls on permanently installed telephones.
- An accessible electrical outlet within 48" of a telephone connection to allow use of a TTY.

Design Suggestions

If a high threshold or change in level is necessary for any patios, terraces, and balconies for water and/or wind protection, some sort of equivalent facilitation, such as ramping, may be required. In as much as this would be

an alteration to an existing condition, steeper slopes are allowed: 1:8 for 3" or less vertical rise, and 1:10 for 6" or less.

Extra storage space for a wheelchair is useful. (Otherwise, a wheelchair user is forced to leave the wheelchair in the route of travel.) Corner guards and door kick plates help protect walls from damage. Crank or lever hardware on a window is easier to operate than locks, which require pinching, and blinds on a continuous pull chain (or on a machine-operated push button) are easier to use than blinds on a standard string. Clear floor space is necessary to reach window drape or blind controls.

Accessible rooms sometimes have one single-person bed instead of a double to create additional maneuvering space. However, double beds are a standard feature in lodging rooms for many reasons, and are strongly recommended for all accessible rooms. Lamps should be within reach range of the bed (swing-arm lamps are useful), with easily operated controls (touch-switches are ideal). Furniture should not block environmental controls (thermostats, air conditioning vents, etc.).

Key Items

Standard room partition materials, finishes, and features installed to compliant dimensions; accessible door and cabinet hardware; either visual alarms, phone alerts, and door alerts or outlets for the same; accessible bathroom features.

Level of Difficulty

Varies. Carpentry, wiring, finish work and possibly plumbing involved.

Estimates

Modify bedroom area of hotel room

Description	Quantity	Unit	Labor-Hours	Material
Remove interior door	1.000	Ea.	0.400	0.00
Remove door frame	2.000	Ea.	1.000	0.00
Remove metal stud/gypsum board partition	28.000	S.F.	1.292	0.00
Re-frame door opening	96.000	S.F.	4.388	64.32
Interior door frame	2.000	Ea.	2.000	150.00
Hollow metal flush door, 3'-0" x 6'-8"	2.000	Ea.	1.882	332.00
Hinges	3.000	Pr.	0.000	133.50
Lever-handled lockset	1.000	Ea.	0.800	121.00
Threshold	1.000	Ea.	0.400	30.00
Paint door	2.000	Ea.	3.200	20.60
Painting	136.000	S.F.	3.454	29.92
Remove closet hardware (labor minimum)	1.000	Job	2.000	0.00
Install D-pull closet handles	4.000	Ea.	0.200	13.92
Adjustable closet rod and shelf, 12" wide, 3' long	1.000	Ea.	0.400	8.50
Cutout demolition of partition	1.000	Ea.	0.333	0.00
Conductor	0.100	C.L.F.	0.296	1.23
Install junction box	1.000	Ea.	0.400	7.75
20-amp rocker switch	1.000	Ea.	0.296	11.70
Install plate	1.000	Ea.	0.100	1.80
Cutout demolition of partition	1.000	Ea.	0.333	0.00
Conductor	0.100	C.L.F.	0.296	1.23
Install junction box	1.000	Ea.	0.400	7.75
Install outlet	1.000	Ea.	0.296	7.05
Install plate	1.000	Ea.	0.100	1.80
Repair gypsum board	1.000	Job	4.000	0.00
Paint gypsum board - minimum	1.000	Job	2.000	0.00
Fire alarm horn	1.000	Ea.	1.194	36.50
#18 fire alarm conductor	0.100	C.L.F.	0.100	6.00
Fire alarm light	1.000	Ea.	1.509	95.00
#18 fire alarm conductor	0.100	C.L.F.	0.100	6.00
Telephone company labor minimum	1.000	Job	2.286	0.00
Totals			**35.455**	**1,087.57**

Total per each including general contractor's overhead and profit	**$4,377**

Modify bathroom in hotel room

Description	Quantity	Unit	Labor-Hours	Material
Remove interior door	1.000	Ea.	0.400	0.00
Remove door frame	2.000	Ea.	1.000	0.00
Remove metal stud/gypsum board partition	28.000	S.F.	1.292	0.00
Re-frame door opening	96.000	S.F.	4.388	64.32
Interior door frame	2.000	Ea.	2.000	150.00
Hollow metal flush door, 3'-0" x 6'-8"	2.000	Ea.	1.882	332.00
Hinges	3.000	Pr.	0.000	133.50
Lever-handled lockset	1.000	Ea.	0.800	121.00
Paint door	2.000	Ea.	3.200	20.60
Painting	136.000	S.F.	3.454	29.92
Remove bathtub	1.000	Ea.	2.000	0.00
Remove partition finishes	66.000	S.F.	0.587	0.00
Water-resistant 5/8" gypsum board	110.000	S.F.	0.880	29.70
Copper shower pan	18.000	S.F.	1.440	55.98
Ceramic tile floor (pitched to drain)	15.000	S.F.	1.311	134.25
Ceramic tile walls, thin-set 4-1/4" x 4-1/4"	110.000	S.F.	9.263	229.90
Ceramic bath accessories	2.000	Ea.	0.390	19.00
Tub grab bar	2.000	Ea.	1.143	157.00
Grab bar vertical arms	4.000	Ea.	2.667	346.00
Bar-mounted hand-held shower head	1.000	Ea.	0.364	63.50
Rough in supply, waste and vent for shower	1.000	Ea.	13.445	141.00
Rough in supply, waste and vent for sink	1.000	Ea.	9.639	220.00
Wall-hung porcelain enamel lavatory (22" x 19")	1.000	Ea.	2.000	405.00
Faucet, handles and drain	1.000	Ea.	0.800	107.00
Mirror with stainless steel shelf	1.000	Ea.	0.400	90.50
Labor minimum for toilet relocation	1.000	Job	4.000	0.00
Rough in supply, waste and vent	1.000	Ea.	5.634	139.00
Labor minimum to repair gypsum board and ceramic tile	1.000	Job	0.080	0.13
Totals			**74.459**	**2,989.30**

Total per each including general contractor's overhead and profit   **$10,418**

71
Install Accessible Pathways to Play Area

ACCESSIBLE ROUTE TO PLAYGROUND

INTERLOCKING RUBBER MATS

It is vital that children with disabilities have access to play areas—and as much equipment as possible. Playgrounds present particular challenges to a designer. How does one provide an environment that stimulates the imagination of all children, who have such a broad range of abilities and skills? How does one provide a ground surface that doesn't cost a fortune and is forgiving to falls, yet firm enough for a wheelchair? What types of play equipment are challenging and interesting, yet within reach of children with limited use of their arms or legs?

ADAAG References
105.2.3, Advisory
204.1, Protruding Objects, Exception 2
206.2.17, Accessible Routes in Play Areas
240, Play Areas
303, Changes in Level
304, Turning Space

305, Clear Floor or Ground Space
308, Reach Ranges
403, Walking Surfaces
405, Ramps
1008, Play Areas
ASTM F 1951-99, Standard Specification for the Determination of Accessibility of Surface Systems Under and Around Playground Equipment
ASTM F 1292-04, Standard Specification for Impact Attenuation of Surface Systems within the Use Zone of Playground Equipment
ASTM 1487-01, Standard Consumer Safety Performance Specifications for Playground Equipment

Where Applicable

All public and common use play areas in day care centers, school yards, etc. Note that relocating existing equipment to a resilient surface does not require one to make the equipment accessible. Similarly, one can alter equipment without installing a resilient surface provided the surface is not being altered.

Design Requirements

- Route at least 36" wide to the play area, with a 2% maximum cross slope and 5% maximum running slope (stable, firm surface, and slip-resistant).
- Accessible route at least 60" wide connecting *all* ground level accessible components, with 2% maximum cross slope and 5% maximum running slope. Constrictions down to 36" wide are allowed for distances up to 60", provided they are at least 60" apart. Play areas 1,000 square feet or less can have 44" wide routes if there are passing spaces every 30' or less. Transfer systems can serve as an accessible route between elevated play components. (See 1008.2 for other special considerations for elevated components.)

- Ramps not steeper than 1:16 (6.25%) and without handrails may connect ground level components. Ramps up to a vertical rise of 12" with handrails may connect elevated components.
- Handrails between 20" and 28" high and a diameter between 0.95" and 1.55".
- Resilient surfaces within play equipment use zones (area around play equipment) meeting ASTM F 1292-04, and other accessible surfaces meeting ASTM F 1951-99.
- Roughly one-third of ground level play components adjacent to a 30" by 48" clear area on an accessible route. (See table 240.2.1.2 for actual number and proportion of different types of components.)
- One-half of elevated play components with a transfer system adjacent to a 30" x 48" transfer area on an accessible route. (Ramps, transfer systems, steps, and decks are not considered elevated play components.) Transfer platforms at least 14" deep, 24" wide, and between 11" and 18" above the ground surface, with the 48" side of the transfer space centered on the 24" side of the transfer platform. Successive platforms can be utilized as steps, provided their risers are no more than 8".
- At least one transfer support, such as rope loops, loop handles, slots in flat surfaces, poles, bars, D-rings, and so forth, at a transfer systems.

Design Suggestions

The accessible route to a play area should be the same as the general route of travel. Try to connect all play components with a pathway meeting the requirements, even if a piece of equipment is not one of the accessible ones. Children with limited abilities may want to be near others using a piece of equipment, or the person monitoring a child might use a mobility aid.

For new play areas, there are manufacturers that design accessible equipment who can help with the layout and specifications. Layout and surface treatment is more of a problem when up-grading existing playgrounds. Consider consulting a landscape architect to design the entire layout, even if you plan to keep a number of the existing pieces and do the work in stages. The strongest recommendation is to discuss proposed playground designs and renovations with disability organizations, users, and parents to determine the optimal design.

The playground surface must not only be "stable, firm, and slip-resistant," but also to "impact attenuating." Child safety guidelines require a safe fall area around all equipment. Some materials seem to comply with both: interlocking rubber matting (which is expensive), plastic matting that allows grass to grow between the holes, poured-in-place rubber surfacing, and specially-bonded wood fibers. Traditional materials, such as sand and loose wood chips, are not accessible, but could be used in conjunction with rubber or plastic mats forming the route to and around all play equipment.

The design and location of benches is important for people supervising children. Locate benches off the path so as not to interfere with children's play, 20' to 30' away from the play area. Benches should allow users to sit together with space for a wheelchair or baby carriage beside the bench and off the path.

Equipment

Many challenging types of designs can accommodate children with a wide range of abilities, such as climbing structures, play structures that allow transfer from a wheelchair, sand tables and water tables that allow children using wheelchairs to roll up to and under them, and others. Wheelchair access is the first issue to consider, but it is only one type of access that must be provided. Consider installing play equipment that is usable by people with differing abilities. Input from children with disabilities, their parents, teachers, and others who have experience in accessible playground design is vital in designing an accessible play area.

Also, when purchasing equipment, don't forget that a monitor might have to retrieve a child from hard-to-reach spot. Inspect the design to make sure that adults—with or without a wheelchair—can reach an elevated structure and come to the rescue.

Key Items

Accessible route surface material, resilient surfacing material, equipment modification.

Level of Difficulty

Moderate to high. Installing pre-fabricated play equipment is fairly easy, but an accessible play area can require professional landscape planning and contractors.

Estimates

Install accessible route to play area, 36″ wide asphalt pathway

Description	Quantity	Unit	Labor-Hours	Material
Excavating	0.100	C.Y.	0.018	0.00
Asphalt sidewalk base	3.000	S.F.	0.029	1.17
2-1/2″ asphalt sidewalk	0.340	S.Y.	0.025	1.44
Install topsoil, 4″ deep	0.230	S.Y.	0.002	0.58
Install sod	0.002	M.S.F.	0.004	0.43
Totals			**0.078**	**3.62**

Total per linear foot including general contractor's overhead and profit **$11**

Install accessible route in play area, rubber mats, interlocking over concrete

Description	Quantity	Unit	Labor-Hours	Material
Excavating	0.030	C.Y.	0.005	0.00
Concrete sidewalk base	1.000	S.F.	0.010	0.39
4″ sidewalk, broom finish	1.000	S.F.	0.040	1.21
Install topsoil, 4″ deep	0.230	S.Y.	0.002	0.58
Install sod	0.002	M.S.F.	0.004	0.43
Interlocking rubber mats	1.000	S.F.	0.038	2.85
Totals			**0.099**	**5.46**

Total per square foot including general contractor's overhead and profit **$16**

Install accessible route in play area, woven rubber mats on existing sand

Description	Quantity	Unit	Labor-Hours	Material
Woven rubber mats	1.000	S.F.	0.052	14.00
Totals			0.052	14.00

Total per square foot including general contractor's overhead and profit **$27**

Install wood mulch on existing sand

Description	Quantity	Unit	Labor-Hours	Material
Wood mulch in place	1.000	S.Y.	0.080	1.63
Totals			0.080	1.63

Total per square yard of 3-inch-deep mulch including general contractor's overhead and profit **$7**

Provide woven plastic mats on existing grass

Description	Quantity	Unit	Labor-Hours	Material
Woven plastic mats	1.000	S.F.	0.052	18.00
Totals			0.052	18.00

Total per square foot including general contractor's overhead and profit **$34**

72
Install Swimming Pool Access

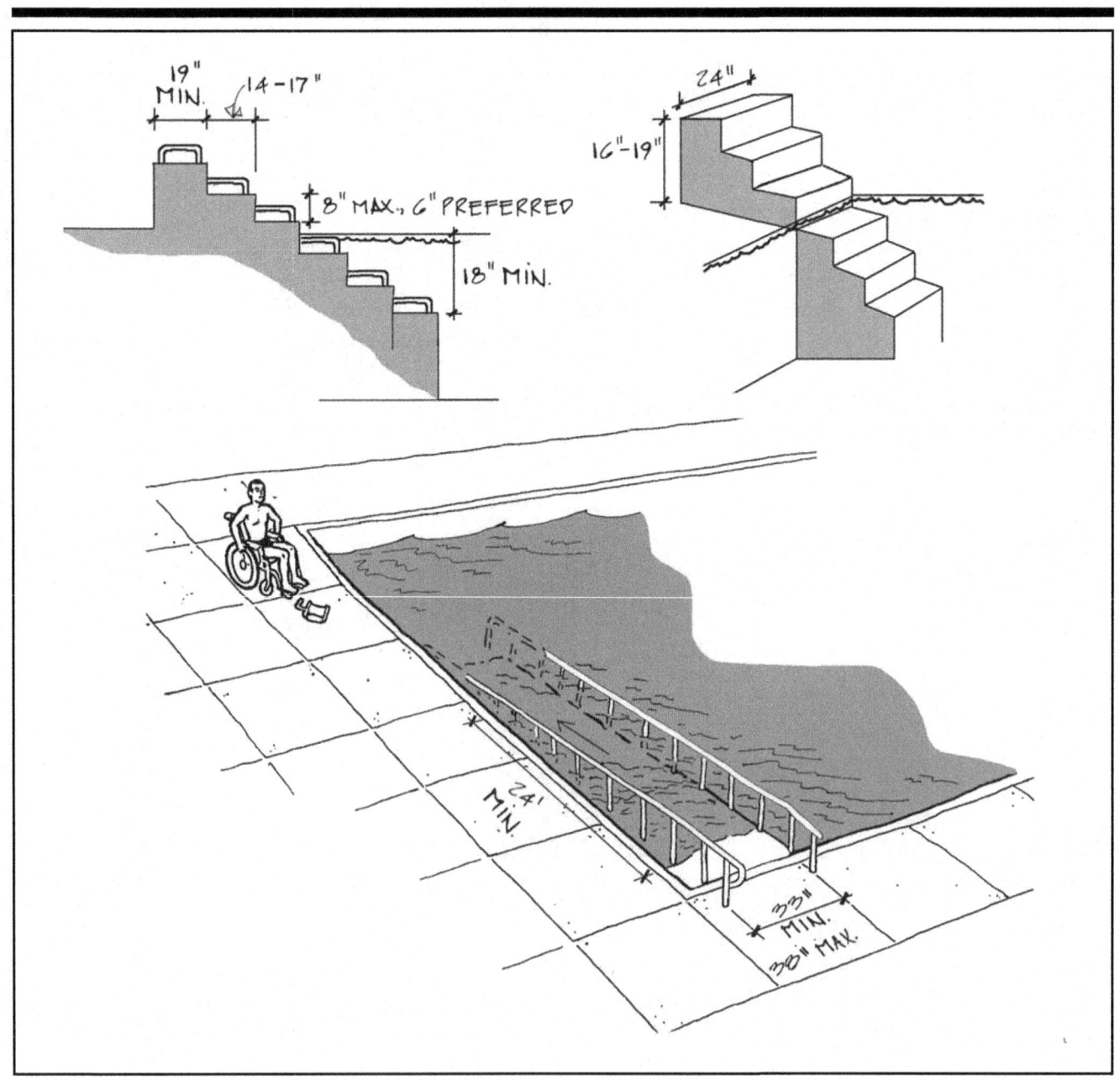

Swimming pool access solutions have been evolving for many years, and in some cases, minimal retrofit is necessary. A wide range of lift, transfer wall, and ramp solutions are available to help a swimming pool become accessible. Not surprisingly, accessible features, particularly ramps, transfer walls, and stairs, have proven to be quite popular with the general population.

ADAAG References

206.2, Accessible Routes
242, Swimming Pools, Wading Pools, and Spas
302, Floor and Ground Surfaces
305, Clear Floor or Ground Space
405, Ramps
504, Stairs
505, Handrails
1009, Swimming Pools, Wading Pools, and Spas

Where Applicable

Public swimming pools, hotel pools, spas, health clubs.

Design Requirements

- On an accessible route.
- A sloped entrance to pool. Two accessible entrances for pools with perimeters 300' or more, one of which may be a lift, transfer wall, or transfer system. (ADAAG requires that at least one means of access be either a ramp or lift.)
- Firm, sloped entrances extending

between 24" and 30" below the water surface, 36" wide minimum with two handrails meeting handrail requirements 505 (but without extension at bottom), spaced between 33" and 38" apart.
- Handrail exception: No handrails required in wading pools. No handrail spacing requirement for sand bottom pools, wave action pools, or rivers.
- Handrails at stairs: between 20" and 24" apart, not required to have an extension at the bottom.

Design Suggestions

Access can be achieved without making the pool look like a "rehab center" by using ramps or transfer systems.

Consider providing an aquatic wheelchair made of non-corrosive materials and designed for access into the water. It will protect the water from contamination and avoid damage to personal wheelchairs or other mobility aids.

Extend the pool to create an accessible area, including a ramp or a roll-in area. Many pool surroundings have the area to expand, but this solution can be expensive.

As an alternative, it is possible to create a ramp to a depth of at least 24", which complies with all standard ramp dimensions in the shallow end. This, too, can be expensive, even without the excavation required.

If a second entry is desired, and there is no problem raising the water level, consider building a transfer wall 16" to 19" high around the edge of the pool.

Another entrance option is a transfer stair that has a 24" by 19" seat, 16" to 19" high, with treads stepping down into the water to a depth of 18" minimum below the surface. There should be grab bars along the edge of each tread.

For stairs leading into the pool, install handrails on both sides to help people who use a wheelchair but have some walking ability.

For people with visual impairments, it is recommended that a 24" wide detectable warning strip be installed around the edge of the pool if there is no continuous wall or lip. (The ADA specifically requires detectable warning strips only at reflecting pools.) This is a useful safety feature, and the usual drawback of snow removal (trying to shovel over the bumps in the materials) will not be a factor.

Key Items

If the pool configuration is not altered, involves installation of a concrete or prefab ramp, and/or low wall, and/or transfer device.

Level of Difficulty

Moderate to high. Involves concrete and/or concrete block work.

72. Install Swimming Pool Access *(continued)*

Estimate

Install transfer device for swimming pool access

Description	Quantity	Unit	Labor-Hours	Material
Saw cutting, concrete per inch of depth	32.000	L.F.	0.523	11.20
Remove concrete deck	0.450	S.Y.	0.071	0.00
Hand excavation	0.500	C.Y.	1.000	0.00
Spread footing	0.500	C.Y.	1.471	52.50
Anchor bolts	4.000	Ea.	0.213	5.80
Lifting device	1.000	Ea.	0.000	1,800.00
Labor to install lifting device	1.000	Job	8.000	0.00
Totals			**11.278**	**1,869.50**

Total per each including general contractor's overhead and profit **$4,080**

Notes

73
Create Accessible Beaches

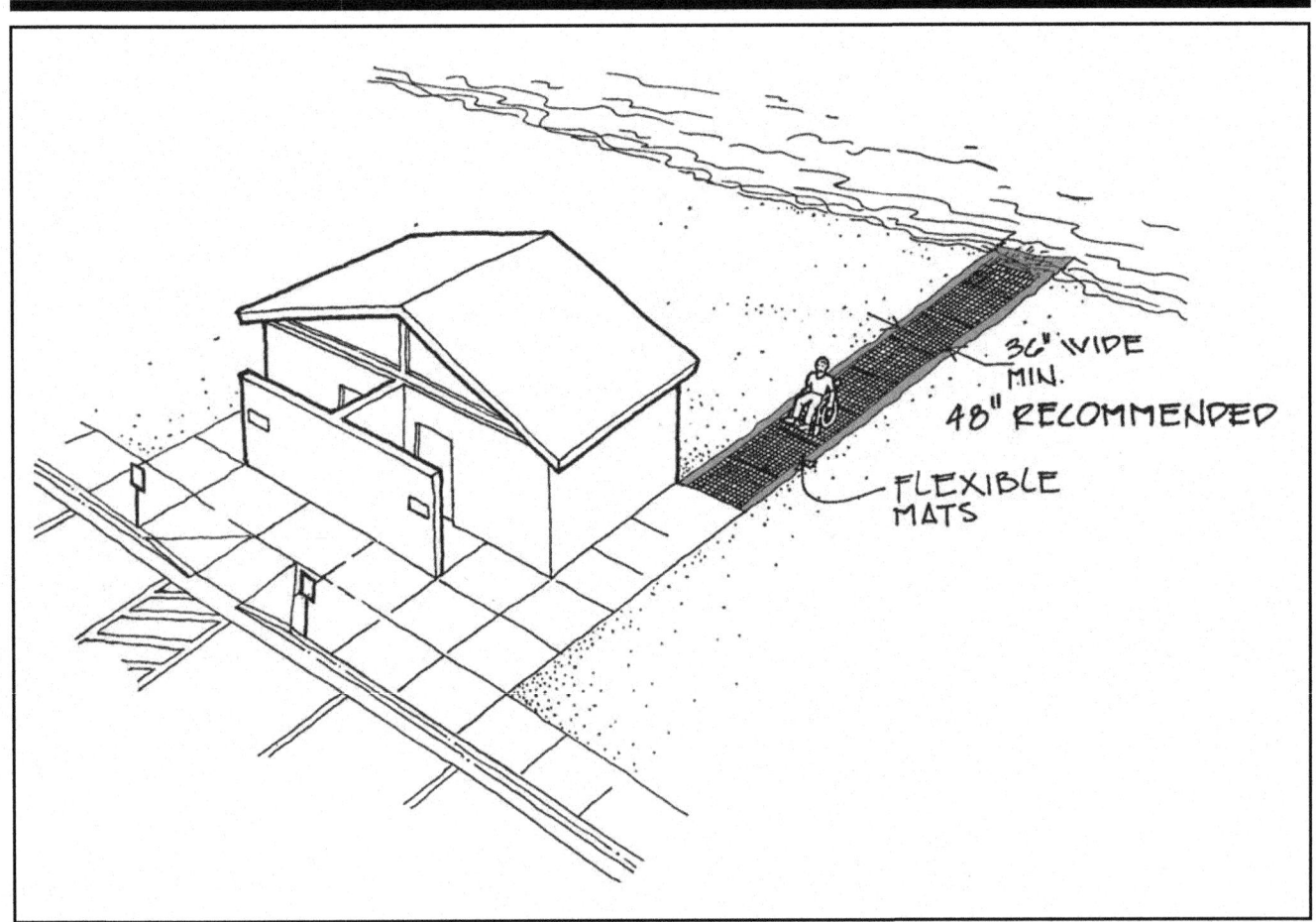

Although the U.S. Access Board has not yet established guidelines for beaches, many beaches are owned by state or local governments, and hence require compliance with the program access section of Title II of the ADA. Beaches at public accommodations (at resorts, hotels, and the like) must comply with Title III. Not surprisingly, the accessible features for beaches have proven to be quite popular with the general population.

ADAAG References
103, Equivalent Facilitation
206.2.1 and 206.2.2, Accessible Routes
302, Floor and Ground Surfaces
303, Changes in Level

Where Applicable

Public beaches at rivers, lakes, and oceans.

Design Requirements

- Sufficient number of all common facilities accessible (bathhouses, showers, changing rooms, concession stands, telephones, etc.).
- Accessible route of travel from parking to the beach free of steps. 36" wide minimum (48" recommended) with 60" x 60" passing spots. No protruding objects; surface stable, firm, and slip-resistant.

Design Suggestions

Several design solutions create access to the water, although sand a difficult ground surface to make accessible, and many of the traditional materials used for creating an accessible route conflict with environmental regulations.

People sometimes use wheelchairs to enter the water at both fresh and saltwater beaches. Other people walk with crutches or canes. Accordingly, provide several accessible paths of travel without violating environmental concerns.

Some types of beach access involve placing movable rubber or plastic mats over the sand, or rolling out a material similar to snow fencing over the sand. These paths are often 36" wide, but 48" is recommended, since rolling off the edge is easy when the material is placed over an unstable surface such as sand. Some localities extend one length down to the high tide line; others to the low tide water line. Others form a "T" to allow access to different sections of the beach. The mats can be set daily in place during a given time period, when the beach is open, but some beaches make them available on an as-needed basis. Clearly, the path systems work best for beaches along rivers and lakes.

While it is possible to install permanent pathways, these are subject to limitations placed by environmental concerns and ordinances. Concrete or wood pathways can inhibit natural movement of sand, and wood walkways placed over sand can block sunlight and adversely affect the growth of grasses needed to stabilize sand dunes. Generally, temporary materials placed in areas where the public is allowed to walk will not disturb the beach any more than typical foot traffic, but local environmental legislation and advisers should be consulted.

There are special beach wheelchairs for individual users, featuring large wheels that can roll on sand and even float. Have one or more available for use on an as-needed basis. For maximum accessibility, these wheelchairs should be used in conjunction with an accessible pathway to ensure that people who prefer to use (or must use) their own wheelchairs have access to the beach.

Key Items

Flexible mats, wood boardwalks, or concrete pathways.

Level of Difficulty

Low to Moderate.

Estimates

Provide plastic mats for beach access (3′ wide, 40′ long)

Description	Quantity	Unit	Labor-Hours	Material
Woven plastic mats	120.000	S.F.	6.193	2,160.00
Totals			6.193	2,160.00

Total per each including general contractor's overhead and profit **$4,052**

73. Create Beach Access *(continued)*

Provide plastic mats for beach access

Description	Quantity	Unit	Labor-Hours	Material
Woven plastic mats	32.000	S.F.	1.652	576.00
Totals			1.652	576.00

Total per 4' x 8' section including general contractor's overhead and profit **$1,080**

Provide wood duckboards for beach access

Description	Quantity	Unit	Labor-Hours	Material
Hardwood strips on rubber backing (3 foot wide)	3.000	S.F.	0.155	35.85
Totals			0.155	35.85

Total per linear foot including general contractor's overhead and profit **$71**

Install concrete pathway to high water line, 36" wide

Description	Quantity	Unit	Labor-Hours	Material
Excavating	0.120	C.Y.	0.021	0.00
Concrete sidewalk base	2.000	S.F.	0.019	0.78
4" sidewalk, broom finish	3.000	S.F.	0.120	3.63
Totals			0.160	4.41

Total per linear foot including general contractor's overhead and profit **$18**

Notes

74
Create Accessible Hiking Trails

ACCESSIBLE TRAIL

Creating access is often thought of only in terms of modifying buildings, but outdoor areas are used by people with disabilities as well as others. Even though accessible trail guidelines are in the works but not final as of the publishing of this book, state and local governments have an obligation under Title II to provide "program access." Public accommodations must also provide an equal opportunity for people with disabilities to participate.

ADAAG References

302, Floor or Ground Surfaces
303, Changes in Level
307, Protruding Objects
402, Accessible Routes
403, Walking Surfaces
Final Report, Regulatory Negotiation Committee on Accessibility Guidelines for Outdoor Developed Areas, U.S. Access Board, Section 16.2

Where Applicable

Public trails.

Design Requirements

As of this book's publication date, the U.S. Access Board was developing guidelines for outdoor developed areas including trails.

Design Suggestions

- 36" minimum clear width (60" recommended) with 60" by 60" or "T" passing spots less than 1,000' apart. Width may be reduced to 32" because of conditions a, b, c, or d above.
- 3" edging, if provided.
- No protruding objects below 80" (unless otherwise impossible because of conditions a, b, c, or d above).
- Obstacles in path: 3" max if running and cross slope is less than 1:20, 2" maximum otherwise.
- Openings should be small enough to stop a 1/2" diameter sphere with elongations perpendicular to the direction of travel. If this is not possible, meet the other specifications.
- Maximum running slope 5%, maximum cross-slope 1:33 (3%) for drainage. However, if absolutely necessary, running slope 1:12 maximum for up to 200', provide resting spots at least at the beginning and end. Slopes of 1:10 for up to 30' and 1:8 for up to 10', provide resting spots at the beginning and end.
- Frequent spots to leave the path and resting spots.

- Accessible trails should be representative of all trails in the recreational area. If a nature park has some trails in wooded areas and other sin open grassland, accessible trails should be provided in both areas, making as much of the program accessible as possible.

If sections are longer than 30' and steeper than 1:18, consider installing a sign at the head of the trail indicating that the slope could be difficult to negotiate. This also applies to cross slopes greater than 1:33 (3%) or other difficult situations. Where the slope is steeper than 1:12 for more than 20', consider installing handrails.

As always, consultation with people with disabilities is necessary to gain data and ideas for modifications.

Key Item

Landscaping equipment. Paving materials, such as stone dust on a base course or wood decking.

Level of Difficulty

Moderate. Basic landscaping.

Estimate

Create accessible trails (no heavy trees)

Description	Quantity	Unit	Labor-Hours	Material
Clear trail	0.070	Acre	3.360	0.00
Stump removal (assume ten 6" diameter stumps per 100 L.F.)	10.000	Ea.	4.000	0.00
Fine grading by hand	11.120	S.Y.	0.381	0.00
Gravel and cinders over stone base	11.120	S.Y.	1.779	130.66
Compaction	10.000	C.Y.	0.400	0.00
Totals			**9.920**	**130.66**

Total per square foot including general contractor's overhead and profit	**$4**
Total per 100 linear feet of 3 foot wide trail including general contractor's overhead and profit	**$1,058**

Notes

Part Three
Case Studies

Two case studies are included in this section as examples of the access design and construction process. Every case is different, but all projects—no matter how small—involve an initial survey to identify existing barriers, an examination of design options, a design process, and construction planning. Deciding what is the best solution is where much of the expense, and the headaches, of access modifications come into play. To say the least, the design process is not always smooth, especially if it is not possible to make a facility fully in compliance with ADAAG. The headaches can be reduced if the intended users are included in the design process as early as possible. There is always an interplay between need and cost, but with more information in the design program, a more successful balance may be achieved.

Each case study is organized by the existing conditions before the project begins, a description of the needed modification and requirements, and the final design solution. This section is useful as an example of realistic challenges and solutions to achieve long-term ADA compliance.

Existing Bathroom Rehab

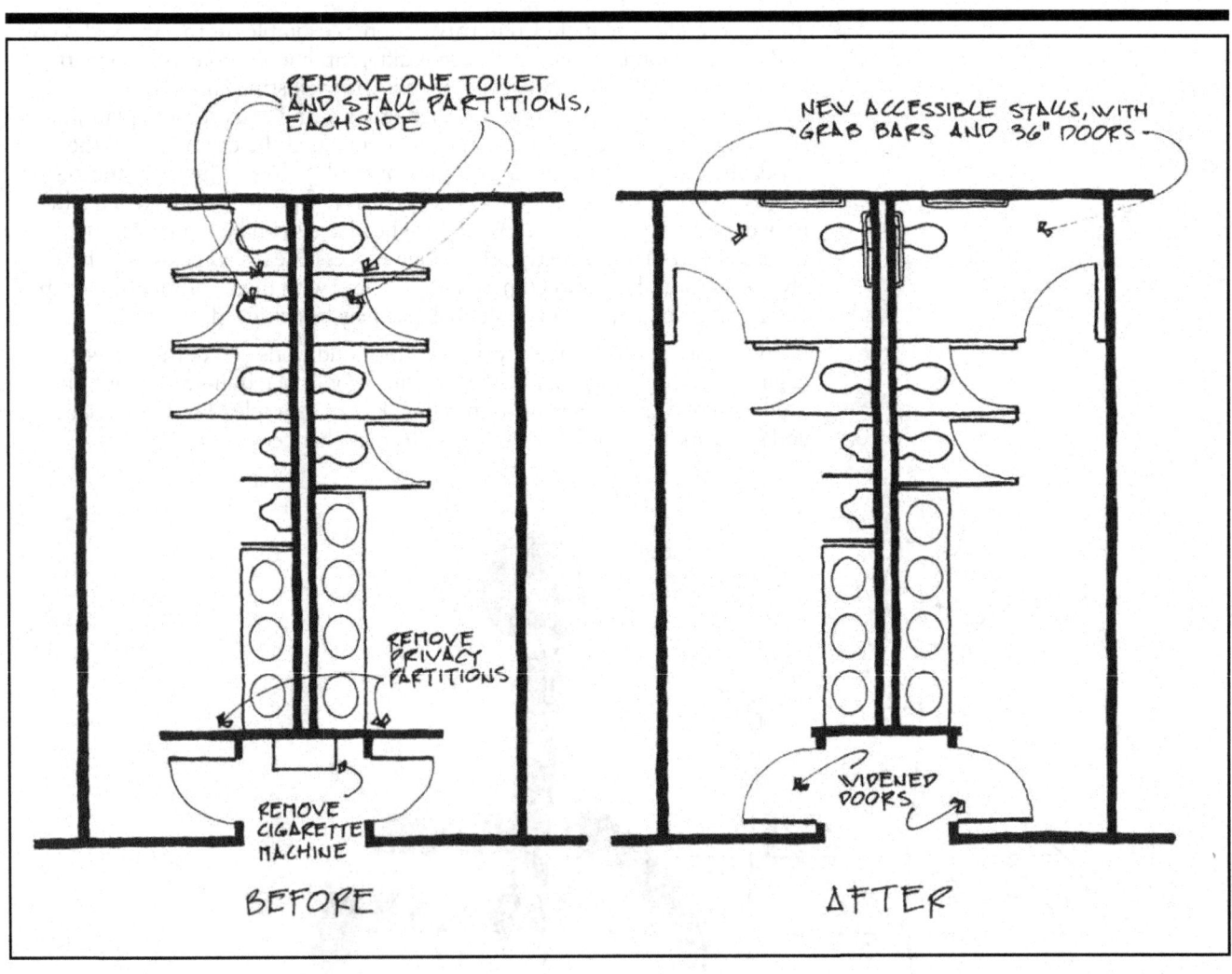

REMOVE ONE TOILET AND STALL PARTITIONS, EACH SIDE

NEW ACCESSIBLE STALLS, WITH GRAB BARS AND 36" DOORS

REMOVE PRIVACY PARTITIONS

REMOVE CIGARETTE MACHINE

WIDENED DOORS

BEFORE

AFTER

Facility

A restaurant with accessible entrance on the first floor of a 1920s-era Art Deco office building.

Space Being Modified

Men's and women's restrooms.

Level of Accessibility

The restrooms are inaccessible, due to narrow entrance; lack of maneuvering space; and lack of accessible toilet stalls, sinks, and dispensers.

Reason for Modification

The restaurant is undergoing renovation due to a change of ownership, with an estimated construction cost of $100,000. The restrooms are in good condition, and the owners were only planning to make minimal modifications for reasons unrelated to ADA. The restrooms serve a primary function area, and the route to them needs to be modified, up to 20% of the total cost of the renovation (in this case about 7.6%). Bringing the restrooms into compliance is triggered by ADAAG.

Existing Conditions

Inaccessible entrances to both restrooms: Narrow doorways (30" wide with 28" clear opening), knob hardware, 3/4" marble thresholds, 30" wide entry vestibules created by vision screens, tiled floors.

Men's Room: Three 36" toilet stalls with enameled metal partitions, toilets centered within stalls, seats 15" a.f.f., two urinals with rims 22" a.f.f. in 32" wide spaces, four sinks set in a plastic laminate countertop at 35"

a.f.f. with 10" apron, ball faucets, mirror 44" a.f.f., paper towel dispenser 60" a.f.f., coat hooks on toilet doors 68" a.f.f.

Women's Room: Four 36" toilet stalls, toilets centered within stall, seats 15" a.f.f., four sinks set in a plastic laminate countertop at 35" a.f.f., coat hooks on toilet doors 68" a.f.f., sanitary napkin dispenser 60" a.f.f.

Design Options

There are three possible design alternatives for bringing the restrooms into compliance with the ADA, with various advantages, disadvantages, and costs.

1. Leave the existing restrooms intact, and add two new single-use accessible restrooms. This option was rejected by the owners, since it would take up too much space in the restaurant, and there is no room on the site for exterior expansion.
2. Leave the existing restrooms intact, and add one new single unisex restroom. This option was rejected by the plumbing inspector. While this solution is permitted by ADAAG, it is prohibited by the local plumbing code, which mandates separate accessible restrooms for men and women.
3. Modify both restrooms for compliance. This was approved and accepted by the owners and plumbing inspector. (Approval by the plumbing inspector was needed since one fixture would be lost in each restroom.)

Preferred Solution

Modify both restrooms. Estimated cost: $10,000. The plans were

prepared by the architect, and reviewed by the building inspector, plumbing inspector, and owners.

Modifications

Entrances: Remove cigarette machine at entrances to create more space, widen restroom doors to 36", reverse door swing to create sufficient clearance on the latch side of the door, replace with a door that has level handles, add raised character/Braille signage to the door jamb to the right of the door, and remove existing thresholds.

Entry vestibules: Remove vision screens to allow more maneuvering space.

Toilet stalls: Remove one toilet in each restroom, combine two stalls to create one accessible stall with a wider door. (The other existing toilet partitions are in good enough condition to remain intact.) Add grab bars. The route of travel is narrow (34"), but allowed to remain as is, since widening the path of travel would require decreasing the other stalls, making them unusable. Existing toilet allowed to remain, but the seat is replaced with a higher model, and the reverse flush valve will face the open side. One urinal in men's room to remain as is, with the second lowered. Stalls around urinals can remain, since there is more than 30" clearance between the protective partitions.

Sinks: Sink heights allowed to remain; 10" aprons replaced with 3" aprons to allow knee space below. All pipes wrapped and faucets replaced with blade handles. Mirrors lowered to 38" a.f.f.

Dispensers: All dispensers lowered to 48" a.f.f.

Historic Entry Modification

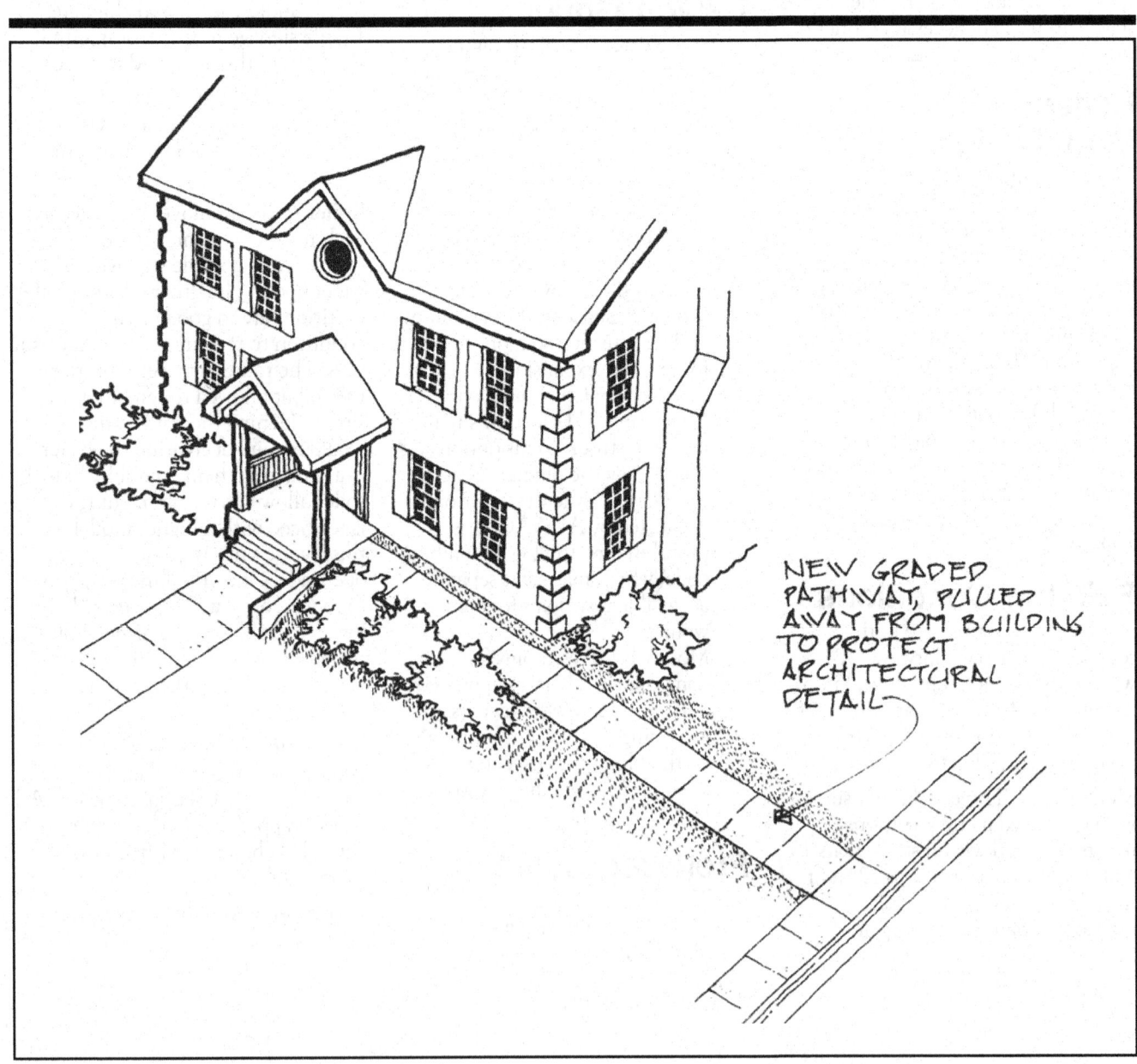

NEW GRADED PATHWAY, PULLED AWAY FROM BUILDING TO PROTECT ARCHITECTURAL DETAIL

Existing Conditions

A two-story library with basement, town-owned, 1840s Federalist design, brick and limestone construction, on local and state historic registers.

Level of Accessibility

The building is set back 50' from the sidewalk. There are two entrances: front and back. Access to the front entrance is by means of four steps, with 6" risers, that extend to a porch. The rear entrance is on grade, but opens into stairhall.

Reason for Modification

Title II compliance. Program access is not possible via a non-structural means, since the town has no other library and no means of providing alternative forms of program accessibility.

Design Options

There are five possible design alternatives for bringing the restrooms into compliance with the ADA, with various advantages, disadvantages, and costs.

1. Add a stair lift on front stairs. This option was rejected by the historic commission.
2. Add a stair lift on rear stairs. This option was accepted by the historic commission, but rejected by the library committee, because it did not meet the intent of the ADA. It would be clumsy to use and would call for leaving the door unlocked, which would compromise security. It was also rejected by the fire marshal, because designated fire egress stair cannot be blocked by a lift.
3. Add an exterior elevator. This was rejected by the historic commission and the town board, since cost would require floating a bond issue (estimated cost, with two-stop elevator with brick shaft, $120,000).
4. Install a ramp at the front entrance. This was deemed acceptable if design would conform to the existing style of the building. It would require a concrete ramp with limestone facing, and metal rails attached to painted fluted posts (estimated cost: $30,000).
5. Install a graded pathway with rough-finish stone pavers up to the front entrance at a slope of 1:20, covering a portion of the existing steps. This option was also found acceptable, with approval from historic commission (estimated cost: $20,000, with landscaping and re-grading, including an extra wide berm to make the lawn area flush with the graded pathway).

Preferred Solution

The graded pathway materials was changed to broom-finish concrete, as it was the only affordable solution that satisfied the historic commission. Concern was raised about necessary snow removal and drainage, since the pathway is flush with the entrance level. This issue was addressed by the landscape contractor. The plans were reviewed and approved by the library manager, town architect, local historic commission, and town disability rights committee.

Part Four
Unit Costs

This section provides over 3,000 construction line items that can be used to adapt the projects in Part Two to your specific requirements. If your project calls for special materials or extra work not described in a project, the cost data here will help you make the necessary customization. For experienced estimators, this section is also helpful for developing additional modification project estimates. For example, if your project involves not only installing a new pathway, but also removing an existing path, curb, or fence, you can find the cost for the added demolition here, under Division 2–Site Construction, and choose the appropriate item. These costs can be tailored to your location by using Part Five, "Location Adjustment Factors."

How to Use
the Unit Costs

The Unit Cost Section is organized according to the 16-Division Construction Specifications Institute (CSI) MasterFormat, the most commonly accepted system for classifying construction information in the U.S. and Canada.

For a listing of these divisions and an outline of their subdivisions, see the Table of Contents of this Unit Cost Section, following this introduction.

The following is a detailed explanation of a sample entry in the Unit Cost Section. Next to each bold number is a definition of the item being described, with the appropriate component of the sample entry following in parenthesis.

1 Division Number/Title (02700/Bases, Ballasts, Pavements, & Appurtenances)

Use the Unit Cost Section Table of Contents to locate specific items. The sections are classified according to the CSI MasterFormat.

2 Line Number (02775-275-0020)

Each unit price line item has been assigned a unique 12-digit code based on the CSI MasterFormat classification.

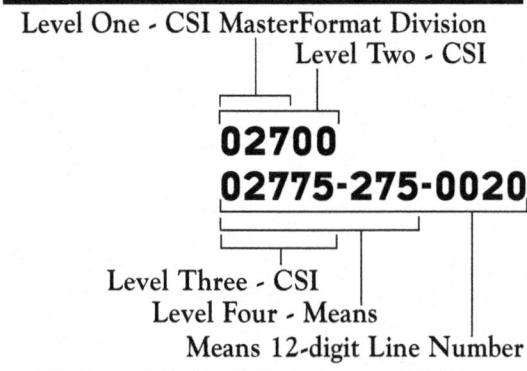

Level One - CSI MasterFormat Division
Level Two - CSI

02700
02775-275-0020

Level Three - CSI
Level Four - Means
Means 12-digit Line Number

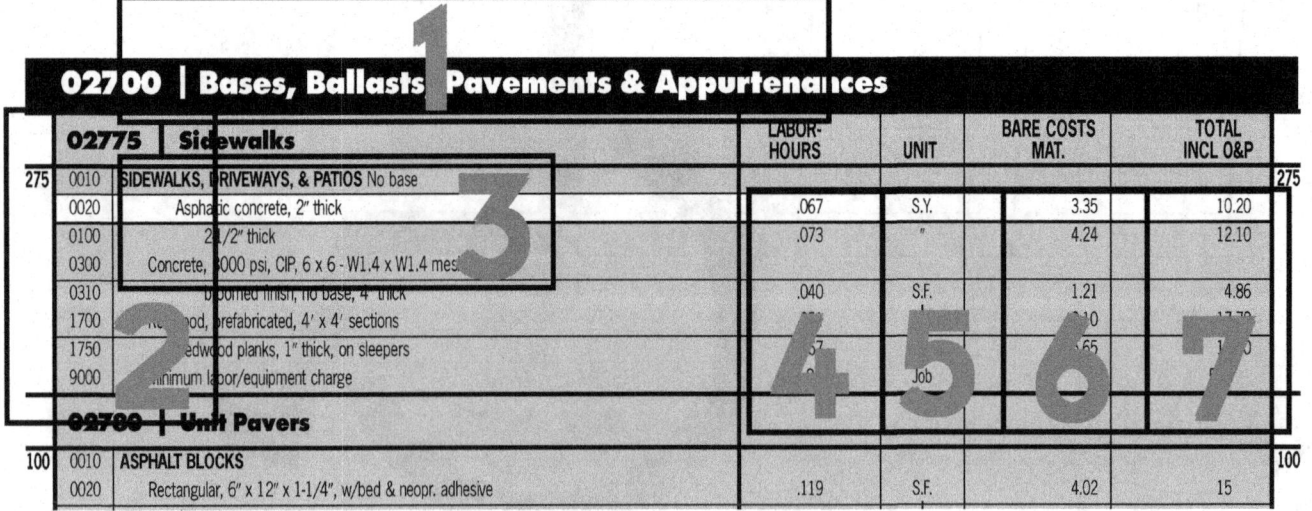

02700	Bases, Ballasts, Pavements & Appurtenances					
02775	**Sidewalks**	LABOR-HOURS	UNIT	BARE COSTS MAT.	TOTAL INCL O&P	
275 0010	SIDEWALKS, DRIVEWAYS, & PATIOS No base					275
0020	Asphaltic concrete, 2" thick	.067	S.Y.	3.35	10.20	
0100	2 1/2" thick	.073	"	4.24	12.10	
0300	Concrete, 3000 psi, CIP, 6 x 6 - W1.4 x W1.4 mesh					
0310	broomed finish, no base, 4" thick	.040	S.F.	1.21	4.86	
1700	Redwood, prefabricated, 4' x 4' sections			10	17.70	
1750	Redwood planks, 1" thick, on sleepers	.67		.65	1.0	
9000	Minimum labor/equipment charge		Job			
02780	**Unit Pavers**					
100 0010	ASPHALT BLOCKS					100
0020	Rectangular, 6" x 12" x 1-1/4", w/bed & neopr. adhesive	.119	S.F.	4.02	15	

3 Description
(Asphaltic Concrete, 2" Thick)

Each line item is described in detail. Sub-items and additional sizes are indented beneath the appropriate line items. The first line or two after the main item (in boldface) may contain descriptive information that pertains to all line items beneath this boldface listing.

4 Labor-Hours (.067)

The "Labor-Hours" figure represents the number of hours required to install one unit of work. To find out the number of work-hours required for your particular task, multiply the quantity of item times the number of work-hours shown. For example:

Quantity	x	Productivity Rate	=	Duration
1,000 S.Y.	x	.067 S.Y.	=	67 Labor-Hours

5 Unit (S.Y.)

The abbreviated designation indicates the unit of measure upon which the price and productivity are based (S.Y. = Square Yard). See "Abbreviations" at the back of the book for a complete list.

6 Bare Costs: Bare Mat.
(Bare Material Cost) (3.35)

This figure for the unit material cost for the line item is the "bare" material cost with no markups included. *Costs shown reflect national average material prices for January 2004 and include delivery to the job site. No sales taxes are included.*

7 Total Costs Including O & P
(10.20)

The figure in this column represents the total cost of the work and materials described in the line item. This figure is the sum of the bare material cost plus 20% (compensation for the contractor's risk in the purchase and handling of materials), plus the marked-up costs for labor and equipment as determined by RSMeans. (Labor is marked up by an average of 55.5% to cover labor burdens, such as Workers' Compensation and employee benefits, and the subcontractor's overhead and profit. Equipment is marked up 10% to cover the contractor's risk related to this item.) This sum is then increased by 15% to cover general conditions items such as permits, trash removal fees, and site supervision, 7.5% for the general contractor coordinating the subcontractors, and 15% for the general contractor's overhead and profit.

*Each trade as tracked by the RSMeans has a different markup. The markups range from a high of 84.1% to a low of 47.0% due to variations in Workers' Compensation rates for different trades.

Note: The line items in the Unit Cost Section were selected for their appropriateness to ADA modification projects. RSMeans produces 27 annual cost data books covering a wide range of construction disciplines.

Unit Costs
Table of Contents

02050 | Basic Site Materials & Methods

02055 | Soils

			LABOR-HOURS	UNIT	BARE COSTS MAT.	TOTAL INCL O&P	
150	0010	**BORROW**					150
	0020	And spread, with 200 H.P. dozer, no compaction					
	0200	Common borrow	.047	C.Y.	5.30	16.75	
	0700	Screened loam	.047		20	42	
	0800	Topsoil, weed free	.047		20	42	
	0900	For 5 mile haul, add	.040	↓		6.17	

02060 | Aggregate

			LABOR-HOURS	UNIT	BARE COSTS MAT.	TOTAL INCL O&P	
150	0010	**BORROW**					150
	0020	And spread, with 200 H.P. dozer, no compaction					
	0100	Bank run gravel	.047	C.Y.	15.40	34	
	0300	Crushed stone, (1.40 tons per CY) 1-1/2"	.047		24.50	49.50	
	0320	3/4"	.047		27	53.50	
	0340	1/2"	.047		17.75	38	
	0360	3/8"	.047		17.90	38	
	0400	Sand, washed, concrete	.047		25	50.50	
	0500	Dead or bank sand	.047		3.93	14.50	
	0600	Select structural fill	.047		7.50	20.50	
	0900	For 5 mile haul, add	.040	↓		6.15	

02200 | Site Preparation

02220 | Site Demolition

			LABOR-HOURS	UNIT	BARE COSTS MAT.	TOTAL INCL. O&P	
130	0010	**BLDG. FOOTINGS AND FOUNDATIONS DEMOLITION**					130
	0200	Floors, concrete slab on grade,					
	0240	4" thick, plain concrete	.080	S.F.		5.45	
	0280	Reinforced, wire mesh	.085			5.80	
	0300	Rods	.100			6.85	
	0400	6" thick, plain concrete	.107			7.30	
	0420	Reinforced, wire mesh	.118			8.05	
	0440	Rods	.133	↓		9.15	
	1000	Footings, concrete, 1' thick, 2' wide	.187	L.F.		17.15	
	1080	1'-6" thick, 2' wide	.224			20.50	
	1120	3' wide	.280			25.50	
	1140	2' thick, 3' wide	.320			29	
	1200	Average reinforcing, add				14.20%	
	1220	Heavy reinforcing, add		↓		28.40%	
	2000	Walls, block, 4" thick	.044	S.F.		2.73	
	2040	6" thick	.047			2.89	
	2080	8" thick	.053			3.27	
	2100	12" thick	.053			3.27	
	2200	For horizontal reinforcing, add				14.20%	
	2220	For vertical reinforcing, add				28.40%	
	2400	Concrete, plain concrete, 6" thick	.250			17.15	
	2420	8" thick	.286			19.55	
	2440	10" thick	.333			23	
	2500	12" thick	.400			27.50	
	2600	For average reinforcing, add				14.20%	
	2620	For heavy reinforcing, add		↓		28.40%	

		02220	Site Demolition	LABOR-HOURS	UNIT	BARE COSTS MAT.	TOTAL INCL O&P	
130	4000		For congested sites or small quantities, add up to		S.F.		284%	130
	9000		Minimum labor/equipment charge	4	Job		279	
220	0010		**FENCING DEMOLITION**					220
	1600		Fencing, barbed wire, 3 strand	.037	L.F.		2.27	
	1650		5 strand	.057			3.50	
	1700		Chain link, posts & fabric, remove only, 8' to 10' high	.054	↓		4.22	
240	0010		**MINOR SITE DEMOLITION**					240
	0015		No hauling, abandon catch basin or manhole	3.429	Ea.		269	
	0020		Remove existing catch basin or manhole, masonry	6			470	
	0030		Catch basin or manhole frames and covers, stored	1.846			145	
	0040		Remove and reset	3.429			269	
	0100		Roadside delineators, remove only	.183			16.40	
	0110		Remove and reset	.320	↓		29	
	0400		Minimum labor/equipment charge	6	Job		470	
	0800		Guiderail, corrugated steel, remove only	.240	L.F.		17.20	
	0850		Remove and reset	.600	"		43.50	
	0860		Guide posts, remove only	.267	Ea.		19.70	
	0870		Remove and reset	.480	"		53.50	
	0890		Minimum labor/equipment charge	4	Job		245	
	0900		Hydrants, fire, remove only	8	Ea.		730	
	0950		Remove and reset	20	"		1,850	
	0990		Minimum labor/equipment charge	8	Job		695	
	1000		Masonry walls, block or tile, solid, remove	.031	C.F.		2.84	
	1100		Cavity wall	.025			2.33	
	1200		Brick, solid	.062			5.70	
	1300		With block back-up	.050			4.55	
	1400		Stone, with mortar	.062			5.70	
	1500		Dry set	.037	↓		3.43	
	1600		Median barrier, precast concrete, remove and store	.112	L.F.		13.45	
	1610		Remove and reset	.123	"		14.80	
	1650		Minimum labor/equipment charge	2	Job		139	
	2900		Pipe removal, sewer/water, no excavation, 12" diameter	.137	L.F.		10.75	
	2930		15" diameter	.160			12.50	
	2960		24" diameter	.200			15.65	
	3000		36" diameter	.267			21	
	3200		Steel, welded connections, 4" diameter	.150			11.80	
	3300		10" diameter	.300	↓		23.50	
	3390		Minimum labor/equipment charge	8	Job		625	
	3500		Railroad track removal, ties and track	.170	L.F.		14	
	3600		Ballast	.096	C.Y.		6.80	
	3700		Remove and re-install, ties & track using new bolts & spikes	.960	L.F.		67.50	
	3800		Turnouts using new bolts and spikes	48	Ea.		3,375	
	3890		Minimum labor/equipment charge	9.600	Job		675	
	4000		Sidewalk removal, bituminous, 2-1/2" thick	.074	S.Y.		5.80	
	4050		Brick, set in mortar	.130			10.15	
	4100		Concrete, plain, 4"	.150			11.80	
	4200		Mesh reinforced	.160	↓		12.50	
	4290		Minimum labor/equipment charge	4	Job		277	
250	0010		**DEMOLISH, REMOVE PAVEMENT AND CURB**					250
	5010		Pavement removal, bituminous roads, 3" thick	.058	S.Y.		5.75	
	5050		4" to 6" thick	.095			9.40	
	5100		Bituminous driveways	.063			6.15	
	5200		Concrete to 6" thick, hydraulic hammer, mesh reinforced	.157			15.50	
	5300		Rod reinforced	.200	↓		19.75	
	5400		Concrete, 7" to 24" thick, plain	1.212	C.Y.		120	
	5500		Reinforced	1.667	"		165	

02220	Site Demolition	LABOR-HOURS	UNIT	BARE COSTS MAT.	TOTAL INCL O&P	
250	5590	Minimum labor/equipment charge	6.667	Job		660	**250**
	5600	With hand held air equipment, bituminous, to 6" thick	.025	S.F.		1.73	
	5700	Concrete to 6" thick, no reinforcing	.030			2.08	
	5800	Mesh reinforced	.034			2.37	
	5900	Rod reinforced	.063	↓		4.34	
	5990	Minimum labor/equipment charge	6.667	Job		660	
	6000	Curbs, concrete, plain	.067	L.F.		5.20	
	6100	Reinforced	.087			6.85	
	6200	Granite	.067			5.20	
	6300	Bituminous	.045	↓		3.55	
	6390	Minimum labor/equipment charge	4	Job		315	
310	0010	**SELECTIVE DEMOLITION, CUTOUT**					**310**
	0020	Concrete, elev. slab, light reinforcement, under 6 CF	.615	C.F.		41.94	
	0050	Light reinforcing, over 6 C.F.	.533	"		37	
	0200	Slab on grade to 6" thick, not reinforced, under 8 S.F.	.471	S.F.		32	
	0250	Not reinforced, over 8 S.F.	.229	"		15.65	
	0600	Walls, not reinforced, under 6 C.F.	.667	C.F.		45.50	
	0650	Not reinforced, over 6 C.F.	.615			42	
	1000	Concrete, elevated slab, bar reinforced, under 6 C.F.	.889			61	
	1050	Bar reinforced, over 6 C.F.	.800	↓		54.50	
	1200	Slab on grade to 6" thick, bar reinforced, under 8 S.F.	.533	S.F.		37	
	1250	Bar reinforced, over 8 S.F.	.381	"		26	
	1400	Walls, bar reinforced, under 6 C.F.	.800	C.F.		54.50	
	1450	Bar reinforced, over 6 C.F.	.727	"		50	
	2000	Brick, to 4 S.F. opening, not including toothing					
	2040	4" thick	1.333	Ea.		91.50	
	2060	8" thick	2.222			152	
	2080	12" thick	4			274	
	2400	Concrete block, to 4 S.F. opening, 2" thick	1.143			78	
	2420	4" thick	1.333			91.50	
	2440	8" thick	1.481			102	
	2460	12" thick	1.667			114	
	2600	Gypsum block, to 4 S.F. opening, 2" thick	.500			34	
	2620	4" thick	.571			39	
	2640	8" thick	.727			50	
	2800	Terra cotta, to 4 S.F. opening, 4" thick	.571			39	
	2840	8" thick	.615			42	
	2880	12" thick	.800	↓		54.50	
	4000	For toothing masonry, see Division 04910-800					
	6000	Walls, interior, not including re-framing,					
	6010	openings to 5 S.F.					
	6100	Drywall to 5/8" thick	.333	Ea.		20.50	
	6200	Paneling to 3/4" thick	.400			24.50	
	6300	Plaster, on gypsum lath	.400			24.50	
	6340	On wire lath	.571	↓		35	
	7000	Wood frame, not including re-framing, openings to 5 S.F.					
	7200	Floors, sheathing and flooring to 2" thick	1.600	Ea.		98	
	7310	Roofs, sheathing to 1" thick, not including roofing	1.333			82	
	7410	Walls, sheathing to 1" thick, not including siding	1.143	↓		70.50	
	8500	Minimum labor/equipment charge	2	Job		122	
320	0010	**SELECTIVE DEMOLITION, DISPOSAL ONLY**					**320**
	0015	Urban bldg w/salvage value allowed					
	0020	Including loading and 5 mile haul to dump					
	0200	Steel frame	.112	C.Y.		13.45	

02200 | Site Preparation

02220 | Site Demolition

			LABOR-HOURS	UNIT	BARE COSTS MAT.	TOTAL INCL O&P	
320	0300	Concrete frame	.132	C.Y.		15.80	**320**
	0400	Masonry construction	.108			12.95	
	0500	Wood frame	.194	↓		23.50	
330	0010	**SELECTIVE DEMOLITION, DUMP CHARGES**					**330**
	0020	Dump charges, typical urban city, tipping fees only					
	0100	Building construction materials		Ton		99.40	
	0200	Trees, brush, lumber				71	
	0300	Rubbish only				85.20	
	0500	Reclamation station, usual charge		↓		120.70	
350	0010	**SELECTIVE DEMOLITION, RUBBISH HANDLING**					**350**
	0020	The following are to be added to the demolition prices					
	0400	Chute, circular, prefabricated steel, 18" diameter	.600	L.F.	28	85.50	
	0440	30" diameter	.800	"	37.50	114	
	0600	Dumpster, weekly rental, 1 dump/week, 6 C.Y. capacity (2 Tons)		Week		447.30	
	0700	10 C.Y. capacity (4 Tons)				532.50	
	0800	30 C.Y. capacity (10 Tons)				944.30	
	0840	40 C.Y. capacity (13 Tons)		↓		1,143.10	
	1000	Dust partition, 6 mil polyethylene, 1" x 3" frame	.008	S.F.	.16	.89	
	1080	2" x 4" frame	.008	"	.26	1.07	
	2000	Load, haul, and dump, 50' haul	.667	C.Y.		40.50	
	2040	100' haul	.970			59.50	
	2080	Over 100' haul, add per 100 L.F.	.451			27.50	
	2120	In elevators, per 10 floors, add	.114			7	
	3000	Loading & trucking, including 2 mile haul, chute loaded	.711			60.50	
	3040	Hand loading truck, 50' haul	.667			57.50	
	3080	Machine loading truck	.267			24	
	3120	Wheeled 50' and ramp dump loaded	.667			40.50	
	5000	Haul, per mile, up to 8 C.Y. truck	.007			1.07	
	5100	Over 8 C.Y. truck	.005	↓		.80	
360	0010	**SELECTIVE DEMOLITION, SAW CUTTING**					**360**
	0015	Asphalt, up to 3" deep	.015	L.F.	.26	1.84	
	0020	Each additional inch of depth	.009		.06	.92	
	0400	Concrete slabs, mesh reinforcing, up to 3" deep	.016		.35	2.10	
	0420	Each additional inch of depth	.010		.12	1.11	
	0800	Concrete walls, hydraulic saw, plain, per inch of depth	.064		.32	7.45	
	0820	Rod reinforcing, per inch of depth	.107		.44	12.20	
	1200	Masonry walls, hydraulic saw, brick, per inch of depth	.053		.32	6.25	
	1220	Block walls, solid, per inch of depth	.064		.33	7.45	
	5000	Wood sheathing to 1" thick, on walls	.040			3.11	
	5020	On roof	.032	↓		2.49	
	9000	Minimum labor/equipment charge	4	Job		279	

02300 | Earthwork

02315 | Excavation and Fill

			LABOR-HOURS	UNIT	BARE COSTS MAT.	TOTAL INCL. O&P	
110	0010	**BACKFILL, GENERAL**					**110**
	0015	By hand, no compaction, light soil	.571	C.Y.		35	
	0100	Heavy soil	.727			45	
	0300	Compaction in 6" layers, hand tamp, add to above	.388	↓		24	

271

02300 | Earthwork

02315 | Excavation and Fill

			LABOR-HOURS	UNIT	BARE COSTS MAT.	TOTAL INCL. O&P	
110	0400	Roller compaction operator walking, add	.120	C.Y.		10.50	110
	0500	Air tamp, add	.211			14.55	
	0600	Vibrating plate, add	.133			8.90	
	0800	Compaction in 12" layers, hand tamp, add to above	.235			14.45	
	1000	Air tamp, add	.140			9.60	
	1100	Vibrating plate, add	.089			6.15	
	1200	Trench, dozer, no compaction, 60 HP	.028			3.10	
	1300	Dozer backfilling, bulk, up to 300' haul, no compaction	.010			1.83	
	1400	Air tamped, add	.200	E.C.Y.		17.15	
	1900	Dozer backfilling, trench, up to 300' haul, no compaction	.013	C.Y.		2.45	
	2000	Air tamped, add	.200	E.C.Y.		17.15	
	2350	Spreading in 8" layers, small dozer	.011	C.Y.		2.08	
	2450	Compacting with vibrating plate, 8" lifts	.110	"		7.30	
462	0010	**EXCAVATION, STRUCTURAL**					462
	0015	Hand, pits to 6' deep, sandy soil	1	C.Y.		61	
	0100	Heavy soil or clay	2			122	
	0300	Pits 6' to 12' deep, sandy soil	1.600			98	
	0500	Heavy soil or clay	2.667			163	
	0700	Pits 12' to 18' deep, sandy soil	2			122	
	0900	Heavy soil or clay	4			245	
	1100	Hand loading trucks from stock pile, sandy soil	.667			40.50	
	1300	Heavy soil or clay	1			61	
	1500	For wet or muck hand excavation, add to above		%		71%	
	9000	Minimum labor/equipment charge	2	Job		122	
520	0010	**FILL**, spread dumped material, no compaction					520
	0020	By dozer, no compaction	.012	C.Y.		2.20	
	0100	By hand	.667	"		40.50	
	9000	Minimum labor/equipment charge	2	Job		122	

02700 | Bases, Ballasts, Pavements & Appurtenances

02760 | Paving Specialties

			LABOR-HOURS	UNIT	BARE COSTS MAT.	TOTAL INCL. O&P	
300	0010	**PAINTED TRAFFIC LINES AND MARKINGS**					300
	0020	Acrylic waterborne, white or yellow, 4" wide	.002	L.F.	.15	.43	
	0200	6" wide	.004	"	.14	.56	
	0760	Arrows	.061	S.F.	1.71	8.70	
500	0010	**PAVEMENT MARKINGS**					500
	0800	Parking stall, paint, white	.109	Stall	3.03	13.25	
	1000	Street letters and numbers	.030	S.F.	.58	3.20	

02770 | Curbs and Gutters

			LABOR-HOURS	UNIT	BARE COSTS MAT.	TOTAL INCL. O&P	
100	0010	**BITUMINOUS CONCRETE CURBS**					100
	0012	Curbs, asphaltic, machine formed, 8" wide, 6" high, 40 L.F./ton	.032	L.F.	.59	3.31	
	0100	8" wide, 8" high, 30 L.F. per ton	.036		.68	3.71	
	0150	Asphaltic berm, 12" W, 3"-6" H, 35 L.F./ton, before pavement	.046		.91	4.84	
	0200	12" W, 1-1/2" to 4" H, 60 L.F. per ton, laid with pavement	.038		.55	3.32	

		**02770	Curbs and Gutters**	LABOR-HOURS	UNIT	BARE COSTS MAT.	TOTAL INCL O&P	
300	0010	**CEMENT CONCRETE CURBS**					**300**
	0300	Concrete, wood forms, 6″ x 18″, straight	.096	L.F.	2.10	10.75	
	0400	6″ x 18″, radius	.240		2.20	21.50	
	0550	Precast, 6″ x 18″, straight	.080		7	19	
	0600	6″ x 18″, radius	.172	▼	8	29	
500	0010	**STONE CURBS**					**500**
	1000	Granite, split face, straight, 5″ x 16″	.096	L.F.	9.35	24.50	
	1100	6″ x 18″	.107		12.25	30	
	1300	Radius curbing, 6″ x 18″, over 10′ radius	.215	▼	15	45	
	1400	Corners, 2′ radius	.700	Ea.	50.50	148	
	1600	Edging, 4-1/2″ x 12″, straight	.187	L.F.	4.67	24.50	
	1800	Curb inlets, (guttermouth) straight	1.366	Ea.	112	310	
	2000	Indian granite (belgian block)					
	2100	Jumbo, 10-1/2″ x 7-1/2″ x 4″, grey	.107	L.F.	1.54	10.05	
	2150	Pink	.107		2.06	10.90	
	2200	Regular, 9″ x 4-1/2″ x 4-1/2″, grey	.100		1.42	9.30	
	2250	Pink	.100		2	10.30	
	2300	Cubes, 4″ x 4″ x 4″, grey	.091		1.40	8.75	
	2350	Pink	.091		1.48	8.80	
	2400	6″ x 6″ x 6″, pink	.103	▼	3.60	13.25	

		**02775	Sidewalks**					
275	0010	**SIDEWALKS, DRIVEWAYS, & PATIOS** No base					**275**
	0020	Asphaltic concrete, 2″ thick	.067	S.Y.	3.35	10.20	
	0100	2-1/2″ thick	.073	″	4.24	12.10	
	0300	Concrete, 3000 psi, CIP, 6 x 6 - W1.4 x W1.4 mesh,					
	0310	broomed finish, no base, 4″ thick	.040	S.F.	1.21	4.86	
	1700	Redwood, prefabricated, 4′ x 4′ sections	.051		8.10	17.70	
	1750	Redwood planks, 1″ thick, on sleepers	.067	▼	5.65	14.90	
	9000	Minimum labor/equipment charge	8	Job		555	

		**02780	Unit Pavers**					
100	0010	**ASPHALT BLOCKS**					**100**
	0020	Rectangular, 6″ x 12″ x 1-1/4″, w/bed & neopr. adhesive	.119	S.F.	4.02	15	
	0100	3″ thick	.123		5.65	18.15	
	0300	Hexagonal tile, 8″ wide, 1-1/4″ thick	.119		4.02	15	
	0400	2″ thick	.123		5.65	18.15	
	0500	Square, 8″ x 8″, 1-1/4″ thick	.119		4.02	15	
	0600	2″ thick	.123	▼	5.65	18.15	
	9000	Minimum labor/equipment charge	4	Job		315	
200	0010	**BRICK PAVING** 4″ x 8″ x 1-1/2″, without joints (4.5 brick/S.F.)	.145	S.F.	2.20	13.80	**200**
	0100	Grouted, 3/8″ joint (3.9 brick/S.F.)	.178		2.55	16.65	
	0200	4″ x 8″ x 2-1/4″, without joints (4.5 bricks/S.F.)	.145		2.82	14.85	
	0300	Grouted, 3/8″ joint (3.9 brick/S.F.)	.178		2.60	16.70	
	1500	Brick on 1″ thick sand bed laid flat, 4.5 per S.F.	.160		2.38	15.05	
	2000	Brick pavers, laid on edge, 7.2 per S.F.	.229		2.16	19.45	
	2500	For 4″ thick concrete bed and joints, add	.027		.96	3.51	
	2800	For steam cleaning, add	.008	▼	.05	.69	
	9000	Minimum labor/equipment charge	4	Job		315	
600	0010	**STONE PAVERS**					**600**
	1300	Slate, natural cleft, irregular, 3/4″ thick	.174	S.F.	5.15	21	
	1350	Random rectangular, gauged, 1/2″ thick	.152		11.15	29.50	
	1400	Random rectangular, butt joint, gauged, 1/4″ thick	.107	▼	12	28	

02700 | Bases, Ballasts, Pavements & Appurtenances

		02780	Unit Pavers	LABOR-HOURS	UNIT	BARE COSTS MAT.	TOTAL INCL O&P	
600	1450		For sand rubbed finish, add		S.F.	5.60	9.55	600
	1550		Granite blocks, 3-1/2″ x 3-1/2″ x 3-1/2″	.174		6.25	22.50	
	1600		4″ to 12″ long, 3″ to 5″ wide, 3″ to 5″ thick	.163	↓	5.20	20	

02900 | Planting

		02910	Plant Preparation	LABOR-HOURS	UNIT	BARE COSTS MAT.	TOTAL INCL. O&P	
810	0010		**LOAM & TOPSOIL**					810
	0700		Furnish and place, truck dumped, screened, 4″ deep	.009	S.Y.	2.53	5.25	
	0800		6″ deep	.015	″	3.23	7	
	0810		Minimum labor/equipment charge	6	Ea.		610	
	0900		Fine grading and seeding, incl. lime, fertilizer & seed,					
	1000		With equipment	.048	S.Y.	.34	3.97	
	2000		Minimum labor/equipment charge	2	Job		122	

03050 | Basic Concrete Materials & Methods

		03055	Selective Demolition	LABOR-HOURS	UNIT	BARE COSTS MAT.	TOTAL INCL. O&P	
110	0010		**SELECTIVE DEMOLITION, CONCRETE FRAMING**					110
	1020		Concrete, average reinforcing, beams, 8″ x 10″	.333	L.F.		23	
	1040		10″ x 12″	.364			25	
	1060		12″ x 14″	.444			30.50	
	1200		Columns, 8″ x 8″	.333			23	
	1240		10″ x 10″	.333			23	
	1280		12″ x 12″	.364			25	
	1320		14″ x 14″	.400			27.50	
	1400		Girders, 14″ x 16″	.727			50	
	1440		16″ x 18″	1	↓		69	
	1600		Slabs, elevated, 6″ thick	.067	S.F.		4.56	
	1640		8″ thick	.089			6.10	
	1680		10″ thick	.111			7.60	
	1900		Add for heavy reinforcement		↓		35.50%	
	1910		Minimum labor/equipment charge	20	Job		1,375	

			03310	Structural Concrete	LABOR-HOURS	UNIT	BARE COSTS MAT.	TOTAL INCL O&P	
240	0010			CONCRETE IN PLACE Including forms (4 uses), reinforcing					240
	0050			steel and finishing unless otherwise indicated					
	0300			Beams, 5 kip per L.F., 10' span	12.804	C.Y.	233	1,500	
	0350			25' span	10.782		214	1,275	
	0500			Chimney foundations, industrial, minimum	3.476		145	510	
	0510			Maximum	4.724		168	640	
	0700			Columns, square, 12" x 12", minimum reinforcing	16.722		212	1,775	
	0720			Average reinforcing	19.743		310	2,200	
	0740			Maximum reinforcing	22.148		390	2,550	
	0800			16" x 16", minimum reinforcing	12.330		197	1,375	
	0820			Average reinforcing	15.911		310	1,850	
	0840			Maximum reinforcing	19.512		425	2,375	
	3800			Footings, spread under 1 C.Y.	2.942		105	400	
	3850			Over 5 C.Y.	1.382		96.50	268	
	3900			Footings, strip, 18" x 9", unreinforced	2.800		91	365	
	3920			18" x 9", reinforced	3.200		105	420	
	3925			20" x 10", unreinforced	2.489		88	340	
	3930			20" x 10", reinforced	2.800		99.50	380	
	3935			24" x 12", unreinforced	2.036		87	300	
	3940			24" x 12", reinforced	2.333		98.50	345	
	3945			36" x 12", unreinforced	1.600		83.50	262	
	3950			36" x 12", reinforced	1.867		94	299	
	4000			Foundation mat, under 10 C.Y.	2.896		129	440	
	4050			Over 20 C.Y.	1.986		115	345	
	4200			Grade walls, 8" thick, 8' high	4.364		126	580	
	4250			14' high	7.337		158	885	
	4260			12" thick, 8' high	3.109		112	450	
	4270			14' high	4.999		126	630	
	4300			15" thick, 8' high	2.499		106	390	
	4350			12' high	3.902	↓	112	515	
	4751			Slab on grade, incl. troweled finish, not incl. forms					
	4760			or reinforcing, over 10,000 S.F., 4" thick	.021	S.F.	.96	3.08	
	4820			6" thick	.021		1.41	3.88	
	4840			8" thick	.023		1.93	4.84	
	4900			12" thick	.026		2.89	6.75	
	4950			15" thick	.029	↓	3.64	8.20	
	5000			Slab on grade, incl. textured finish, not incl. forms					
	5001			or reinforcing, 4" thick	.019	S.F.	.94	2.93	
	5010			6" thick	.022		1.47	3.98	
	5020			8" thick	.024	↓	1.92	4.91	
	5200			Lift slab in place above the foundation, incl. forms,					
	5210			reinforcing, concrete and columns, minimum	.098	S.F.	5.10	16.95	
	5250			Average	.126		5.75	20.50	
	5300			Maximum	.139	↓	6.50	22.50	
	5900			Pile caps, incl. forms and reinf., sq. or rect., under 5 C.Y.	2.069	C.Y.	89	310	
	5950			Over 10 C.Y.	1.493		92	269	
	6000			Triangular or hexagonal, under 5 C.Y.	2.113		86.50	305	
	6050			Over 10 C.Y.	1.318		94.50	261	
	6200			Retaining walls, gravity, 4' high see division 02830-100	3.021		97.50	415	
	6250			10' high	1.600		86	280	
	6300			Cantilever, level backfill loading, 8' high	2.857		109	425	
	6350			16' high	2.198	↓	103	355	
	6800			Stairs, not including safety treads, free standing, 3'-6" wide	.578	LF Nose	6.90	56.50	
	6850			Cast on ground	.384	"	4.85	37.50	
	7000			Stair landings, free standing	.240	S.F.	2.65	23	
	7050			Cast on ground	.101	"	1.53	10.45	

03300 | Cast-In-Place Concrete

		03310	Structural Concrete	LABOR-HOURS	UNIT	BARE COSTS MAT.	TOTAL INCL O&P	
240	9000		Minimum labor/equipment charge	16	Job		1,250	240

04050 | Basic Masonry Materials and Methods

		04055	Selective Demolition	LABOR-HOURS	UNIT	BARE COSTS MAT.	TOTAL INCL. O&P	
110	0010		**SELECTIVE DEMOLITION, MASONRY**					110
	1000		Chimney, 16" x 16", soft old mortar	.333	V.L.F.		20.50	
	1020		Hard mortar	.444			27	
	1080		20" x 20", soft old mortar	.667			40.50	
	1100		Hard mortar	.800			49	
	1140		20" x 32", soft old mortar	.800			49	
	1160		Hard mortar	1			61	
	1200		48" x 48", soft old mortar	1.600			98	
	1220		Hard mortar	2			122	
	2000		Columns, 8" x 8", soft old mortar	.167			10.25	
	2020		Hard mortar	.200			12.25	
	2060		16" x 16", soft old mortar	.500			30.50	
	2100		Hard mortar	.571			35	
	2140		24" x 24", soft old mortar	1			61	
	2160		Hard mortar	1.333			82	
	2200		36" x 36", soft old mortar	2			122	
	2220		Hard mortar	2.667	▼		163	
	3000		Copings, precast or masonry, to 8" wide					
	3020		Soft old mortar	.044	L.F.		2.73	
	3040		Hard mortar	.050	"		3.07	
	3100		To 12" wide					
	3120		Soft old mortar	.050	L.F.		3.07	
	3140		Hard mortar	.057	"		3.50	
	4000		Fireplace, brick, 30" x 24" opening					
	4020		Soft old mortar	4	Ea.		245	
	4040		Hard mortar	6.400			390	
	4100		Stone, soft old mortar	5.333			325	
	4120		Hard mortar	8	▼		490	
	5000		Veneers, brick, soft old mortar	.057	S.F.		3.50	
	5020		Hard mortar	.064			3.92	
	5100		Granite and marble, 2" thick	.044			2.73	
	5120		4" thick	.047			2.89	
	5140		Stone, 4" thick	.044			2.73	
	5160		8" thick	.046	▼		2.80	
	5400		Alternate pricing method, stone, 4" thick	.133	C.F.		8.15	
	5420		8" thick	.094	"		5.75	
	9000		Minimum labor/equipment charge	4	Job		245	

		04810	Unit Masonry Assemblies	LABOR-HOURS	UNIT	BARE COSTS MAT.	TOTAL INCL O&P	
100	0010		**BRICK VENEER** Scaffolding not included, truck load lots					100
	0015		Material costs incl. 3% brick and 25% mortar waste					
	2000		Standard, sel. common, 4" x 2-2/3" x 8", (6.75/S.F.)	.174	S.F.	2.76	17.10	
	2020		Standard, red, 4" x 2-2/3" x 8", running bond (6.75/SF)	.182		2.76	17.65	
	2050		Full header every 6th course (7.88/S.F.)	.216		3.21	21	
	2100		English, full header every 2nd course (10.13/S.F.)	.286		4.12	27.50	
	2150		Flemish, alternate header every course (9.00/S.F.)	.267		3.67	25.50	
	2200		Flemish, alt. header every 6th course (7.13/S.F.)	.195		2.91	18.80	
	2250		Full headers throughout (13.50/S.F.)	.381		5.50	36.50	
	2300		Rowlock course (13.50/S.F.)	.400		5.50	38	
	2350		Rowlock stretcher (4.50/S.F.)	.129		1.85	12.35	
	2400		Soldier course (6.75/S.F.)	.200		2.76	18.95	
	2450		Sailor course (4.50/S.F.)	.138		1.85	13	
	2600		Buff or gray face, running bond, (6.75/S.F.)	.182		2.93	17.90	
	2700		Glazed face brick, running bond	.190		7.95	27	
	2750		Full header every 6th course (7.88/S.F.)	.235		9.30	32.50	
	3000		Jumbo, 6" x 4" x 12" running bond (3.00/S.F.)	.092		3.61	12.65	
	3050		Norman, 4" x 2-2/3" x 12" running bond, (4.5/S.F.)	.125		3.83	15.40	
	3100		Norwegian, 4" x 3-1/5" x 12" (3.75/S.F.)	.107		2.94	12.65	
	3150		Economy, 4" x 4" x 8" (4.50/S.F.)	.129		3.39	15	
	3200		Engineer, 4" x 3-1/5" x 8" (5.63/S.F.)	.154		2.86	15.85	
	3250		Roman, 4" x 2" x 12" (6.00/S.F.)	.160		4.62	19.30	
	3300		SCR, 6" x 2-2/3" x 12" (4.50/S.F.)	.129		4.23	16.40	
	3350		Utility, 4" x 4" x 12" (3.00/S.F.)	.089	↓	3.31	12	
	9000		Minimum labor/equipment charge	8	Job		555	
172	0010		**CONCRETE BLOCK, BACK-UP,** C90, 2000 psi					172
	0020		Normal weight, 8" x 16" units, tooled joint 1 side					
	0050		Not-reinforced, 2000 psi, 2" thick	.084	S.F.	.82	7.35	
	0200		4" thick	.087		.97	7.80	
	0300		6" thick	.091		1.42	8.85	
	0350		8" thick	.100		1.54	9.75	
	0400		10" thick	.121		2.16	12.25	
	0450		12" thick	.155	↓	2.22	14.45	
	9000		Minimum labor/equipment charge	8	Job		555	
184	0010		**CONCRETE BLOCK, EXTERIOR** C90, 2000 psi					184
	0020		Reinforced alt courses, tooled joints 2 sides					
	0100		Normal weight, 8" x 16" x 6" thick	.101	S.F.	1.65	9.95	
	0200		8" thick	.111		2.46	12.10	
	0250		10" thick	.138		2.93	14.80	
	0300		12" thick	.192	↓	2.99	18.40	
	9000		Minimum labor/equipment charge	8	Job		555	
186	0010		**CONCRETE BLOCK FOUNDATION WALL** C90/C145					186
	0050		Normal-weight, cut joints, horiz joint reinf, no vert reinf					
	0200		Hollow, 8" x 16" x 6" thick	.088	S.F.	1.63	9.05	
	0250		8" thick	.094		1.76	9.70	
	0300		10" thick	.114		2.38	12.15	
	0350		12" thick	.160	↓	2.46	15.20	
	1000		Reinforced, #4 vert @ 48"					
	1100		Hollow, 8" x 16" block, 4" thick	.088	S.F.	1.57	8.90	
	1125		6" thick	.090		2.27	10.25	
	1150		8" thick	.096		2.67	11.40	
	1200		10" thick	.118		3.56	14.45	
	1250		12" thick	.166	↓	3.91	18.10	

04800 | Masonry Assemblies

		04810	Unit Masonry Assemblies	LABOR-HOURS	UNIT	BARE COSTS MAT.	TOTAL INCL. O&P	
186	9000		Minimum labor/equipment charge	8	Job		555	186
210	0010		**CONCRETE BLOCK, PARTITIONS**, scaffolding not included					210
	1000		Lightweight block, tooled joints, 2 sides, hollow					
	1100		Not reinforced, 8" x 16" x 4" thick	.091	S.F.	1.11	8.35	
	1150		6" thick	.098		1.51	9.55	
	1200		8" thick	.104		1.86	10.55	
	1250		10" thick	.108		2.45	11.85	
	1300		12" thick	.137	↓	2.87	14.35	
	4000		Regular block, tooled joints, 2 sides, hollow					
	4100		Not reinforced, 8" x 16" x 4" thick	.093	S.F.	.93	8.15	
	4150		6" thick	.100		1.38	9.45	
	4200		8" thick	.107		1.50	10.15	
	4250		10" thick	.111		2.12	11.55	
	4300		12" thick	.141	↓	2.19	13.45	
	9000		Minimum labor/equipment charge	8	Job		555	

04900 | Masonry Restoration and Cleaning

		04910	Unit Masonry Restoration	LABOR-HOURS	UNIT	BARE COSTS MAT.	TOTAL INCL. O&P	
800	0010		**TOOTHING MASONRY**					800
	0500		Brickwork, soft old mortar	.200	V.L.F.		12.25	
	0520		Hard mortar	.267			16.35	
	0700		Blockwork, soft old mortar	.114			7	
	0720		Hard mortar	.160	↓		9.80	
	9000		Minimum labor/equipment charge	2	Job		122	

05050 | Basic Metal Materials & Methods

		05060	Selective Demolition	LABOR-HOURS	UNIT	BARE COSTS MAT.	TOTAL INCL. O&P	
110	0010		**SELECTIVE DEMOLITION, STEEL FRAMING**					110
	2000		Steel framing, beams, 4" x 6"	.112	L.F.		9.15	
	2020		4" x 8"	.140			11.50	
	2080		8" x 12"	.224			18.50	
	2200		Columns, 6" x 6"	.140			11.50	
	2240		8" x 8"	.160			13.15	
	2280		10" x 10"	.175			14.45	
	2400		Girders, 10" x 12"	.249			20.50	
	2440		10" x 14"	.280			23	
	2480		10" x 16"	.339			28	
	2520		10" x 24"	.448	↓		37	
	2950		Minimum labor/equipment charge	28	Job		2,300	
		05090	**Metal Fastenings**					
340	0010		**DRILLING** For anchors, up to 4" deep, incl. bit and layout					340
	0050		in concrete or brick walls and floors, no anchor					

05090	Metal Fastenings	LABOR-HOURS	UNIT	BARE COSTS MAT.	TOTAL INCL O&P	
340	0100	Holes, 1/4" diameter	.107	Ea.	.08	8.45	**340**
	0150	For each additional inch of depth, add	.019		.02	1.48	
	0200	3/8" diameter	.127		.07	10.05	
	0250	For each additional inch of depth, add	.024		.02	1.87	
	0300	1/2" diameter	.160		.07	12.60	
	0350	For each additional inch of depth, add	.032		.02	2.52	
	0400	5/8" diameter	.167		.13	13.15	
	0450	For each additional inch of depth, add	.033		.03	2.65	
	0500	3/4" diameter	.178		.16	14.10	
	0550	For each additional inch of depth, add	.036		.04	2.89	
	0600	7/8" diameter	.186		.19	14.80	
	0650	For each additional inch of depth, add	.038		.05	3.04	
	0700	1" diameter	.200		.22	15.95	
	0750	For each additional inch of depth, add	.042		.06	3.36	
	0800	1-1/4" diameter	.211		.31	16.90	
	0850	For each additional inch of depth, add	.044		.08	3.59	
	0900	1-1/2" diameter	.229		.48	18.60	
	0950	For each additional inch of depth, add	.048	▼	.12	3.97	
	1100	Drilling & layout for drywall/plaster walls, up to 1" deep, no anchor					
	1200	Holes, 1/4" diameter	.053	Ea.	.01	4.17	
	1300	3/8" diameter	.057		.01	4.47	
	1400	1/2" diameter	.062		.01	4.81	
	1500	3/4" diameter	.067		.02	5.20	
	1600	1" diameter	.073		.03	5.70	
	1700	1-1/4" diameter	.080		.04	6.30	
	1800	1-1/2" diameter	.089		.06	7	
	1920	Holes, 1/4" diameter	.071		.10	7.20	
	1925	For each additional 1/4" depth, add	.024		.10	2.52	
	1930	3/8" diameter	.077		.12	7.75	
	1935	For each additional 1/4" depth, add	.026		.12	2.72	
	1940	1/2" diameter	.083		.13	8.40	
	1945	For each additional 1/4" depth, add	.028		.13	2.95	
	1950	5/8" diameter	.091		.21	9.35	
	1955	For each additional 1/4" depth, add	.030		.21	3.34	
	1960	3/4" diameter	.100		.24	10.20	
	1965	For each additional 1/4" depth, add	.033		.24	3.67	
	1970	7/8" diameter	.111		.28	11.40	
	1975	For each additional 1/4" depth, add	.037		.28	4.12	
	1980	1" diameter	.125		.32	12.85	
	1985	For each additional 1/4" depth, add	.042	▼	.32	4.64	
	2000	Minimum labor/equipment charge	2	Job		155	
380	0010	**EXPANSION ANCHORS** & shields					**380**
	0100	Bolt anchors for concrete, brick or stone, no layout and drilling					
	0200	Expansion shields, zinc, 1/4" diameter, 1-5/16" long, single	.089	Ea.	.96	8.50	
	0300	1-3/8" long, double	.094		1.06	9.10	
	0400	3/8" diameter, 1-1/2" long, single	.094		1.58	10.05	
	0500	2" long, double	.100		1.95	11.10	
	0600	1/2" diameter, 2-1/16" long, single	.100		2.62	12.25	
	0700	2-1/2" long, double	.107		2.52	12.60	
	0800	5/8" diameter, 2-5/8" long, single	.107		3.74	14.70	
	0900	2-3/4" long, double	.114		3.74	15.25	
	1000	3/4" diameter, 2-3/4" long, single	.114		5.55	18.35	
	1100	3-15/16" long, double	.123		7.40	22	
	1500	Self drilling anchor, snap-off, for 1/4" diameter bolt	.308		.72	25	
	1600	3/8" diameter bolt	.348	▼	1.04	28.50	

		05090	**Metal Fastenings**	LABOR-HOURS	UNIT	BARE COSTS MAT.	TOTAL INCL O&P	
380	1700		1/2" diameter bolt	.400	Ea.	1.60	34.50	**380**
	1800		5/8" diameter bolt	.444		2.68	39.50	
	1900		3/4" diameter bolt	.500	▼	4.50	47	
	2100		Hollow wall anchors for gypsum wall board, plaster or tile					
	2300		1/8" diameter, short	.050	Ea.	.23	4.28	
	2400		Long	.053		.27	4.62	
	2500		3/16" diameter, short	.053		.48	4.97	
	2600		Long	.057		.52	5.35	
	2700		1/4" diameter, short	.057		.59	5.45	
	2800		Long	.062		.66	5.90	
	3000		Toggle bolts, bright steel, 1/8" diameter, 2" long	.094		.21	7.70	
	3100		4" long	.100		.32	8.30	
	3200		3/16" diameter, 3" long	.100		.36	8.35	
	3300		6" long	.107		.50	9.15	
	3400		1/4" diameter, 3" long	.107		.41	9	
	3500		6" long	.114		.57	9.90	
	3600		3/8" diameter, 3" long	.114		.78	10.20	
	3700		6" long	.133		1.37	12.70	
	3800		1/2" diameter, 4" long	.133		2.03	13.85	
	3900		6" long	.160	▼	3.35	18.20	
	4000		Nailing anchors					
	4100		Nylon nailing anchor, 1/4" diameter, 1" long	2.500	C	16.80	224	
	4200		1-1/2" long	2.857		21.50	259	
	4300		2" long	3.333		36	320	
	4400		Metal nailing anchor, 1/4" diameter, 1" long	2.500		25	238	
	4500		1-1/2" long	2.857		34	281	
	4600		2" long	3.333	▼	43.50	335	
	8000		Wedge anchors, not including layout or drilling					
	8050		Carbon steel, 1/4" diameter, 1-3/4" long	.053	Ea.	.36	4.77	
	8100		3 1/4" long	.057		.48	5.25	
	8150		3/8" diameter, 2-1/4" long	.055		.55	5.20	
	8200		5" long	.057		.96	6.10	
	8250		1/2" diameter, 2-3/4" long	.057		.84	5.90	
	8300		7" long	.064		1.43	7.40	
	8350		5/8" diameter, 3-1/2" long	.062		1.65	7.65	
	8400		8-1/2" long	.070		3.52	11.45	
	8450		3/4" diameter, 4-1/4" long	.070		1.99	8.80	
	8500		10" long	.084		4.53	14.30	
	8550		1" diameter, 6" long	.080		6.70	17.65	
	8575		9" long	.094		8.70	22	
	8600		12" long	.107		9.35	24.50	
	8650		1-1/4" diameter, 9" long	.114		12.15	29.50	
	8700		12" long	.133	▼	15.60	37	
	8750		For type 303 stainless steel, add			350%		
	8800		For type 316 stainless steel, add			450%		
	9000		Minimum labor/equipment charge	2	Job		155	
900	0010		**WELDING STRUCTURAL**					**900**
	0020		Field welding, 1/8" E6011, cost per welder, no oper. engr	1	Hr.	3	123	
	0200		With 1/2 operating engineer	1.500		3	160	
	0300		With 1 operating engineer	2	▼	3	197	
	0500		With no operating engineer, 2# weld rod per ton	1	Ton	3	123	
	0600		8# E6011 per ton	4		12	490	
	0800		With one operating engineer per welder, 2# E6011 per ton	2		3	197	
	0900		8# E6011 per ton	8	▼	12	785	
	1200		Continuous fillet, stick welding, incl. equipment					
	1300		Single pass, 1/8" thick, 0.1#/L.F.	.053	L.F.	.15	6.55	

05050 | Basic Metal Materials & Methods

05090 | Metal Fastenings

			LABOR-HOURS	UNIT	BARE COSTS MAT.	TOTAL INCL O&P	
900	1400	3/16" thick, 0.2#/L.F.	.107	L.F.	.30	13.20	900
	1500	1/4" thick, 0.3#/L.F.	.160		.45	19.70	
	1610	5/16" thick, 0.4#/L.F.	.211		.60	26	
	1800	3 passes, 3/8" thick, 0.5#/L.F.	.267		.75	33	
	2010	4 passes, 1/2" thick, 0.7#/L.F.	.364		1.05	45	
	2200	5 to 6 passes, 3/4" thick, 1.3#/L.F.	.667		1.95	82	
	2400	8 to 11 passes, 1" thick, 2.4#/L.F.	1.333	↓	3.60	164	
	9000	Minimum labor/equipment charge	2	Job		237	

05500 | Metal Fabrications

05517 | Metal Stairs

			LABOR-HOURS	UNIT	BARE COSTS MAT.	TOTAL INCL. O&P	
700	0010	**STAIR** Steel, safety nosing, steel stringers					700
	0020	Grating tread and pipe railing, 3'-6" wide	.914	Riser	111	284	
	0100	4'-0" wide	1.067		145	360	
	0200	Cement fill metal pan, picket rail, 3'-6" wide	.914		167	380	
	0300	4'-0" wide	1.067		189	435	
	0350	Wall rail, both sides, 3'-6" wide	.604		128	280	
	0500	Checkered plate tread, industrial, 3'-6" wide	1.143		111	310	
	0550	Circular, for tanks, 3'-0" wide	.970		123	310	
	0800	Custom steel stairs, 3'-6" wide, minimum	.914		167	380	
	0810	Average	1.067		223	495	
	0900	Maximum	1.600		279	635	
	1100	For 4' wide stairs, add			5%		
	1300	For 5' wide stairs, add		↓	10%		
	1500	Landing, steel pan, conventional	.200	S.F.	22.50	58.50	
	1810	Spiral aluminum, 5'-0" diameter, stock units	.711	Riser	202	420	
	1820	Custom units	.711		380	720	
	1900	Spiral, cast iron, 4'-0" diameter, ornamental, minimum	.711		180	380	
	1920	Maximum	1.280		246	555	
	2000	Spiral, steel, industrial checkered plate, 4' diameter	.711		180	380	
	2200	Stock units, 6'-0" diameter	.800	↓	218	455	
	3110	Spiral steel, stock units, primed, flat metal tread, 3'-6" dia	10	Flight	880	2,250	
	3120	4'-0" dia	11.034		1,000	2,550	
	3130	4'-6" dia	11.852		1,100	2,825	
	3140	5'-0" dia	12.800		1,200	3,050	
	3210	Galvanized, 3'-6" dia	10		1,550	3,425	
	3220	4'-0" dia	11.034		1,725	3,800	
	3230	4'-6" dia	11.852		1,875	4,150	
	3240	5'-0" dia	12.800		2,025	4,450	
	3310	Checkered plate tread, 3'-6" dia	11.034		1,075	2,700	
	3320	4'-0" dia	11.852		1,225	3,025	
	3330	4'-6" dia	12.800		1,350	3,275	
	3340	5'-0" dia	13.913		1,450	3,500	
	3410	Galvanized, 3'-6" dia	11.034		1,750	3,850	
	3420	4'-0" dia	11.852		1,975	4,300	
	3430	4'-6" dia	12.800		2,150	4,650	
	3440	5'-0" dia	13.913		2,300	5,025	
	3510	Red oak tread on flat metal, 3'-6" dia	11.852		1,475	3,400	
	3520	4'-0" dia	12.800	↓	1,625	3,775	

05500 | Metal Fabrications

05517 | Metal Stairs

		LABOR-HOURS	UNIT	BARE COSTS MAT.	TOTAL INCL O&P	
700						**700**
3530	4'-6" dia	13.913	Flight	1,775	4,100	
3540	5'-0" dia	15.238	↓	1,900	4,425	
3900	Industrial ships ladder, 3' W, grating treads, 2 line pipe rail	1.067	Riser	72.50	234	
4000	Aluminum	1.067	"	111	300	
9000	Minimum labor/equipment charge	16	Job		1,675	

05520 | Handrails & Railings

		LABOR-HOURS	UNIT	BARE COSTS MAT.	TOTAL INCL O&P	
700	0010	**RAILING, PIPE**				**700**
0020	Aluminum, 2 rail, satin finish, 1-1/4" diameter	.200	L.F.	13.90	44.50	
0030	Clear anodized	.200		17.20	50	
0040	Dark anodized	.200		19.45	54	
0080	1-1/2" diameter, satin finish	.200		16.65	49.50	
0090	Clear anodized	.200		18.60	52.50	
0100	Dark anodized	.200		20.50	55.50	
0140	Aluminum, 3 rail, 1-1/4" diam., satin finish	.234		21.50	60.50	
0150	Clear anodized	.234		26.50	70	
0160	Dark anodized	.234		29.50	74.50	
0200	1-1/2" diameter, satin finish	.234		25.50	67.50	
0210	Clear anodized	.234		29	74	
0220	Dark anodized	.234		32	78.50	
0500	Steel, 2 rail, on stairs, primed, 1-1/4" diameter	.200		10.45	38.50	
0520	1-1/2" diameter	.200		11.45	40	
0540	Galvanized, 1-1/4" diameter	.200		14.45	45.50	
0560	1-1/2" diameter	.200		16.20	48.50	
0580	Steel, 3 rail, primed, 1-1/4" diameter	.234		15.55	50.50	
0600	1-1/2" diameter	.234		16.50	52	
0620	Galvanized, 1-1/4" diameter	.234		22	61.50	
0640	1-1/2" diameter	.234		25.50	67.50	
0700	Stainless steel, 2 rail, 1-1/4" diam. #4 finish	.234		37.50	88.50	
0720	High polish	.234		60.50	127	
0740	Mirror polish	.234		76	153	
0760	Stainless steel, 3 rail, 1-1/2" diam., #4 finish	.267		56.50	125	
0770	High polish	.267		93.50	187	
0780	Mirror finish	.267		114	222	
0900	Wall rail, alum. pipe, 1-1/4" diam., satin finish	.150		7.95	29	
0905	Clear anodized	.150		9.70	32	
0910	Dark anodized	.150		11.75	36	
0915	1-1/2" diameter, satin finish	.150		8.80	30.50	
0920	Clear anodized	.150		11.10	34.50	
0925	Dark anodized	.150		13.70	39	
0930	Steel pipe, 1-1/4" diameter, primed	.150		6.35	26.50	
0935	Galvanized	.150		9.15	31	
0940	1-1/2" diameter	.182		6.50	30	
0945	Galvanized	.150		9.20	31	
0955	Stainless steel pipe, 1-1/2" diam., #4 finish	.299		30	82.50	
0960	High polish	.299		61	136	
0965	Mirror polish	.299	↓	72	154	
9000	Minimum labor/equipment charge	4	Job		395	

05700 | Ornamental Metal

		05720	Ornamental Handrails & Railings	LABOR-HOURS	UNIT	BARE COSTS MAT.	TOTAL INCL O&P	
700	0010		**RAILINGS, ORNAMENTAL**					700
	0020		Aluminum, bronze or stainless, minimum	.333	L.F.	25	75.50	
	0100		Maximum	.889		216	455	
	0200		Aluminum ornamental rail, minimum	.533		25	95.50	
	0300		Maximum	1		75	227	
	0400		Hand-forged wrought iron, minimum	.667		71	186	
	0500		Maximum	1		213	465	
	0600		Composite metal/wood/glass, minimum	1.333		150	385	
	0700		Maximum	1.600	▼	299	670	
	9000		Minimum labor/equipment charge	4	Job		395	

06050 | Basic Wood / Plastic Materials / Methods

		06052	Selective Demolition	LABOR-HOURS	UNIT	BARE COSTS MAT.	TOTAL INCL. O&P	
110	0010		**SELECTIVE DEMOLITION, WOOD FRAMING**					110
	3000		6" x 8"	.145	L.F.		9.05	
	3040		6" x 10"	.182			11.30	
	3080		6" x 12"	.216			13.45	
	3120		8" x 12"	.286			17.75	
	3160		10" x 12"	.364			22.50	
	3400		Fascia boards, 1" x 6"	.016			.98	
	3440		1" x 8"	.018			1.09	
	3480		1" x 10"	.020	▼		1.22	
	3520		For trim boards, see division 06052-120					
	3800		Headers over openings, 2 @ 2" x 6"	.073	L.F.		4.45	
	3840		2 @ 2" x 8"	.080			4.90	
	3880		2 @ 2" x 10"	.089			5.45	
	4230		2" x 6"	.016			1.01	
	4240		2" x 8"	.017			1.04	
	4250		2" x 10"	.018			1.08	
	4280		2" x 12"	.018			1.11	
	5400		Posts, 4" x 4"	.020			1.22	
	5440		6" x 6"	.040			2.45	
	5480		8" x 8"	.053			3.27	
	5500		10" x 10"	.067			4.08	
	5800		2" x 6" (alternate method)	.019			1.15	
	5840		2" x 8" (alternate method)	.019			1.17	
	5900		Hip & valley rafters, 2" x 6"	.032			1.96	
	5940		2" x 8"	.038	▼		2.33	
	6200		Stairs and stringers, minimum	.400	Riser		24.50	
	6240		Maximum	.615	"		37.50	
	6600		2" x 4"	.008	L.F.		.48	
	6640		2" x 6"	.010	"		.61	
	9000		Minimum labor/equipment charge	2	Job		122	
	9500		See Div. 02220-350 for rubbish handling					
120	0010		**SELECTIVE DEMOLITION, MILLWORK AND TRIM**					120
	1000		Cabinets, wood, base cabinets, per L.F.	.200	L.F.		12.25	
	1020		Wall cabinets, per L.F.	.200	"		12.25	
	1060		Remove and reset, base cabinets	.889	Ea.		69	

283

06050 | Basic Wood / Plastic Materials / Methods

		06052	Selective Demolition	LABOR-HOURS	UNIT	BARE COSTS MAT.	TOTAL INCL O&P	
120	1070		Wall cabinets	.800	Ea.		62.50	120
	1100		Steel, painted, base cabinets	.267	L.F.		16.35	
	1120		Wall cabinets	.267	"		16.35	
	1200		Casework, large area	.050	S.F.		3.07	
	1220		Selective	.080	"		4.90	
	1500		Counter top, minimum	.080	L.F.		4.90	
	1510		Maximum	.133			8.15	
	1550		Remove and reset, minimum	.320			25	
	1560		Maximum	.400	↓		31.50	
	2000		Paneling, 4' x 8' sheets	.008	S.F.		.48	
	2100		Boards, 1" x 4"	.023			1.41	
	2120		1" x 6"	.021			1.31	
	2140		1" x 8"	.020	↓		1.22	
	3000		Trim, baseboard, to 6" wide	.013	L.F.		.81	
	3040		Greater than 6" and up to 12" wide	.016			.98	
	3080		Remove and reset, minimum	.040			3.11	
	3090		Maximum	.053			4.15	
	3100		Ceiling trim	.016			.98	
	3120		Chair rail	.013			.81	
	3140		Railings with balusters	.067	↓		4.08	
	3160		Wainscoting	.023	S.F.		1.41	
	9000		Minimum labor/equipment charge	2	Job		122	

06100 | Rough Carpentry

		06110	Wood Framing	LABOR-HOURS	UNIT	BARE COSTS MAT.	TOTAL INCL. O&P	
100	0010	**BLOCKING**						100
	2600		Miscellaneous, to wood construction					
	2620		2" x 4"	47.059	M.B.F.	465	4,450	
	2625		Pneumatic nailed	38.095		465	3,725	
	2660		2" x 8"	29.630		600	3,325	
	2665		Pneumatic nailed	24.242	↓	600	2,900	
	2720		To steel construction					
	2740		2" x 4"	57.143	M.B.F.	465	5,225	
	2780		2" x 8"	38.095	"	600	3,950	
	9000		Minimum labor/equipment charge	2	Job		155	
515	0010	**FRAMING, COLUMNS**						515
	0100		4" x 4"	.041	L.F.	1.16	5.15	
	0150		4" x 6"	.058		2.51	8.80	
	0200		4" x 8"	.073		3.34	11.35	
	0250		6" x 6"	.074		5.05	14.35	
	0300		6" x 8"	.091		7.10	19.20	
	0350		6" x 10"	.107	↓	9.20	24	
	9000		Minimum labor/equipment charge	4	Job		310	
530	0010	**FRAMING, JOISTS**						530
	2002		Joists, 2" x 4"	.013	L.F.	.31	1.52	
	2007		Pneumatic nailed	.011		.31	1.39	
	2100		2" x 6"	.013		.52	1.88	
	2105		Pneumatic nailed	.011		.52	1.75	
	2152		2" x 8"	.015	↓	.80	2.50	

06110	Wood Framing	LABOR-HOURS	UNIT	BARE COSTS MAT.	TOTAL INCL O&P	
530	2157	Pneumatic nailed	.013	L.F.	.80	2.35	**530**
	2202	2" x 10"	.018		1.08	3.22	
	2207	Pneumatic nailed	.015		1.08	3.05	
	2252	2" x 12"	.018		1.47	3.93	
	2257	Pneumatic nailed	.016		1.47	3.75	
	2302	2" x 14"	.021		1.77	4.64	
	2307	Pneumatic nailed	.018		1.77	4.43	
	2352	3" x 6"	.017		1.64	4.14	
	2402	3" x 10"	.021		2.73	6.25	
	2452	3" x 12"	.027		3.28	7.65	
	2502	4" x 6"	.020		2.51	5.85	
	2552	4" x 10"	.027		4.18	9.20	
	2602	4" x 12"	.036		5	11.30	
	2607	Sister joist, 2" x 6"	.020		.52	2.43	
	2608	Pneumatic nailed	.017		.52	2.18	
	2612	2" x 8"	.025		.80	3.31	
	2613	Pneumatic nailed	.021		.80	2.99	
	2617	2" x 10"	.030		1.08	4.18	
	2618	Pneumatic nailed	.025		1.08	3.78	
	2622	2" x 12"	.035		1.47	5.25	
	2627	Pneumatic nailed	.029	▼	1.47	4.79	
	3000	Composite wood joist 9-1/2" deep	17.778	M.L.F.	1,600	4,100	
	3010	11-1/2" deep	18.182		1,700	4,325	
	3020	14" deep	19.512		1,875	4,700	
	3030	16" deep	20.513		2,550	5,950	
	4000	Open web joist 12" deep	18.182		1,675	4,300	
	4010	14" deep	19.512		1,950	4,850	
	4020	16" deep	20.513		2,025	5,050	
	4030	18" deep	21.622	▼	2,050	5,150	
	9000	Minimum labor/equipment charge	2	Job		155	
545	0010	**FRAMING, MISCELLANEOUS**					**545**
	2002	Firestops, 2" x 4"	.021	L.F.	.31	2.12	
	2007	Pneumatic nailed	.017		.31	1.84	
	2102	2" x 6"	.027		.52	2.96	
	2107	Pneumatic nailed	.022		.52	2.59	
	5002	Nailers, treated, wood construction, 2" x 4"	.020		.42	2.26	
	5007	Pneumatic nailed	.017		.42	2.01	
	5102	2" x 6"	.021		.76	2.97	
	5107	Pneumatic nailed	.018		.76	2.68	
	5122	2" x 8"	.023		.95	3.40	
	5127	Pneumatic nailed	.019		.95	3.11	
	5202	Steel construction, 2" x 4"	.021		.42	2.38	
	5222	2" x 6"	.023		.76	3.08	
	5242	2" x 8"	.025		.95	3.55	
	7002	Rough bucks, treated, for doors or windows, 2" x 6"	.040		.76	4.42	
	7007	Pneumatic nailed	.033		.76	3.89	
	7102	2" x 8"	.042		.95	4.90	
	7107	Pneumatic nailed	.035		.95	4.36	
	8000	Stair stringers, 2" x 10"	.123		1.08	11.45	
	8100	2" x 12"	.123		1.47	12.10	
	8150	3" x 10"	.128		2.73	14.60	
	8200	3" x 12"	.128	▼	3.28	15.55	
	9000	Minimum labor/equipment charge	2	Job		155	
550	0010	**PARTITIONS** Wood stud with single bottom plate and					**550**
	0020	double top plate, no waste, std. & better lumber					

		06110	Wood Framing	LABOR-HOURS	UNIT	BARE COSTS MAT.	TOTAL INCL O&P	
550	0182		2" x 4" studs, 8' high, studs 12" O.C.	.200	L.F.	3.74	22	**550**
	0187		12" O.C., pneumatic nailed	.167		3.74	19.30	
	0202		16" O.C.	.160		3.06	17.65	
	0207		16" O.C., pneumatic nailed	.133		3.06	15.60	
	0302		24" O.C.	.128		2.38	14	
	0307		24" O.C., pneumatic nailed	.107		2.38	12.35	
	0382		10' high, studs 12" O.C.	.200		4.42	23	
	0387		12" O.C., pneumatic nailed	.167		4.42	20.50	
	0402		16" O.C.	.160		3.57	18.55	
	0407		16" O.C., pneumatic nailed	.133		3.57	16.50	
	0502		24" O.C.	.128		2.72	14.60	
	0507		24" O.C., pneumatic nailed	.107		2.72	12.95	
	0582		12' high, studs 12" O.C.	.246		5.10	28	
	0587		12" O.C., pneumatic nailed	.205		5.10	24.50	
	0602		16" O.C.	.200		4.08	22.50	
	0607		16" O.C., pneumatic nailed	.167		4.08	19.90	
	0701		24" O.C.	.160		3.06	17.65	
	0706		24" O.C., pneumatic nailed	.133		3.06	15.60	
	0782		2" x 6" studs, 8' high, studs 12" O.C.	.229		6.30	28.50	
	0787		12" O.C., pneumatic nailed	.190		6.30	25.50	
	0802		16" O.C.	.178		5.15	22.50	
	0807		16" O.C., pneumatic nailed	.148		5.15	20.50	
	0902		24" O.C.	.139		4.01	17.65	
	0907		24" O.C., pneumatic nailed	.116		4.01	15.85	
	0982		10' high, studs 12" O.C.	.229		7.45	30	
	0987		12" O.C., pneumatic nailed	.190		7.45	27.50	
	1002		16" O.C.	.178		6	24	
	1007		16" O.C., pneumatic nailed	.148		6	22	
	1102		24" O.C.	.139		4.59	18.65	
	1107		24" O.C., pneumatic nailed	.116		4.59	16.85	
	1182		12' high, studs 12" O.C.	.291		8.60	37.50	
	1187		12" O.C., pneumatic nailed	.242		8.60	33	
	1202		16" O.C.	.229		6.90	29.50	
	1207		16" O.C., pneumatic nailed	.190		6.90	26.50	
	1302		24" O.C.	.178		5.15	22.50	
	1307		24" O.C., pneumatic nailed	.148		5.15	20.50	
	1402		For horizontal blocking, 2" x 4", add	.027		.34	2.65	
	1502		2" x 6", add	.027		.57	3.05	
	1600		For openings, add	.064	↓		4.98	
	1702		Headers for above openings, material only, add		B.F.	.66	1.13	
	9000		Minimum labor/equipment charge	2	Job		155	
575	0010		**FRAMING, TREATED LUMBER**					**575**
	0020		Water-borne salt, C.C.A., A.C.A., wet, .40 P.C.F. retention					
	0100		2" x 4"		M.B.F.	630	1,075	
	0110		2" x 6"			765	1,300	
	0120		2" x 8"			715	1,225	
	0130		2" x 10"			710	1,200	
	0140		2" x 12"			905	1,550	
	0200		4" x 4"			965	1,625	
	0210		4" x 6"			1,025	1,750	
	0220		4" x 8"		↓	1,050	1,775	
	0250		Add for .60 P.C.F. retention			40%		
	0260		Add for 2.5 P.C.F. retention			200%		
	0270		Add for K.D.A.T.			20%		

06110 | Wood Framing

			LABOR-HOURS	UNIT	BARE COSTS MAT.	TOTAL INCL O&P	
590	0010	**FRAMING, WALLS**					590
	2002	Headers over openings, 2" x 6"	.044	L.F.	.52	4.34	
	2007	2" x 6", pneumatic nailed	.037		.52	3.77	
	2052	2" x 8"	.047		.80	5	
	2057	2" x 8", pneumatic nailed	.039		.80	4.42	
	2100	2" x 10"	.050		1.08	5.75	
	2105	2" x 10", pneumatic nailed	.042		1.08	5.10	
	2152	2" x 12"	.053		1.47	6.65	
	2157	2" x 12", pneumatic nailed	.044		1.47	5.95	
	2202	4" x 12"	.084		5	15.05	
	2207	4" x 12", pneumatic nailed	.070		5	14	
	2252	6" x 12"	.114		10.85	27.50	
	5002	Plates, untreated, 2" x 3"	.019		.24	1.88	
	5007	2" x 3", pneumatic nailed	.016		.24	1.64	
	5022	2" x 4"	.020		.31	2.08	
	5027	2" x 4", pneumatic nailed	.017		.31	1.82	
	5041	2" x 6"	.021		.52	2.55	
	5046	2" x 6", pneumatic nailed	.018		.52	2.26	
	5122	Studs, 8' high wall, 2" x 3"	.013		.24	1.46	
	5127	2" x 3", pneumatic nailed	.011		.24	1.29	
	5142	2" x 4"	.015		.31	1.66	
	5147	2" x 4", pneumatic nailed	.012		.31	1.47	
	5162	2" x 6"	.016		.52	2.14	
	5167	2" x 6", pneumatic nailed	.013		.52	1.92	
	5182	3" x 4"	.020		1.09	3.41	
	5187	3" x 4", pneumatic nailed	.017		1.09	3.15	
	9000	Minimum labor/equipment charge	2	Job		155	

06160 | Sheathing

			LABOR-HOURS	UNIT	BARE COSTS MAT.	TOTAL INCL O&P	
800	0010	**SHEATHING** Plywood on roof, CDX					800
	0032	5/16" thick	.010	S.F.	.42	1.50	
	0037	Pneumatic nailed	.008		.42	1.35	
	0052	3/8" thick	.010		.42	1.52	
	0057	Pneumatic nailed	.009		.42	1.38	
	0102	1/2" thick	.011		.39	1.56	
	0103	Pneumatic nailed	.009		.39	1.39	
	0202	5/8" thick	.012		.45	1.73	
	0207	Pneumatic nailed	.010		.45	1.56	
	0302	3/4" thick	.013		.57	2.01	
	0307	Pneumatic nailed	.011		.57	1.83	
	0502	Plywood on walls with exterior CDX, 3/8" thick	.013		.42	1.75	
	0507	Pneumatic nailed	.011		.42	1.55	
	0602	1/2" thick	.014		.39	1.78	
	0607	Pneumatic nailed	.011		.39	1.56	
	0702	5/8" thick	.015		.45	1.95	
	0707	Pneumatic nailed	.012		.45	1.73	
	0802	3/4" thick	.016		.57	2.26	
	0807	Pneumatic nailed	.013		.57	2	
	1200	For structural 1 exterior plywood, add			10%		
	1402	With boards, on roof 1" x 6" boards, laid horizontal	.022		.82	3.12	
	1502	Laid diagonal	.025		.82	3.32	
	1702	1" x 8" boards, laid horizontal	.018		.80	2.79	
	1802	Laid diagonal	.022		.80	3.09	
	2000	For steep roofs, add					
	2200	For dormers, hips and valleys, add			5%		

		06160	Sheathing	LABOR-HOURS	UNIT	BARE COSTS MAT.	TOTAL INCL O&P	
800	2402	Boards on walls, 1" x 6" boards, laid regular	.025	S.F.	.82	3.32	800
	2502	Laid diagonal	.027		.82	3.53	
	2702	1" x 8" boards, laid regular	.021		.80	2.99	
	2802	Laid diagonal	.025		.80	3.28	
	2852	Gypsum, weatherproof, 1/2" thick	.015		.28	1.66	
	2902	Sealed, 4/10" thick	.015		.44	1.88	
	3000	Wood fiber, regular, no vapor barrier, 1/2" thick	.013		.53	1.94	
	3100	5/8" thick	.013		.71	2.25	
	3300	No vapor barrier, in colors, 1/2" thick	.013		.77	2.36	
	3400	5/8" thick	.013		.95	2.67	
	3600	With vapor barrier one side, white, 1/2" thick	.013		.54	1.95	
	3700	Vapor barrier 2 sides, 1/2" thick	.013		.75	2.32	
	3800	Asphalt impregnated, 25/32" thick	.013		.34	1.61	
	3850	Intermediate, 1/2" thick	.013		.27	1.50	
	9000	Minimum labor/equipment charge	4	Job		310	
850	0010	**SUBFLOOR** Plywood, CDX, 1/2" thick	.011	SF Flr.	.39	1.49	850
	0017	Pneumatic nailed	.009	"	.39	1.33	
	0102	5/8" thick	.012	SF Flr.	.45	1.70	
	0107	Pneumatic nailed	.010		.45	1.51	
	0202	3/4" thick	.013		.57	1.97	
	0207	Pneumatic nailed	.010		.57	1.77	
	0302	1-1/8" thick, 2-4-1 including underlayment	.015		1.80	4.25	
	0452	1" x 8" S4S, laid regular	.016		.80	2.62	
	0462	Laid diagonal	.019		.80	2.83	
	0502	1" x 10" S4S, laid regular	.015		1.06	2.94	
	0602	Laid diagonal	.018		1.06	3.18	
	9000	Minimum labor/equipment charge	2	Job		155	

		06220	Millwork	LABOR-HOURS	UNIT	BARE COSTS MAT.	TOTAL INCL. O&P	
500	0010	**MOLDINGS, EXTERIOR**					500
	1500	Cornice, boards, pine, 1" x 2"	.024	L.F.	.32	2.43	
	1700	1" x 6"	.032		1.15	4.46	
	2000	1" x 12"	.044		1.72	6.40	
	2200	Three piece, built-up, pine, minimum	.100		2.13	11.40	
	2300	Maximum	.123		4.71	17.65	
	3000	Corner board, sterling pine, 1" x 4"	.040		.66	4.25	
	3100	1" x 6"	.040		.97	4.77	
	3350	Fascia, sterling pine, 1" x 6"	.032		.97	4.15	
	3370	1" x 8"	.036		1.51	5.35	
	3400	Trim, exterior, sterling pine, back band	.032		.66	3.62	
	3500	Casing	.032		1.75	5.50	
	3600	Crown	.032		1.68	5.35	
	3700	Porch rail with balusters	.364		13.70	51.50	
	3800	Screen	.020		1.06	3.39	
	3850						
	4100	Verge board, sterling pine, 1" x 4"	.040	L.F.	.64	4.20	
	4200	1" x 6"	.040		.96	4.76	

		06220	Millwork	LABOR-HOURS	UNIT	BARE COSTS MAT.	TOTAL INCL O&P	
500	4300		2" x 6"	.048	L.F.	1.56	6.45	**500**
	4400		2" x 8"	.048	↓	2.07	7.30	
	4700		For redwood trim, add			200%		
	9000		Minimum labor/equipment charge	2	Job		155	
700	0010		**MOLDINGS, TRIM**					**700**
	0200		Astragal, stock pine, 11/16" x 1-3/4"	.031	L.F.	.96	4.09	
	0250		1-5/16" x 2-3/16"	.033		2.94	7.60	
	0800		Chair rail, stock pine, 5/8" x 2-1/2"	.030		.85	3.76	
	0900		5/8" x 3-1/2"	.033		1.41	4.99	
	1000		Closet pole, stock pine, 1-1/8" diameter	.040		.80	4.48	
	1100		Fir, 1-5/8" diameter	.040		1.19	5.15	
	3300		Half round, stock pine, 1/4" x 1/2"	.030		.16	2.58	
	3350		1/2" x 1"	.031	↓	.39	3.11	
	3400		Handrail, fir, single piece, stock, hardware not included					
	3450		1-1/2" x 1-3/4"	.100	L.F.	1.22	9.85	
	3470		Pine, 1-1/2" x 1-3/4"	.100		1.07	9.60	
	3500		1-1/2" x 2-1/2"	.105		1.28	10.35	
	3600		Lattice, stock pine, 1/4" x 1-1/8"	.030		.23	2.69	
	3700		1/4" x 1-3/4"	.032		.35	3.09	
	3800		Miscellaneous, custom, pine, 1" x 1"	.030		.31	2.83	
	3900		1" x 3"	.033		.63	3.66	
	4100		Birch or oak, nominal 1" x 1"	.033		.49	3.42	
	4200		Nominal 1" x 3"	.037		1.66	5.75	
	4400		Walnut, nominal 1" x 1"	.037		.80	4.27	
	4500		Nominal 1" x 3"	.040		2.39	7.20	
	4700		Teak, nominal 1" x 1"	.037		1.13	4.82	
	4800		Nominal 1" x 3"	.040		3.22	8.65	
	4900		Quarter round, stock pine, 1/4" x 1/4"	.029		.15	2.52	
	4950		3/4" x 3/4"	.031	↓	.36	3.07	
	5600		Wainscot moldings, 1-1/8" x 9/16", 2' high, minimum	.105	S.F.	8.75	23	
	5700		Maximum	.123	"	18.05	40	
	9000		Minimum labor/equipment charge	2	Job		155	
800	0010		**MOLDINGS, WINDOW AND DOOR**					**800**
	2800		Door moldings, stock, decorative, 1-1/8" wide, plain	.471	Set	29	86	
	2900		Detailed	.471	"	76.50	167	
	3150		Door trim set, 1 head and 2 sides, pine, 2-1/2 wide	1.356	Opng.	14.45	130	
	3170		3-1/2" wide	1.509	"	22.50	157	
	3250		Glass beads, stock pine, 3/8" x 1/2"	.029	L.F.	.36	2.88	
	3270		3/8" x 7/8"	.030		.40	2.99	
	4850		Parting bead, stock pine, 3/8" x 3/4"	.029		.26	2.71	
	4870		1/2" x 3/4"	.031		.31	2.97	
	5000		Stool caps, stock pine, 11/16" x 3-1/2"	.040		1.53	5.70	
	5100		1-1/16" x 3-1/4"	.053	↓	2.67	8.70	
	5300		Threshold, oak, 3' long, inside, 5/8" x 3-5/8"	.250	Ea.	7.60	32.50	
	5400		Outside, 1-1/2" x 7-5/8"	.500	"	33	95.50	
	5900		Window trim sets, including casings, header, stops,					
	5910		stool and apron, 2-1/2" wide, minimum	.615	Opng.	19.90	81.50	
	5950		Average	.800		23.50	102	
	6000		Maximum	1.333	↓	55	198	
	9000		Minimum labor/equipment charge	2	Job		155	

		06415	Countertops	LABOR-HOURS	UNIT	BARE COSTS MAT.	TOTAL INCL O&P	
100	0010	**COUNTER TOP** Stock, plastic lam., 24" wide w/backsplash, min.		.267	L.F.	8.30	35.50	100
	0100	Maximum		.320		15.85	52	
	0300	Custom plastic, 7/8" thick, aluminum molding, no splash		.267		17	50	
	0400	Cove splash		.267		22	58.50	
	0600	1-1/4" thick, no splash		.286		20	56	
	0700	Square splash		.286		25	64.50	
	0900	Square edge, plastic face, 7/8" thick, no splash		.267		21.50	57	
	1000	With splash		.267		28	68	
	1200	For stainless channel edge, 7/8" thick, add				2.27	3.88	
	1300	1-1/4" thick, add				2.66	4.54	
	1500	For solid color suede finish, add				2.12	3.61	
	1700	For end splash, add			Ea.	13.35	23	
	1900	For cut outs, standard, add, minimum		.250		2.71	24	
	2000	Maximum		1		3.23	83	
	2100	Postformed, including backsplash and front edge		.267	L.F.	8.80	36	
	2110	Mitred, add		.667	Ea.		52	
	2200	Built-in place, 25" wide, plastic laminate		.320	L.F.	11.70	45	
	2300	Ceramic tile mosaic		.320		25.50	68.50	
	2500	Marble, stock, with splash, 1/2" thick, minimum		.471		31	89.50	
	2700	3/4" thick, maximum		.615		78.50	183	
	2900	Maple, solid, laminated, 1-1/2" thick, no splash		.286		52.50	111	
	3000	With square splash		.286		62	128	
	3200	Stainless steel		.333	S.F.	94	185	
	3400	Recessed cutting block with trim, 16" x 20" x 1"		1	Ea.	61	182	
	9000	Minimum labor/equipment charge		2.133	Job		166	

		06430	Stairs & Railings					
500	0010	**RAILING** Custom design, architectural grade, hardwood, minimum		.211	L.F.	5.60	26	500
	0100	Maximum		.267		45.50	98	
	0300	Stock interior railing with spindles 6" O.C., 4' long		.200		29	64.50	
	0400	8' long		.167		27	58.50	
	9000	Minimum labor/equipment charge		2.667	Job		208	
620	0010	**STAIRS, PREFABRICATED**						620
	0100	Box stairs, prefabricated, 3'-0" wide						
	0110	Oak treads, up to 14 risers		.410	Riser	71.50	154	
	0400	8' floor to floor		5.333	Flight	870	1,900	
	0600	With pine treads for carpet, up to 14 risers		.410	Riser	46	110	
	1100	For 4' wide stairs, add			Flight	25%		
	1550	Stairs, prefabricated stair handrail with balusters		.267	L.F.	49	104	
	1700	Basement stairs, prefabricated, pine treads						
	1710	Pine risers, 3' wide, up to 14 risers		.308	Riser	46	103	
	1920	3'-6" wide		.410	"	71.50	154	
	2300	8' floor to floor		5.333	Flight	1,000	2,125	
	2900	Prefabricated stair rail with balusters, 13 risers		1.333	Ea.	585	1,100	
	4000	Residential, wood, oak treads, prefabricated		10.667	Flight	930	2,400	
	4200	Built in place		36.364	"	1,675	5,700	
	4400	Spiral, oak, 4'-6" diameter, unfinished, prefabricated,						
	4500	incl. railing, 9' high		10.667	Flight	4,400	8,350	
	9000	Minimum labor/equipment charge		5.333	Job		415	
630	0010	**STAIR PARTS** Balusters, turned, 30" high, pine, minimum		.286	Ea.	4.10	29	630
	0100	Maximum		.308	"	19	56.50	
	0300	30" high birch balusters, minimum		.286	Ea.	6.30	33	
	0400	Maximum		.308		25.50	67.50	

06430		Stairs & Railings	LABOR-HOURS	UNIT	BARE COSTS MAT.	TOTAL INCL O&P	
630	0600	42" high, pine balusters, minimum	.296	Ea.	5.30	32	630
	0700	Maximum	.320		27.50	71.50	
	0900	42" high birch balusters, minimum	.296		5.30	32	
	1000	Maximum	.320	↓	37	87.50	
	1050	Baluster, stock pine, 1-1/4" x 1-1/4"	.033	L.F.	1.55	5.25	
	1100	1-3/4" x 1-3/4"	.036	"	7.60	15.80	
	1200	Newels, 3-1/4" wide, starting, minimum	1.143	Ea.	37	153	
	1300	Maximum	1.333		425	835	
	1500	Landing, minimum	1.600		98	293	
	1600	Maximum	2	↓	460	940	
	1800	Railings, oak, built-up, minimum	.133	L.F.	30	62	
	1900	Maximum	.145		43.50	86	
	2100	Add for sub rail	.073		5	14.20	
	2300	Risers, beech, 3/4" x 7-1/2" high	.125		5.80	19.60	
	2400	Fir, 3/4" x 7-1/2" high	.125		1.60	12.45	
	2600	Oak, 3/4" x 7-1/2" high	.125		6	19.95	
	2800	Pine, 3/4" x 7-1/2" high	.121		3.05	14.65	
	2850	Skirt board, pine, 1" x 10"	.145		2.90	16.25	
	2900	1" x 12"	.154	↓	3.45	17.85	
	3000	Treads, oak, 1-1/4" x 10" wide, 3' long	.444	Ea.	74	161	
	3100	4' long, oak	.471		98.50	205	
	3300	1-1/4" x 11-1/2" wide, 3' long, oak	.444		26	79	
	3400	6' long, oak	.571		155	310	
	3600	Beech treads, add		↓	40%		
	3800	For mitered return nosings, add		L.F.	3.01	5.15	
	9000	Minimum labor/equipment charge	2.667	Job		208	

07060		Selective Demolition	LABOR-HOURS	UNIT	BARE COSTS MAT.	TOTAL INCL. O&P	
110	0010	**SELECTIVE DEMOLITION, ROOFING AND SIDING**					110
	1000	Deck, roof, concrete plank	.033	S.F.		2.74	
	1100	Gypsum plank	.014			1.19	
	1150	Metal decking	.016			1.32	
	1200	Wood, boards, tongue and groove, 2" x 6"	.017			1.02	
	1220	2" x 10"	.015			.94	
	1280	Standard planks, 1" x 6"	.015			.91	
	1320	1" x 8"	.014			.84	
	1340	1" x 12"	.013			.81	
	1350	Plywood, to 1" thick	.008	↓		.48	
	2000	Gutters, aluminum or wood, edge hung	.033	L.F.		2.05	
	2100	Built-in	.080	"		4.90	
	2500	Roof accessories, plumbing vent flashing	.571	Ea.		35	
	2600	Adjustable metal chimney flashing	.889	"		54.50	
	3000	Roofing, built-up, 5 ply roof, no gravel	.025	S.F.		1.55	
	3100	Gravel removal, minimum	.008			.50	
	3120	Maximum	.020			1.25	
	3400	Roof insulation board, up to 2" thick	.010			.64	
	4000	Shingles, asphalt strip, 1 layer	.011			.71	
	4100	Slate	.016	↓		1	

07050 | Basic Thermal & Moisture Protection Materials & Methods

07060 | Selective Demolition

			LABOR-HOURS	UNIT	BARE COSTS MAT.	TOTAL INCL. O&P	
110	4300	Wood	.018	S.F.		1.14	110
	4500	Skylight to 10 S.F.	1	Ea.		61	
	5000	Siding, metal, horizontal	.018	S.F.		1.11	
	5020	Vertical	.020			1.22	
	5200	Wood, boards, vertical	.020			1.22	
	5220	Clapboards, horizontal	.021			1.29	
	5240	Shingles	.023			1.41	
	5260	Textured plywood	.011	↓		.68	
	9000	Minimum labor/equipment charge	4	Job		245	

07100 | Dampproofing and Waterproofing

07110 | Dampproofing

			LABOR-HOURS	UNIT	BARE COSTS MAT.	TOTAL INCL. O&P	
100	0010	**BITUMINOUS ASPHALT COATING** For foundation					100
	0030	Brushed on, below grade, 1 coat	.012	S.F.	.07	.99	
	0100	2 coat	.016		.10	1.34	
	0300	Sprayed on, below grade, 1 coat, 25.6 S.F./gal.	.010		.07	.82	
	0400	2 coat, 20.5 S.F./gal.	.016		.14	1.41	
	0600	Troweled on, asphalt with fibers, 1/16" thick	.016		.16	1.44	
	0700	1/8" thick	.020		.29	1.95	
	1000	1/2" thick	.023	↓	.94	3.25	
	9000	Minimum labor/equipment charge	2.667	Job		193	

07300 | Shingles, Roof Tiles and Roof Coverings

07310 | Shingles

			LABOR-HOURS	UNIT	BARE COSTS MAT.	TOTAL INCL. O&P	
980	0010	**WOOD** 16" No. 1 red cedar shingles, 5" exposure, on roof	3.200	Sq.	163	530	980
	0015	Pneumatic nailed	2.462		163	470	
	0200	7-1/2" exposure, on walls	3.902		108	485	
	0205	Pneumatic nailed	2.996		108	420	
	0300	18" No. 1 red cedar perfections, 5-1/2" exposure, on roof	2.909		160	500	
	0305	Pneumatic nailed	2.241		160	450	
	0600	Resquared, and rebutted, 5-1/2" exposure, on roof	2.667		199	545	
	0605	Pneumatic nailed	2.051		199	495	
	0900	7-1/2" exposure, on walls	3.265		146	505	
	0905	Pneumatic nailed	2.516		146	445	
	1000	Add to above for fire retardant shingles, 16" long			30	51	
	1050	18" long		↓	28.50	49	
	1060	Preformed ridge shingles	.020	L.F.	1.65	4.37	
	1100	Hand-split red cedar shakes, 1/2" thick x 24" long, 10" exp. on roof	3.200	Sq.	138	480	
	1105	Pneumatic nailed	2.462		138	430	
	1110	3/4" thick x 24" long, 10" exp. on roof	3.556	↓	138	510	

		07310	Shingles	LABOR-HOURS	UNIT	BARE COSTS MAT.	TOTAL INCL O&P	
980	1115		Pneumatic nailed	2.740	Sq.	138	445	980
	1200		1/2″ thick, 18″ long, 8-1/2″ exp. on roof	4		97.50	475	
	1205		Pneumatic nailed	3.077		97.50	405	
	1210		3/4″ thick x 18″ long, 8 1/2″ exp. on roof	4.444		97.50	510	
	1215		Pneumatic nailed	3.419		97.50	430	
	1700		Add to above for fire retardant shakes, 24″ long			30	51	
	1800		18″ long			30	51	
	1810		Ridge shakes	.023	L.F.	2.35	5.80	
	2000		White cedar shingles, 16″ long, extras, 5″ exposure, on roof	3.333	Sq.	118	460	
	2005		Pneumatic nailed	2.564		118	400	
	2100		7-1/2″ exposure, on walls	4		84.50	455	
	2105		Pneumatic nailed	3.077		84.50	385	
	2300		For 15# organic felt underlayment on roof, 1 layer, add	.125		2.68	14.30	
	2400		2 layers, add	.250		5.35	28.50	
	2700		Panelized systems, No.1 cedar shingles on 5/16″ CDX plywood					
	2800		On walls, 8′ strips, 7″ or 14″ exposure	.023	S.F.	3.20	7.25	
	3500		On roofs, 8′ strips, 7″ or 14″ exposure	2.667	Sq.	320	750	
	3505		Pneumatic nailed	2	″	320	700	
	9000		Minimum labor/equipment charge	2.667	Job		208	

		07920	Joint Sealants	LABOR-HOURS	UNIT	BARE COSTS MAT.	TOTAL INCL. O&P	
800	0010	**CAULKING AND SEALANTS**						800
	0020	Acoustical sealant, elastomeric, cartridges			Ea.	2.05	3.50	
	0032	Backer rod, polyethylene, 1/4″ diameter		.017	L.F.	.02	1.41	
	0052	1/2″ diameter		.017		.03	1.44	
	0072	3/4″ diameter		.017		.06	1.47	
	0092	1″ diameter		.017		.09	1.53	
	0100	Acrylic latex caulk, white						
	0200	11 fl. oz cartridge			Ea.	1.82	3.10	
	0500	1/4″ x 1/2″		.032	L.F.	.15	2.79	
	0600	1/2″ x 1/2″		.032		.30	3.04	
	0800	3/4″ x 3/4″		.035		.67	3.89	
	0900	3/4″ x 1″		.040		.89	4.68	
	1000	1″ x 1″		.044		1.12	5.40	
	1400	Butyl based, bulk			Gal.	22	38	
	1500	Cartridges			″	27	46	
	1700	Bulk, in place 1/4″ x 1/2″, 154 L.F./gal.		.035	L.F.	.14	2.99	
	1800	1/2″ x 1/2″, 77 L.F./gal.		.044	″	.29	4.01	
	2000	Latex acrylic based, bulk			Gal.	23	39.50	
	2100	Cartridges			″	26	44	
	2200	Bulk in place, 1/4″ x 1/2″, 154 L.F./gal.		.035	L.F.	.15	3.01	
	2300	Polysulfide compounds, 1 component, bulk			Gal.	44	74.50	
	2400	Cartridges			″	46.50	80	
	2600	1 or 2 component, in place, 1/4″ x 1/4″, 308 L.F./gal.		.055	L.F.	.14	4.61	
	2700	1/2″ x 1/4″, 154 L.F./gal.		.059		.28	5.15	
	2900	3/4″ x 3/8″, 68 L.F./gal.		.062		.64	5.95	
	3000	1″ x 1/2″, 38 L.F./gal.		.062		1.15	6.85	

07900 | Joint Sealers

		07920	Joint Sealants	LABOR-HOURS	UNIT	BARE COSTS MAT.	TOTAL INCL O&P	
800	3200		Polyurethane, 1 or 2 component		Gal.	47.50	81.50	800
	3300		Cartridges		"	46.50	80	
	3500		Bulk, in place, 1/4" x 1/4"	.053	L.F.	.15	4.49	
	3600		1/2" x 1/4"	.055		.31	4.89	
	3800		3/4" x 3/8", 68 L.F./gal.	.062		.70	6.05	
	3900		1" x 1/2"	.073	↓	1.24	7.85	
	4100		Silicone rubber, bulk		Gal.	34.50	59	
	4200		Cartridges		"	40.50	69	
	4300		Bulk in place, 1/4" x 1/2", 154 L.F./gal.	.034	L.F.	.22	3.07	
	4400		Neoprene gaskets, closed cell, adhesive, 1/8" x 3/8"	.033		.20	2.97	
	4500		1/4" x 3/4"	.037		.48	3.76	
	4700		1/2" x 1"	.040		1.40	5.55	
	4800		3/4" x 1-1/2"	.048	↓	2.91	8.80	
	5500		Resin epoxy coating, 2 component, heavy duty		Gal.	26	44	
	5800		Tapes, sealant, P.V.C. foam adhesive, 1/16" x 1/4"		L.F.	.05	.08	
	5900		1/16" x 1/2"			.07	.12	
	5950		1/16" x 1"			.11	.19	
	6000		1/8" x 1/2"		↓	.08	.13	
	6200		Urethane foam, 2 component, handy pack, 1 C.F.		Ea.	27.50	47.50	
	6300		50.0 C.F. pack		C.F.	14.05	24	
	9000		Minimum labor/equipment charge	2	Job		158	

08050 | Basic Door and Window Materials and Methods

		08060	Selective Demolition	LABOR-HOURS	UNIT	BARE COSTS MAT.	TOTAL INCL. O&P	
110	0010		**SELECTIVE DEMOLITION, DOORS**					110
	0200		Doors, exterior, 1-3/4" thick, single, 3' x 7' high	.500	Ea.		30.50	
	0220		Double, 6' x 7' high	.667			40.50	
	0500		Interior, 1-3/8" thick, single, 3' x 7' high	.400			24.50	
	0520		Double, 6' x 7' high	.500			30.50	
	0700		Bi-folding, 3' x 6'-8" high	.400			24.50	
	0720		6' x 6'-8" high	.444			27	
	0900		Bi-passing, 3' x 6'-8" high	.500			30.50	
	0940		6' x 6'-8" high	.571			35	
	1500		Remove and reset, minimum	1			77.50	
	1520		Maximum	1.333			104	
	2000		Frames, including trim, metal	1			77.50	
	2200		Wood	.500	↓		39	
	2201		Alternate pricing method	.040	L.F.		3.11	
	2950		Minimum labor/equipment charge	2	Job		122	
	3000		Special doors, counter doors	2.667	Ea.		208	
	3100		Double acting	1.600			124	
	3200		Floor door (trap type)	2			155	
	3300		Glass, sliding, including frames	1.333			104	
	3400		Overhead, commercial, 12' x 12' high	4			310	
	3440		20' x 16' high	5.333			415	
	3500		Residential, 9' x 7' high	2			155	
	3540		16' x 7' high	2.286			178	
	3600		Remove and reset, minimum	4	↓		310	

08050 | Basic Door and Window Materials and Methods

08060 | Selective Demolition

			LABOR-HOURS	UNIT	BARE COSTS MAT.	TOTAL INCL O&P	
110	3620	Maximum	6.400	Ea.		500	110
	3700	Roll-up grille	3.200			249	
	3800	Revolving door	8			625	
	3900	Storefront swing door	5.333	↓		415	
	9000	Minimum labor/equipment charge	2	Job		155	
120	0010	**SELECTIVE DEMOLITION, WINDOWS**					120
	0200	Aluminum, including trim, to 12 S.F.	.500	Ea.		30.50	
	0240	To 25 S.F.	.727			45	
	0280	To 50 S.F.	1.600			98	
	0320	Storm windows, to 12 S.F.	.296			18.15	
	0360	To 25 S.F.	.381			23.50	
	0400	To 50 S.F.	.500	↓		30.50	
	0600	Glass, minimum	.040	S.F.		2.45	
	0620	Maximum	.053	"		3.27	
	1000	Steel, including trim, to 12 S.F.	.615	Ea.		37.50	
	1020	To 25 S.F.	.889			54.50	
	1040	To 50 S.F.	2			122	
	2000	Wood, including trim, to 12 S.F.	.364			22.50	
	2020	To 25 S.F.	.444			27	
	2060	To 50 S.F.	.615			37.50	
	5020	Remove and reset window, minimum	1.333			104	
	5040	Average	2			155	
	5080	Maximum	4	↓		310	
	9000	Minimum labor/equipment charge	2	Job		122	

08100 | Metal Doors and Frames

08110 | Steel Doors and Frames

			LABOR-HOURS	UNIT	BARE COSTS MAT.	TOTAL INCL. O&P	
200	0010	**COMMERCIAL STEEL DOORS**					200
	0015	Flush, full panel, hollow core					
	0020	1-3/8" thick, 20 ga., 2'-0"x 6'-8"	.800	Ea.	159	335	
	0040	2'-8" x 6'-8"	.889		163	345	
	0060	3'-0" x 6'-8"	.941		166	355	
	0100	3'-0" x 7'-0"	.941		161	350	
	0120	For vision lite, add			48.50	83	
	0140	For narrow lite, add			48.50	82	
	0320	Half glass, 20 ga., 2'-0" x 6'-8"	.800		223	445	
	0340	2'-8" x 6'-8"	.889		228	460	
	0360	3'-0" x 6'-8"	.941		230	465	
	0400	3'-0" x 7'-0"	.941		230	465	
	0500	Hollow core, 1-3/4" thick, full panel, 20 ga., 2'-8" x 6'-8"	.889		173	365	
	0520	3'-0" x 6'-8"	.941		166	355	
	0640	3'-0" x 7'-0"	.941		181	385	
	0680	4'-0" x 7'-0"	1.067		259	525	
	0700	4'-0" x 8'-0"	1.231		310	625	
	1000	18 ga., 2'-8" x 6'-8"	.941		191	400	
	1020	3'-0" x 6'-8"	1		186	395	
	1120	3'-0" x 7'-0"	.941	↓	201	415	

08110	Steel Doors and Frames	LABOR-HOURS	UNIT	BARE COSTS MAT.	TOTAL INCL O&P	
200	1180	4'-0" x 7'-0"	1.143	Ea.	259	535	**200**
	1200	4'-0" x 8'-0"	.941		310	600	
	1230	Half glass, 20 ga., 2'-8" x 6'-8"	.800		230	455	
	1240	3'-0" x 6'-8"	.889		230	460	
	1260	3'-0" x 7'-0"	.889		238	475	
	1280	4'-0" x 7'-0"	1		293	575	
	1300	4'-0" x 8'-0"	1.231		370	735	
	1320	18 ga., 2'-8" x 6'-8"	.889		253	500	
	1340	3'-0" x 6'-8"	.941		250	500	
	1360	3'-0" x 7'-0"	.941		257	515	
	1380	4'-0" x 7'-0"	1.067		325	635	
	1400	4'-0" x 8'-0"	1.143		370	730	
	1720	Insulated, 1-3/4" thick, full panel, 18 ga., 3'-0" x 6'-8"	1.067		186	400	
	1740	2'-8" x 7'-0"	1		197	415	
	1760	3'-0" x 7'-0"	1.067		199	425	
	1800	4'-0" x 8'-0"	1.231		310	625	
	1820	Half glass, 18 ga., 3'-0" x 6'-8"	1		250	505	
	1840	2'-8" x 7'-0"	.941		261	520	
	1860	3'-0" x 7'-0"	1		280	560	
	1900	4'-0" x 8'-0"	1.143		370	730	
	2000	For bottom louver, add		▼	88.50	151	
	2020	For baked enamel finish, add			30%		
	2040	For galvanizing, add			15%		
	9000	Minimum labor/equipment charge	2	Job		155	
820	0010	**STEEL FRAMES, KNOCK DOWN**					**820**
	0020	16 ga., up to 5-3/4" jamb depth					
	0025	6'-8" high, 3'-0" wide, single	1	Ea.	75	205	
	0040	6'-0" wide, double	1.143		90.50	243	
	0100	7'-0" high, 3'-0" wide, single	1		77	209	
	0140	6'-0" wide, double	1.143		94.50	251	
	1000	16 ga., up to 4-7/8" deep, 7'-0" H, 3'-0" W, single	1		74	204	
	1140	6'-0" wide, double	1.143		91	245	
	2800	14 ga., up to 3-7/8" deep, 7'-0" high, 3'-0" wide, single	1		76	207	
	2840	6'-0" wide, double	1.143		93	248	
	3600	up to 5-3/4" jamb depth, 7'-0" high, 4'-0" wide, single	1.067		65	194	
	3640	8'-0" wide, double	1.333		86	251	
	3700	8'-0" high, 4'-0" wide, single	1.067		86	230	
	3740	8'-0" wide, double	1.333		107	285	
	4000	6-3/4" deep, 7'-0" high, 4'-0" wide, single	1.067		90	237	
	4020	6'-0" wide, double	1.333		103	281	
	4040	8'-0", wide double	1.333		120	310	
	4100	8'-0" high, 4'-0" wide, single	1.067		102	258	
	4140	8'-0" wide, double	1.333		119	305	
	4400	8-3/4" deep, 7'-0" high, 4'-0" wide, single	1.067		95.50	247	
	4440	8'-0" wide, double	1.333		126	320	
	4500	8'-0" high, 4'-0" wide, single	1.067		112	275	
	4540	8'-0" wide, double	1.333		131	325	
	4900	For welded frames, add			28	48	
	5400	14 ga., "B" label, up to 5-3/4" deep, 7'-0" high, 4'-0" wide, single	1.067		99.50	253	
	5440	8'-0" wide, double	1.333		116	300	
	5800	6-3/4" deep, 7'-0" high, 4'-0" wide, single	1.067		88.50	234	
	5840	8'-0" wide, double	1.333		128	320	
	6200	8-3/4" deep, 7'-0" high, 4'-0" wide, single	1.067		108	267	
	6240	8'-0" wide, double	1.333	▼	138	340	
	6300	For "A" label use same price as "B" label					
	6400	For baked enamel finish, add			30%		

08100 | Metal Doors and Frames

		08110	Steel Doors and Frames	LABOR-HOURS	UNIT	BARE COSTS MAT.	TOTAL INCL O&P	
820	6500		For galvanizing, add			15%		820
	7900		Transom lite frames, fixed, add	.103	S.F.	27	53.50	
	8000		Movable, add	.123	"	32.50	65	
	9000		Minimum labor/equipment charge	2	Job		155	

08200 | Wood and Plastic Doors

		08210	Wood Doors	LABOR-HOURS	UNIT	BARE COSTS MAT.	TOTAL INCL. O&P	
900	0010		**WOOD DOOR, ARCHITECTURAL**					900
	0015		Flush, int., 1-3/8", 7 ply, hollow core,					
	0020		Lauan face, 2'-0" x 6'-8"	.941	Ea.	27	119	
	0040		2'-6" x 6'-8"	.941		31	126	
	0080		3'-0" x 6'-8"	.941		37	136	
	0100		4'-0" x 6'-8"	1		65	188	
	0120		Birch face, 2'-0" x 6'-8"	.941		43.50	148	
	0140		2'-6" x 6'-8"	.941		41.50	145	
	0180		3'-0" x 6'-8"	.941		47	155	
	0200		4'-0" x 6'-8"	1		88.50	229	
	0220		Oak face, 2'-0" x 6'-8"	.941		67.50	189	
	0240		2'-6" x 6'-8"	.941		72.50	197	
	0280		3'-0" x 6'-8"	.941		77.50	206	
	0300		4'-0" x 6'-8"	1		98	246	
	0320		Walnut face, 2'-0" x 6'-8"	.941		137	310	
	0340		2'-6" x 6'-8"	.941		140	315	
	0380		3'-0" x 6'-8"	.941		145	320	
	0400		4'-0" x 6'-8"	1		164	360	
	0430		For 7'-0" high, add			13.90	23.50	
	0440		For 8'-0" high, add			19.45	33.50	
	0480		For prefinishing, clear, add			30.50	52	
	0500		For prefinishing, stain, add			41.50	71.50	
	1320		M.D. overlay on hardboard, 2'-0" x 6'-8"	.941		86	221	
	1340		2'-6" x 6'-8"	.941		86	221	
	1380		3'-0" x 6'-8"	.941		102	248	
	1400		4'-0" x 6'-8"	1		139	315	
	1420		For 7'-0" high, add			7.65	13.05	
	1440		For 8'-0" high, add			20.50	35	
	1720		H.P. plastic laminate, 2'-0" x 6'-8"	1		203	425	
	1740		2'-6" x 6'-8"	1		203	425	
	1780		3'-0" x 6'-8"	1.067		232	480	
	1800		4'-0" x 6'-8"	1.143		325	645	
	1820		For 7'-0" high, add			8.05	13.75	
	1840		For 8'-0" high, add			20.50	35	
	2020		5 ply particle core, lauan face, 2'-6" x 6'-8"	1.067		70	203	
	2040		3'-0" x 6'-8"	1.143		73	214	
	2080		3'-0" x 7'-0"	1.231		80	233	
	2100		4'-0" x 7'-0"	1.333		94.50	265	
	2120		Birch face, 2'-6" x 6'-8"	1.067		80	220	
	2140		3'-0" x 6'-8"	1.143		87.50	238	
	2180		3'-0" x 7'-0"	1.231		89.50	249	
	2200		4'-0" x 7'-0"	1.333		109	290	

		08210	Wood Doors	LABOR-HOURS	UNIT	BARE COSTS MAT.	TOTAL INCL O&P	
900	2220		Oak face, 2'-6" x 6'-8"	1.067	Ea.	88	234	900
	2240		3'-0" x 6'-8"	1.143		97	255	
	2280		3'-0" x 7'-0"	1.231		99.50	267	
	2300		4'-0" x 7'-0"	1.333		122	310	
	2320		Walnut face, 2'-0" x 6'-8"	1.067		98	251	
	2340		2'-6" x 6'-8"	1.143		112	280	
	2380		3'-0" x 6'-8"	1.231		126	310	
	2400		4'-0" x 6'-8"	1.333		164	385	
	2440		For 8'-0" high, add			24	41	
	2460		For 8'-0" high walnut, add			13.15	22.50	
	2480		For solid wood core, add			29.50	50.50	
	2720		For prefinishing, clear, add			19.15	32.50	
	2740		For prefinishing, stain, add			42.50	73	
	3320		M.D. overlay on hardboard, 2'-6" x 6'-8"	1.143		83.50	232	
	3340		3'-0" x 6'-8"	1.231		87.50	246	
	3380		3'-0" x 7'-0"	1.333		89	256	
	3400		4'-0" x 7'-0"	1.600		109	310	
	3440		For 8'-0" height, add			24.50	42	
	3460		For solid wood core, add			31.50	54	
	3720		H.P. plastic laminate, 2'-6" x 6'-8"	1.231		122	305	
	3740		3'-0" x 6'-8"	1.333		139	340	
	3780		3'-0" x 7'-0"	1.455		144	360	
	3800		4'-0" x 7'-0"	2		175	450	
	3840		For 8'-0" height, add			24.50	42	
	3860		For solid wood core, add			30	50.50	
	4000		Exterior, flush, solid wood stave core, birch, 1-3/4" x 7'-0" x 2'-6"	1.067		152	345	
	4020		2'-8" wide	1.067		159	355	
	4040		3'-0" wide	1.143		167	375	
	4100		Oak faced 1-3/4" x 7'-0" x 2'-6" wide	1.067		167	370	
	4120		2'-8" wide	1.067		179	390	
	4140		3'-0" wide	1.143		190	415	
	4200		Walnut faced, 1-3/4" x 7'-0" x 2'-6" wide	1.067		245	505	
	4220		2'-8" wide	1.067		256	520	
	4240		3'-0" wide	1.143		267	545	
	4300		For 6'-8" high door, deduct from 7'-0" door		↓	13.65	23.50	
	9000		Minimum labor/equipment charge	2	Job		155	
960	0010	**WOOD FRAMES**						960
	0400		Exterior frame, incl. ext. trim, pine, 5/4 x 4-9/16" deep	.043	L.F.	4.41	10.85	
	0420		5-3/16" deep	.043		7.35	15.90	
	0440		6-9/16" deep	.043		6.85	15	
	0600		Oak, 5/4 x 4-9/16" deep	.046		8.45	17.90	
	0620		5-3/16" deep	.046		9.50	19.75	
	0640		6-9/16" deep	.046		10.55	21.50	
	0800		Walnut, 5/4 x 4-9/16" deep	.046		9.90	20.50	
	0820		5-3/16" deep	.046		14.40	28	
	0840		6-9/16" deep	.046		16.95	32.50	
	1000		Sills, 8/4 x 8" deep, oak, no horns	.160		11.95	33	
	1020		2" horns	.160		13.30	35.50	
	1040		3" horns	.160		15.35	38.50	
	1100		8/4 x 10" deep, oak, no horns	.178		16.10	41.50	
	1120		2" horns	.178		17.95	44.50	
	1140		3" horns	.178	↓	19.55	47	
	2000		Exterior, colonial, frame & trim, 3' opng., in-swing, minimum	.727	Ea.	279	530	
	2010		Average	.762		415	765	
	2020		Maximum	.800		940	1,650	
	2100		5'-4" opening, in-swing, minimum	.941	↓	315	605	

298

08200 | Wood and Plastic Doors

08210 | Wood Doors

		LABOR-HOURS	UNIT	BARE COSTS MAT.	TOTAL INCL. O&P	
960	2120	Maximum	1.067	Ea.	940	1,650	**960**
	2140	Out-swing, minimum	.941		325	620	
	2160	Maximum	1.067		975	1,750	
	2400	6'-0" opening, in-swing, minimum	1		300	595	
	2420	Maximum	1.600		975	1,800	
	2460	Out-swing, minimum	1		325	630	
	2480	Maximum	1.600	↓	1,200	2,200	
	2600	For two sidelights, add, minimum	.533	Opng.	310	575	
	2620	Maximum	.800	"	995	1,775	
	2700	Custom birch frame, 3'-0" opening	1	Ea.	179	385	
	2750	6'-0" opening	1	"	270	535	
	3000	Interior frame, pine, 11/16" x 3-5/8" deep	.043	L.F.	3.99	10.10	
	3020	4-9/16" deep	.043		5.45	12.65	
	3200	Oak, 11/16" x 3-5/8" deep	.046		3.59	9.70	
	3220	4-9/16" deep	.046		3.87	10.15	
	3240	5-3/16" deep	.046		3.98	10.40	
	3400	Walnut, 11/16" x 3-5/8" deep	.046		6.05	13.85	
	3420	4-9/16" deep	.046		6.40	14.40	
	3440	5-3/16" deep	.046		6.60	14.85	
	3800	Threshold, oak, 5/8" x 3-5/8" deep	.080		2.33	10.20	
	3820	4-5/8" deep	.084		2.96	11.60	
	3840	5-5/8" deep	.089	↓	5	15.45	
	9000	Minimum labor/equipment charge	2	Job		155	

08400 | Entrances and Storefronts

08410 | Metal-Framed Storefronts

			LABOR-HOURS	UNIT	BARE COSTS MAT.	TOTAL INCL. O&P	
140	0010	**STOREFRONT SYSTEMS** Aluminum frame, clear 3/8" plate glass,					**140**
	0020	incl. 3' x 7' door with hardware (400 sq. ft. max. wall)					
	0500	Wall height to 12' high, commercial grade	.107	S.F.	12.60	29.50	
	0600	Institutional grade	.123		15.85	36.50	
	0700	Monumental grade	.139		24	51	
	1000	6' x 7' door with hardware, commercial grade	.119		12.90	31	
	1100	Institutional grade	.139		17.65	40	
	1200	Monumental grade	.160		32.50	68	
	1500	For bronze anodized finish, add			15%		
	1600	For black anodized finish, add			30%		
	1700	For stainless steel framing, add to monumental		↓	75%		
	9000	Minimum labor/equipment charge	16	Job		1,175	
300	0010	**STAINLESS STEEL AND GLASS** Entrance unit, narrow stiles					**300**
	0020	3' x 7' opening, including hardware, minimum	10	Opng.	4,600	8,825	
	0050	Average	11.429		4,975	9,625	
	0100	Maximum	13.333		5,325	10,400	
	1000	For solid bronze entrance units, statuary finish, add		↓	60%		
	1100	Without statuary finish, add			45%		
	2000	Balanced doors, 3' x 7', economy	17.778	Ea.	6,225	12,400	
	2100	Premium	22.857	"	10,700	20,600	
	9000	Minimum labor/equipment charge	8	Job		790	

08400 | Entrances and Storefronts

		08460 Automatic Entrance Doors	LABOR-HOURS	UNIT	BARE COSTS MAT.	TOTAL INCL O&P	
600	0010	**SLIDING ENTRANCE** 12' x 7'-6" opng., 5' x 7' door, 2 way traf.,					600
	0020	mat activated, panic pushout, incl. operator & hardware,					
	0030	not including glass or glazing	22.857	Opng.	5,900	11,800	
	9000	Minimum labor/equipment charge	22.857	Job		1,675	

08700 | Hardware

		08710 Door Hardware	LABOR-HOURS	UNIT	BARE COSTS MAT.	TOTAL INCL. O&P	
100	0010	**AUTOMATIC OPENERS COMMERCIAL**					100
	0020	Pneumatic, incl opener, motion sens, control box, tubing, compressor					
	0050	For single swing door, per opening	20	Ea.	3,550	7,625	
	0100	Pair, per opening	32	Opng.	5,900	12,600	
	1000	For single sliding door, per opening	26.667		3,925	8,800	
	1300	Bi-parting pair	32	↓	5,925	12,600	
	1420	Electronic door opener incl motion sens, 12V control box, motor					
	1450	For single swing door, per opening	20	Opng.	3,200	7,000	
	1500	Pair, per opening	32		5,200	11,400	
	1600	For single sliding door, per opening	26.667		3,450	8,000	
	1700	Bi-parting pair	32	↓	5,300	11,500	
	1750	Handicap actuator buttons, 2, including 12V DC wiring, add	5.333	Pr.	355	1,025	
300	0010	**DOOR CLOSER** Rack and pinion	1.231	Ea.	123	305	300
	0020	Adjustable backcheck, 3 way mount, all sizes, regular arm	1.333		125	315	
	0040	Hold open arm	1.333		141	345	
	0100	Fusible link	1.231		110	284	
	0200	Non sized, regular arm	1.333		123	315	
	0240	Hold open arm	1.333		153	365	
	0400	4 way mount, non sized, regular arm	1.333		168	390	
	0440	Hold open arm	1.333	↓	181	410	
	2000	Backcheck and adjustable power, hinge face mount					
	2010	All sizes, regular arm	1.231	Ea.	155	360	
	2040	Hold open arm	1.231		167	380	
	2400	Top jamb mount, all sizes, regular arm	1.333		155	370	
	2440	Hold open arm	1.333		167	390	
	2800	Top face mount, all sizes, regular arm	1.231		155	360	
	2840	Hold open arm	1.231		166	380	
	4000	Backcheck, overhead concealed, all sizes, regular arm	1.455		164	395	
	4040	Concealed arm	1.600		175	425	
	4400	Compact overhead, concealed, all sizes, regular arm	1.455		298	625	
	4440	Concealed arm	1.600		310	655	
	4800	Concealed in door, all sizes, regular arm	1.455		110	300	
	4840	Concealed arm	1.600		119	330	
	4900	Floor concealed, all sizes, single acting	3.636		140	525	
	4940	Double acting	3.636		181	595	
	5000	For cast aluminum cylinder, deduct			14.85	25.50	
	5040	For delayed action, add			26	44	
	5080	For fusible link arm, add			10.75	18.40	
	5120	For shock absorbing arm, add			32	55	
	5160	For spring power adjustment, add		↓	25	42	

08710		Door Hardware	LABOR-HOURS	UNIT	BARE COSTS MAT.	TOTAL INCL O&P	
300	6000	Closer-holder, hinge face mount, all sizes, exposed arm	1.231	Ea.	114	291	300
	7000	Electronic closer-holder, hinge facemount, concealed arm	1.600		173	420	
	7400	With built-in detector	1.600		520	1,025	
	9000	Minimum labor/equipment charge	2	Job		155	
320	0010	**DEADLOCKS** Mortise, heavy duty, outside key	.889	Ea.	118	270	320
	0020	Double cylinder	.889		130	291	
	0100	Medium duty, outside key	.800		91.50	218	
	0110	Double cylinder	.800		114	258	
	1000	Tubular, standard duty, outside key	.800		49.50	146	
	1010	Double cylinder	.800		63	171	
	1200	Night latch, outside key	.800		62	169	
340	0010	**DOORSTOPS** Holder and bumper, floor or wall	.250	Ea.	28	67.50	340
	1300	Wall bumper, 4" diameter, with rubber pad, aluminum	.250		8.50	34	
	1600	Door bumper, floor type, aluminum	.250		4.44	27	
	1900	Plunger type, door mounted	.250		24	59.50	
	9000	Minimum labor/equipment charge	1.333	Job		104	
400	0010	**ENTRANCE LOCKS** Cylinder, grip handle, deadlocking latch	.889	Ea.	108	254	400
	0020	Deadbolt	1		131	300	
	0100	Push and pull plate, dead bolt	1		125	291	
	0900	For handicapped lever, add			136	233	
520	0011	**HINGES,MAT'L.ONLY,** full mort.,avg.freq., stl.base, 4-1/2"x 4-1/2",USP		Pr.	19.35	33.50	520
	0100	5" x 5", USP			31.50	53.50	
	0200	6" x 6", USP			67	115	
	0400	Brass base, 4-1/2" x 4-1/2", US10			40	68	
	0500	5" x 5", US10			56.50	97	
	0600	6" x 6", US10			96.50	164	
	0800	Stainless steel base, 4-1/2" x 4-1/2", US32			61.50	105	
	0900	For non removable pin, add		Ea.	2.23	3.80	
	0910	For floating pin, driven tips, add			2.50	4.27	
	0930	For hospital type tip on pin, add			10.80	18.45	
	0940	For steeple type tip on pin, add			9.45	16.15	
	0950	Full mortise, high frequency, steel base, 3-1/2" x 3-1/2", US26D		Pr.	17.90	30.50	
	1000	4-1/2" x 4-1/2", USP			40.50	69	
	1100	5" x 5", USP			44.50	76	
	1200	6" x 6", USP			107	182	
	1400	Brass base, 3-1/2" x 3-1/2", US4			37	63	
	1430	4-1/2" x 4-1/2", US10			63.50	108	
	1500	5" x 5", US10			94	160	
	1600	6" x 6", US10			136	233	
	1800	Stainless steel base, 4-1/2" x 4-1/2", US32			100	171	
	1810	Stainless steel base, 5" x 4-1/2", US32			141	242	
	1930	For hospital type tip on pin, add		Ea.	6.70	11.40	
	1950	Full mortise, low frequency, steel base, 3-1/2" x 3-1/2", US26D		Pr.	9.45	16.15	
	2000	4-1/2" x 4-1/2", USP			8.85	15.05	
	2100	5" x 5", USP			23.50	40.50	
	2200	6" x 6", USP			47.50	80.50	
	2300	4-1/2" x 4-1/2", US3			13.95	24	
	2310	5" x 5", US3			34.50	58	
	2400	Brass bass, 4-1/2" x 4-1/2", US10			33.50	56.50	
	2500	5" x 5", US10			50.50	86	
	2800	Stainless steel base, 4-1/2" x 4-1/2", US32			57.50	97.50	
650	0010	**LOCKSET** Standard duty, cylindrical, with sectional trim					650
	0020	Non-keyed, passage	.667	Ea.	40	119	

08710	Door Hardware	LABOR-HOURS	UNIT	BARE COSTS MAT.	TOTAL INCL O&P	
650						**650**
0100	Privacy	.667	Ea.	48	134	
0400	Keyed, single cylinder function	.800		68.50	180	
0420	Hotel	1		95.50	241	
0500	Lever handled, keyed, single cylinder function	.800		121	269	
1000	Heavy duty with sectional trim, non-keyed, passages	.667		111	242	
1100	Privacy	.667		140	291	
1400	Keyed, single cylinder function	.800		166	345	
1420	Hotel	1		247	495	
1600	Communicating	.800		196	400	
1690	For re-core cylinder, add			28	47.50	
1700	Residential, interior door, minimum	.500		13.40	61.50	
1720	Maximum	1		35.50	138	
1800	Exterior, minimum	.571		30	96	
1810	Average	1		63	185	
1820	Maximum	1	↓	125	292	
9000	Minimum labor/equipment charge	1.333	Job		104	
700	**MORTISE LOCKSET** Comm., wrought knobs & full escutcheon trim					**700**
0010						
0020	Non-keyed, passage, minimum	.889	Ea.	143	315	
0030	Maximum	1		232	475	
0040	Privacy, minimum	.889		153	330	
0050	Maximum	1		250	505	
0100	Keyed, office/entrance/apartment, minimum	1		175	375	
0110	Maximum	1.143		300	605	
0120	Single cylinder, typical, minimum	1		150	335	
0130	Maximum	1.143		278	565	
0200	Hotel, minimum	1.143		181	400	
0210	Maximum	1.333		294	610	
0300	Communication, double cylinder, minimum	1		181	385	
0310	Maximum	1.143		234	490	
1000	Wrought knobs and sectional trim, non-keyed, passage, minimum	.800		95	225	
1010	Maximum	.889		188	390	
1040	Privacy, minimum	.800		111	252	
1050	Maximum	.889		200	410	
1100	Keyed, entrance, office/apartment, minimum	.889		165	350	
1110	Maximum	1		239	480	
1120	Single cylinder, typical, minimum	.889		159	340	
1130	Maximum	1	↓	231	475	
2000	Cast knobs and full escutcheon trim					
2010	Non-keyed, passage, minimum	.889	Ea.	202	415	
2020	Maximum	1		330	635	
2040	Privacy, minimum	.889		243	480	
2050	Maximum	1		355	685	
2120	Keyed, single cylinder, typical, minimum	1		243	495	
2130	Maximum	1.143		385	745	
2200	Hotel, minimum	1.143		285	580	
2210	Maximum	1.333		510	975	
3000	Cast knob and sectional trim, non-keyed, passage, minimum	.800		157	330	
3010	Maximum	.800		315	600	
3040	Privacy, minimum	.800		179	370	
3050	Maximum	.800		315	600	
3100	Keyed, office/entrance/apartment, minimum	.889		203	415	
3110	Maximum	.889		320	615	
3120	Single cylinder, typical, minimum	.889		203	415	
3130	Maximum	.889		395	745	
3190	For re-core cylinder, add		↓	27.50	46.50	
3900	Keyless, pushbutton type					

		08710 | Door Hardware	LABOR-HOURS	UNIT	BARE COSTS MAT.	TOTAL INCL O&P	
700	4000	Residential/light commercial, deadbolt, standard	.889	Ea.	91.50	226	700
	4010	Heavy duty	.889		109	256	
	4020	Industrial, heavy duty, with deadbolt	.889		223	450	
	4030	Key override	.889		247	490	
	4040	Lever activated handle	.889		271	530	
	4050	Key override	.889		300	585	
	4060	Double sided pushbutton type	1		495	925	
	4070	Key override	1	↓	535	985	
750	0010	**PANIC DEVICE** For rim locks, single door, exit only	1.333	Ea.	340	690	750
	0020	Outside key and pull	1.600		385	785	
	0200	Bar and vertical rod, exit only	1.600		490	965	
	0210	Outside key and pull	2		585	1,150	
	0400	Bar and concealed rod	2		495	1,000	
	0600	Touch bar, exit only	1.333		390	775	
	0610	Outside key and pull	1.600		470	935	
	0700	Touch bar and vertical rod, exit only	1.600		535	1,025	
	0710	Outside key and pull	2		620	1,225	
	1000	Mortise, bar, exit only	2		440	900	
	1600	Touch bar, exit only	2		500	1,000	
	2000	Narrow stile, rim mounted, bar, exit only	1.333		525	1,000	
	2010	Outside key and pull	1.600		570	1,100	
	2200	Bar and vertical rod, exit only	1.600		545	1,050	
	2210	Outside key and pull	2		545	1,075	
	2400	Bar and concealed rod, exit only	2.667		630	1,275	
	3000	Mortise, bar, exit only	2		455	930	
	3600	Touch bar, exit only	2	↓	660	1,300	
	4000	Double doors, exit only	4	Pr.	685	1,475	
	4500	Exit & entrance	4	"	780	1,650	
	9000	Minimum labor/equipment charge	3.200	Job		249	
780	0010	**PUSH-PULL PLATE**					780
	0100	Push plate, .050 thick, 4" x 16", aluminum	.667	Ea.	6.20	62.50	
	0500	Bronze	.667		15.70	79	
	1500	Pull handle and push bar, aluminum	.727		110	243	
	2000	Bronze	.800		142	305	
	3000	Push plate both sides, aluminum	.571		13.35	67.50	
	3500	Bronze	.615		33	104	
	4000	Door pull, designer style, cast aluminum, minimum	.667		58.50	152	
	5000	Maximum	1		293	580	
	6000	Cast bronze, minimum	.667		68	169	
	7000	Maximum	1		320	620	
	8000	Walnut, minimum	.667		52	141	
	9000	Maximum	1	↓	292	575	
	9800	Minimum labor/equipment charge	1.600	Job		124	
800	0010	**SPECIAL HINGES**					800
	0015	Paumelle, high frequency					
	0020	Steel base, 6" x 4-1/2", US10		Pr.	116	197	
	0100	Bronze base, 5" x 4-1/2", US10			153	262	
	0200	Paumelle, average frequency, steel base, 4-1/2" x 3-1/2", US10			78.50	133	
	0400	Olive knuckle, low frequency, brass base, 6" x 4-1/2", US10		↓	140	239	
	1000	Electric hinge with concealed conductor, average frequency					
	1010	Steel base, 4-1/2" x 4-1/2", US26D		Pr.	255	435	
	1100	Bronze base, 4-1/2" x 4-1/2", US26D		"	268	455	
	1200	Electric hinge with concealed conductor, high frequency					
	1210	Steel base, 4-1/2" x 4-1/2", US26D		Pr.	190	325	
	1600	Double weight, 800 lb., steel base, removable pin, 5" x 6", USP		↓	117	200	

08710		Door Hardware	LABOR-HOURS	UNIT	BARE COSTS MAT.	TOTAL INCL O&P	
800	1700	Steel base-welded pin, 5″ x 6″, USP		Pr.	129	220	800
	1800	Triple weight, 2000 lb., steel base, welded pin, 5″ x 6″, USP			135	230	
	2000	Pivot reinf., high frequency, steel base, 7-3/4″ door plate, USP			155	264	
	2200	Bronze base, 7-3/4″ door plate, US10		↓	181	310	
	3000	Swing clear, full mortise, full or half surface, high frequency,					
	3010	Steel base, 5″ high, USP		Pr.	129	220	
	3200	Swing clear, full mortise, average frequency					
	3210	Steel base, 4-1/2″ high, USP		Pr.	102	175	
	4000	Wide throw, average frequency, steel base, 4-1/2″ x 6″, USP			78.50	133	
	4200	High frequency, steel base, 4-1/2″ x 6″, USP		↓	120	205	
	4600	Spring hinge, single acting, 6″ flange, steel		Ea.	44	75	
	4700	Brass			77.50	133	
	4900	Double acting, 6″ flange, steel			79.50	136	
	4950	Brass		↓	130	222	
	9000	Continuous hinge, steel, full mortise, heavy duty	.250	L.F.	10.80	38	

08720		Weatherstripping & Seals					
800	0010	THRESHOLD 3′ long door saddles, aluminum	.167	L.F.	3.43	18.75	800
	0100	Aluminum, 8″ wide, 1/2″ thick	.667	Ea.	29	102	
	0500	Bronze	.133	L.F.	27	57	
	0600	Bronze, panic threshold, 5″ wide, 1/2″ thick	.667	Ea.	57	149	
	0700	Rubber, 1/2″ thick, 5-1/2″ wide	.400		30	82.50	
	0800	2-3/4″ wide	.400	↓	13.65	54.50	
	9000	Minimum labor/equipment charge	2	Job		155	

09060		Selective Demolition	LABOR-HOURS	UNIT	BARE COSTS MAT.	TOTAL INCL. O&P	
110	0010	SELECTIVE DEMOLITION, CEILINGS					110
	0200	Ceiling dml, drywall, furred and nailed or screwed	.020	S.F.		1.22	
	0220	On metal frame	.021			1.29	
	0240	On suspension system, including system	.022			1.36	
	1000	Plaster, lime and horse hair, on wood lath, incl. lath	.023			1.41	
	1020	On metal lath	.028			1.72	
	1100	Gypsum, on gypsum lath	.022			1.36	
	1120	On metal lath	.032			1.96	
	1200	Suspended ceiling, mineral fiber, 2′x2′ or 2′x4′	.011			.65	
	1250	On suspension system, incl. system	.013			.81	
	1500	Tile, wood fiber, 12″ x 12″, glued	.018			1.09	
	1540	Stapled	.011			.65	
	1580	On suspension system, incl. system	.021			1.29	
	2000	Wood, tongue and groove, 1″ x 4″	.016			.98	
	2040	1″ x 8″	.015			.90	
	2400	Plywood or wood fiberboard, 4′ x 8′ sheets	.013	↓		.81	
	9000	Minimum labor/equipment charge	4	Job		245	
120	0010	SELECTIVE DEMOLITION, FLOORING					120
	0200	Brick with mortar	.034	S.F.		2.06	

		09060	Selective Demolition	LABOR-HOURS	UNIT	BARE COSTS MAT.	TOTAL INCL O&P	
120	0400		Carpet, bonded, including surface scraping	.008	S.F.		.48	**120**
	0480		Tackless	.002			.11	
	0600		Composition, acrylic or epoxy	.040			2.45	
	0700		Concrete, scarify skin	.036			4.69	
	0800		Resilient, sheet goods	.011			.70	
	0820		For gym floors	.018			1.09	
	0900		Vinyl composition tile, 12" x 12"	.016			.98	
	2000		Tile, ceramic, thin set	.024			1.45	
	2020		Mud set	.026			1.56	
	2200		Marble, slate, thin set	.024			1.45	
	2220		Mud set	.026			1.56	
	2600		Terrazzo, thin set	.036			2.18	
	2620		Mud set	.038			2.30	
	2640		Cast in place	.053			3.27	
	3000		Wood, block, on end	.020			1.55	
	3200		Parquet	.018			1.38	
	3400		Strip flooring, interior, 2-1/4" x 25/32" thick	.025			1.92	
	3500		Exterior, porch flooring, 1" x 4"	.036			2.83	
	3800		Subfloor, tongue and groove, 1" x 6"	.025			1.92	
	3820		1" x 8"	.019			1.45	
	3840		1" x 10"	.015			1.19	
	4000		Plywood, nailed	.013			1.04	
	4100		Glued and nailed	.020	▼		1.55	
	9000		Minimum labor/equipment charge	2	Job		122	
130	0010		**SELECTIVE DEMOLITION, WALLS AND PARTITIONS**					**130**
	0100		Brick, 4" to 12" thick	.182	C.F.		12.45	
	0200		Concrete block, 4" thick	.040	S.F.		2.74	
	0280		8" thick	.049			3.38	
	1000		Drywall, nailed	.008			.48	
	1020		Glued and nailed	.009			.54	
	1500		Fiberboard, nailed	.009			.54	
	1520		Glued and nailed	.010			.61	
	2000		Movable walls, metal, 5' high	.027			1.63	
	2020		8' high	.020			1.22	
	2200		Metal or wood studs, finish 2 sides, fiberboard	.046			2.90	
	2250		Lath and plaster	.092			5.80	
	2300		Plasterboard (drywall)	.046			2.90	
	2350		Plywood	.053			3.36	
	3000		Plaster, lime and horsehair, on wood lath	.020			1.22	
	3020		On metal lath	.024			1.46	
	3400		Gypsum or perlite, on gypsum lath	.020	▼		1.19	
	3420		On metal lath	.027			1.63	
	3800		Toilet partitions, slate or marble	1.600	Ea.		98	
	3820		Hollow metal	1	"		61	
	9000		Minimum labor/equipment charge	2	Job		122	

09110	Non-Load Bearing Wall Framing	LABOR-HOURS	UNIT	BARE COSTS MAT.	TOTAL INCL O&P	
100	**0010** **METAL STUDS AND TRACK**					**100**
	1600	Non-load bearing, galv, 8' high, 25 ga. 1-5/8" wide, 16" O.C.	.013	S.F.	.19	1.33
	1610	24" O.C.	.008		.14	.90
	1620	2-1/2" wide, 16" O.C.	.013		.21	1.37
	1630	24" O.C.	.009		.16	.93
	1640	3-5/8" wide, 16" O.C.	.013		.22	1.42
	1650	24" O.C.	.009		.17	.96
	1660	4" wide, 16" O.C.	.013		.25	1.49
	1670	24" O.C.	.009		.19	.99
	1680	6" wide, 16" O.C.	.014		.36	1.66
	1690	24" O.C.	.009		.27	1.13
	1700	20 ga. studs, 1-5/8" wide, 16" O.C.	.016		.29	1.76
	1710	24" O.C.	.010		.22	1.18
	1720	2-1/2" wide, 16" O.C.	.016		.32	1.82
	1730	24" O.C.	.011		.24	1.24
	1740	3-5/8" wide, 16" O.C.	.017		.38	1.94
	1750	24" O.C.	.011		.29	1.33
	1760	4" wide, 16" O.C.	.017		.40	1.99
	1770	24" O.C.	.011		.30	1.35
	1780	6" wide, 16" O.C.	.017		.53	2.22
	1790	24" O.C.	.011		.40	1.54
	2000	Non-load bearing, galv, 10' high, 25 ga. 1-5/8" wide, 16" O.C.	.016		.18	1.56
	2100	24" O.C.	.011		.13	1.06
	2200	2-1/2" wide, 16" O.C.	.016		.20	1.61
	2250	24" O.C.	.011		.15	1.07
	2300	3-5/8" wide, 16" O.C.	.017		.21	1.65
	2350	24" O.C.	.011		.16	1.10
	2400	4" wide, 16" O.C.	.017		.24	1.71
	2450	24" O.C.	.011		.18	1.13
	2500	6" wide, 16" O.C.	.017		.34	1.90
	2550	24" O.C.	.011		.25	1.27
	2600	20 ga. studs, 1-5/8" wide, 16" O.C.	.020		.28	2.04
	2650	24" O.C.	.013		.20	1.38
	2700	2-1/2" wide, 16" O.C.	.021		.30	2.10
	2750	24" O.C.	.013		.23	1.43
	2800	3-5/8" wide, 16" O.C.	.021		.36	2.24
	2850	24" O.C.	.014		.27	1.52
	2900	4" wide, 16" O.C.	.021		.38	2.29
	2950	24" O.C.	.014		.28	1.53
	3000	6" wide, 16" O.C.	.021		.50	2.52
	3050	24" O.C.	.014		.37	1.70
	3060	Non-load bearing, galv, 12' high, 25 ga. 1-5/8" wide, 16" O.C.	.019		.17	1.80
	3070	24" O.C.	.013		.13	1.20
	3080	2-1/2" wide, 16" O.C.	.020		.19	1.85
	3090	24" O.C.	.013		.14	1.23
	3100	3-5/8" wide, 16" O.C.	.020		.20	1.89
	3110	24" O.C.	.013		.15	1.26
	3120	4" wide, 16" O.C.	.020		.23	1.95
	3130	24" O.C.	.013		.17	1.29
	3140	6" wide, 16" O.C.	.020		.33	2.15
	3150	24" O.C.	.013		.24	1.43
	3160	20 ga. studs, 1-5/8" wide, 16" O.C.	.024		.26	2.34
	3170	24" O.C.	.016		.19	1.55
	3180	2-1/2" wide, 16" O.C.	.025		.29	2.42
	3190	24" O.C.	.016		.21	1.61
	3200	3-5/8" wide, 16" O.C.	.025		.35	2.52

09100 | Metal Support Assemblies

09110	Non-Load Bearing Wall Framing	LABOR-HOURS	UNIT	BARE COSTS MAT.	TOTAL INCL O&P	
100 3210	24" O.C.	.016	S.F.	.25	1.70	**100**
3220	4" wide, 16" O.C.	.025		.37	2.58	
3230	24" O.C.	.016		.27	1.72	
3240	6" wide, 16" O.C.	.026		.48	2.81	
3250	24" O.C.	.017	↓	.35	1.88	
9000	Minimum labor/equipment charge	2	Job		155	

09200 | Plaster & Gypsum Board

09250	Gypsum Board	LABOR-HOURS	UNIT	BARE COSTS MAT.	TOTAL INCL. O&P	
300 0010	**BLUEBOARD** For use with thin coat					**300**
0100	plaster application (see division 09210-900)					
1000	3/8" thick, on walls or ceilings, standard, no finish included	.008	S.F.	.18	.96	
1100	With thin coat plaster finish	.018		.25	1.84	
1400	On beams, columns, or soffits, standard, no finish included	.024		.21	2.21	
1450	With thin coat plaster finish	.034		.27	3.08	
3000	1/2" thick, on walls or ceilings, standard, no finish included	.008		.20	1	
3100	With thin coat plaster finish	.018		.27	1.87	
3300	Fire resistant, no finish included	.008		.20	1	
3400	With thin coat plaster finish	.018		.27	1.87	
3450	On beams, columns, or soffits, standard, no finish included	.024		.23	2.24	
3500	With thin coat plaster finish	.034		.30	3.13	
3700	Fire resistant, no finish included	.024		.23	2.24	
3800	With thin coat plaster finish	.034		.30	3.13	
5000	5/8" thick, on walls or ceilings, fire resistant, no finish included	.008		.21	1.01	
5100	With thin coat plaster finish	.018		.28	1.89	
5500	On beams, columns, or soffits, no finish included	.024		.24	2.27	
5600	With thin coat plaster finish	.034		.31	3.14	
6000	For high ceilings, over 8' high, add	.005			.41	
6500	For over 3 stories high, add per story	.003	↓		.20	
9000	Minimum labor/equipment charge	4	Job		310	
700 0010	**DRYWALL** Gypsum plasterboard, nailed or screwed					**700**
0100	to studs unless otherwise noted					
0150	3/8" thick, on walls, standard, no finish included	.008	S.F.	.20	.97	
0200	On ceilings, standard, no finish included	.009		.20	1.04	
0250	On beams, columns, or soffits, no finish included	.024		.20	2.19	
0300	1/2" thick, on walls, standard, no finish included	.008		.19	.95	
0350	Taped and finished (level 4 finish)	.017		.23	1.68	
0390	With compound skim coat (level 5 finish)	.021		.27	2.07	
0400	Fire resistant, no finish included	.008		.20	.97	
0450	Taped and finished (level 4 finish)	.017		.24	1.70	
0490	With compound skim coat (level 5 finish)	.021		.28	2.09	
0500	Water resistant, no finish included	.008		.24	1.03	
0550	Taped and finished (level 4 finish)	.017		.28	1.77	
0590	With compound skim coat (level 5 finish)	.021		.32	2.15	
0600	Prefinished, vinyl, clipped to studs	.018		.56	2.34	
1000	On ceilings, standard, no finish included	.009		.19	1.02	
1050	Taped and finished (level 4 finish)	.021		.23	2.01	
1090	With compound skim coat (level 5 finish)	.026	↓	.27	2.50	

		Description	LABOR-HOURS	UNIT	BARE COSTS MAT.	TOTAL INCL O&P	
700		**09250	Gypsum Board**					**700**
	1100	Fire resistant, no finish included	.009	S.F.	.20	1.04	
	1150	Taped and finished (level 4 finish)	.021		.24	2.02	
	1195	With compound skim coat (level 5 finish)	.026		.28	2.51	
	1200	Water resistant, no finish included	.009		.24	1.10	
	1250	Taped and finished (level 4 finish)	.021		.28	2.10	
	1290	With compound skim coat (level 5 finish)	.026		.32	2.58	
	1500	On beams, columns, or soffits, standard, no finish included	.024		.22	2.22	
	1550	Taped and finished (level 4 finish)	.034		.23	3	
	1590	With compound skim coat (level 5 finish)	.030		.27	2.77	
	1600	Fire resistant, no finish included	.024		.23	2.24	
	1650	Taped and finished (level 4 finish)	.034		.24	3.02	
	1690	With compound skim coat (level 5 finish)	.030		.28	2.78	
	1700	Water resistant, no finish included	.024		.28	2.31	
	1750	Taped and finished (level 4 finish)	.034		.28	3.10	
	1790	With compound skim coat (level 5 finish)	.030		.32	2.85	
	2000	5/8" thick, on walls, standard, no finish included	.008		.22	1	
	2050	Taped and finished (level 4 finish)	.017		.26	1.73	
	2090	With compound skim coat (level 5 finish)	.021		.30	2.12	
	2100	Fire resistant, no finish included	.008		.23	1.01	
	2150	Taped and finished (level 4 finish)	.017		.27	1.74	
	2195	With compound skim coat (level 5 finish)	.021		.31	2.13	
	2200	Water resistant, no finish included	.008		.27	1.09	
	2250	Taped and finished (level 4 finish)	.017		.31	1.82	
	2290	With compound skim coat (level 5 finish)	.021		.35	2.21	
	2300	Prefinished, vinyl, clipped to studs	.018		.65	2.50	
	3000	On ceilings, standard, no finish included	.009		.22	1.07	
	3050	Taped and finished (level 4 finish)	.021		.26	2.06	
	3090	With compound skim coat (level 5 finish)	.026		.30	2.53	
	3100	Fire resistant, no finish included	.009		.23	1.08	
	3150	Taped and finished (level 4 finish)	.021		.27	2.07	
	3190	With compound skim coat (level 5 finish)	.026		.31	2.55	
	3200	Water resistant, no finish included	.009		.27	1.16	
	3250	Taped and finished (level 4 finish)	.021		.31	2.15	
	3290	With compound skim coat (level 5 finish)	.026		.35	2.62	
	3500	On beams, columns, or soffits, no finish included	.024		.25	2.28	
	3550	Taped and finished (level 4 finish)	.034		.30	3.13	
	3590	With compound skim coat (level 5 finish)	.042		.35	3.86	
	3600	Fire resistant, no finish included	.024		.26	2.30	
	3650	Taped and finished (level 4 finish)	.034		.31	3.14	
	3690	With compound skim coat (level 5 finish)	.042		.31	3.80	
	3700	Water resistant, no finish included	.024		.31	2.38	
	3750	Taped and finished (level 4 finish)	.034		.35	3.22	
	3790	With compound skim coat (level 5 finish)	.042		.35	3.87	
	4000	Fireproofing, beams or columns, 2 layers, 1/2" thick, incl finish	.048		.44	4.51	
	4050	5/8" thick	.053		.54	5.05	
	4100	3 layers, 1/2" thick	.071		.64	6.60	
	4150	5/8" thick	.076		.80	7.30	
	5050	For 1" thick coreboard on columns	.033		.45	3.36	
	5100	For foil-backed board, add			.10	.17	
	5200	For work over 8' high, add	.005			.41	
	5270	For textured spray, add	.010		.04	.74	
	5300	For over 3 stories high, add per story	.003	↓		.20	
	5350	For finishing inner corners, add	.017	L.F.	.08	1.45	
	5355	For finishing outer corners, add	.013		.17	1.28	
	5500	For acoustical sealant, add per bead	.016	↓	.03	1.30	
	5550	Sealant, 1 quart tube		Ea.	4.89	8.35	

09200 | Plaster & Gypsum Board

09250 | Gypsum Board

			LABOR-HOURS	UNIT	BARE COSTS MAT.	TOTAL INCL O&P	
700	5600	Sound deadening board, 1/4" gypsum	.009	S.F.	.19	1.02	**700**
	5650	1/2" wood fiber	.009	"	.34	1.27	
	9000	Minimum labor/equipment charge	4	Job		310	

09270 | Drywall Accessories

			LABOR-HOURS	UNIT	BARE COSTS MAT.	TOTAL INCL O&P	
100	0010	**ACCESSORIES, DRYWALL** Casing bead, galvanized steel	2.759	C.L.F.	16.55	243	**100**
	0100	Vinyl	2.667		15.85	234	
	0400	Corner bead, galvanized steel, 1-1/4" x 1-1/4"	2.286		9.50	193	
	0600	Vinyl corner bead	2		20.50	191	
	0900	Furring channel, galv. steel, 7/8" deep, standard	3.077		17.40	268	
	1000	Resilient	3.137		18.50	277	
	1100	J trim, galvanized steel, 1/2" wide	2.667		14.60	232	
	1120	5/8" wide	2.712		14.95	235	
	1500	Z stud, galvanized steel, 1-1/2" wide	3.077	↓	31.50	293	
	9000	Minimum labor/equipment charge	2.667	Job		208	

09300 | Tile

09310 | Ceramic Tile

			LABOR-HOURS	UNIT	BARE COSTS MAT.	TOTAL INCL. O&P	
100	0010	**CERAMIC TILE**					**100**
	0600	Cove base, 4-1/4" x 4-1/4" high, mud set	.176	L.F.	3.07	16.20	
	0700	Thin set	.125		3.09	13.10	
	0900	6" x 4-1/4" high, mud set	.160		2.77	14.70	
	1000	Thin set	.117		2.89	12.20	
	1200	Sanitary cove base, 6" x 4-1/4" high, mud set	.172		3.56	16.80	
	1300	Thin set	.129		3.62	14.25	
	1500	6" x 6" high, mud set	.190		3.28	17.45	
	1600	Thin set	.137		3.32	14.20	
	2400	Bullnose trim, 4-1/4" x 4-1/4", mud set	.195		2.66	16.75	
	2500	Thin set	.125		2.60	12.25	
	2700	6" x 4-1/4" bullnose trim, mud set	.190		2.02	15.30	
	2800	Thin set	.129	↓	2.02	11.45	
	3000	Floors, natural clay, random or uniform, thin set, color group 1	.087	S.F.	3.62	11.60	
	3100	Color group 2	.087		3.90	12.15	
	3300	Porcelain type, 1 color, color group 2, 1" x 1"	.087		4.19	12.60	
	3310	2" x 2" or 2" x 1", thin set	.084		4.01	12.10	
	3350	For random blend, 2 colors, add			.77	1.32	
	3360	4 colors, add			1.10	1.88	
	4300	Specialty tile, 4-1/4" x 4-1/4" x 1/2", decorator finish	.087		8.95	21	
	4500	Add for epoxy grout, 1/16" joint, 1" x 1" tile	.020		.54	2.17	
	4600	2" x 2" tile	.020	↓	.50	2.08	
	4800	Pregrouted sheets, walls, 4-1/4" x 4-1/4", 6" x 4-1/4"					
	4810	and 8-1/2" x 4-1/4", 4 S.F. sheets, silicone grout	.067	S.F.	4.15	11.25	
	5100	Floors, unglazed, 2 S.F. sheets,					
	5110	urethane adhesive	.089	S.F.	4.13	12.60	
	5400	Walls, interior, thin set, 4-1/4" x 4-1/4" tile	.084		2.09	8.85	
	5500	6" x 4-1/4" tile	.084		2.30	9.20	
	5700	8-1/2" x 4-1/4" tile	.084		3.25	10.85	
	5800	6" x 6" tile	.080	↓	2.63	9.45	

09300 | Tile

09310 | Ceramic Tile

		LABOR-HOURS	UNIT	BARE COSTS MAT.	TOTAL INCL O&P	
100	5810	8" x 8" tile	.071	S.F.	3.16	9.85	100
	5820	12" x 12" tile	.053		2.98	8.40	
	5830	16" x 16" tile	.032		3.23	7.50	
	6000	Decorated wall tile, 4-1/4" x 4-1/4", minimum	.059		3.50	9.65	
	6100	Maximum	.089		39	71.50	
	6600	Crystalline glazed, 4-1/4" x 4-1/4", mud set, plain	.160		3.25	15.55	
	6700	4-1/4" x 4-1/4", scored tile	.160		4.05	16.85	
	6900	6" x 6" plain	.172		3.25	16.30	
	7000	For epoxy grout, 1/16" joints, 4-1/4" tile, add	.020		.33	1.81	
	7200	For tile set in dry mortar, add	.009			.57	
	7300	For tile set in portland cement mortar, add	.055	↓		3.44	
	9500	Minimum labor/equipment charge	4.923	Job		305	

09330 | Quarry Tile

		LABOR-HOURS	UNIT	BARE COSTS MAT.	TOTAL INCL O&P	
100	0010	QUARRY TILE Base, cove or sanitary, 2" or 5" high, mud set					100
	0100	1/2" thick	.145	L.F.	3.85	15.70	
	0300	Bullnose trim, red, mud set, 6" x 6" x 1/2" thick	.133		3.81	14.85	
	0400	4" x 4" x 1/2" thick	.145		4.30	16.45	
	0600	4" x 8" x 1/2" thick, using 8" as edge	.123	↓	3.78	14.10	
	0700	Floors, mud set, 1,000 S.F. lots, red, 4" x 4" x 1/2" thick	.133	S.F.	3.81	14.85	
	0900	6" x 6" x 1/2" thick	.114		2.96	12.15	
	1000	4" x 8" x 1/2" thick	.123		3.81	14.20	
	1300	For waxed coating, add			.61	1.04	
	1500	For colors other than green, add			.36	.62	
	1600	For abrasive surface, add			.43	.73	
	1800	Brown tile, imported, 6" x 6" x 3/4"	.133		4.53	16.05	
	1900	8" x 8" x 1"	.145		5.15	17.85	
	2100	For thin set mortar application, deduct	.023			1.42	
	2700	Stair tread, 6" x 6" x 3/4", plain	.320		4.20	27	
	2800	Abrasive	.340		4.68	29	
	3000	Wainscot, 6" x 6" x 1/2", thin set, red	.152		3.49	15.50	
	3100	Colors other than green	.152	↓	3.89	16.20	
	3300	Window sill, 6" wide, 3/4" thick	.178	L.F.	4.47	18.70	
	3400	Corners	.200	Ea.	4.91	21	
	9000	Minimum labor/equipment charge	4.923	Job		305	

09500 | Ceilings

09510 | Acoustical Ceilings

		LABOR-HOURS	UNIT	BARE COSTS MAT.	TOTAL INCL O&P	
700	0010	SUSPENDED ACOUSTIC CEILING TILES, Not including					700
	0100	suspension system					
	0300	Fiberglass boards, film faced, 2' x 2' or 2' x 4', 5/8" thick	.013	S.F.	.51	1.86	
	0400	3/4" thick	.013		1.14	2.98	
	0500	3" thick, thermal, R11	.018		1.26	3.53	
	0600	Glass cloth faced fiberglass, 3/4" thick	.016		1.62	4.01	
	0700	1" thick	.016		1.79	4.33	
	0820	1-1/2" thick, nubby face	.017		2.23	5.10	
	0900	Mineral fiber boards, 5/8" thick, aluminum faced, 24" x 24"	.013		1.47	3.55	
	0930	24" x 48"	.012	↓	.99	2.64	

09500 | Ceilings

09510 | Acoustical Ceilings

			LABOR-HOURS	UNIT	BARE COSTS MAT.	TOTAL INCL O&P	
700	0960	Standard face	.012	S.F.	.62	1.98	**700**
	1000	Plastic coated face	.020		.96	3.19	
	1100	3/4" thick, plain faced	.012		.73	2.17	
	1200	Mineral fiber, 2 hour rating, 5/8" thick	.012		.84	2.35	
	1300	Mirror faced panels, 15/16" thick, 2' x 2'	.016		9.80	18.05	
	1900	Eggcrate, acrylic, 1/2" x 1/2" x 1/2" cubes	.016		1.42	3.67	
	2100	Polystyrene eggcrate, 3/8" x 3/8" x 1/2" cubes	.016		1.19	3.25	
	2200	1/2" x 1/2" x 1/2" cubes	.016	↓	1.60	3.98	
	2210						
	2400	Luminous panels, prismatic, acrylic	.020	S.F.	1.73	4.50	
	2500	Polystyrene	.020		.88	3.05	
	2700	Flat white acrylic	.020		3.01	6.70	
	2800	Polystyrene	.020		2.06	5.05	
	3000	Drop pan, white, acrylic	.020		4.41	9.10	
	3100	Polystyrene	.020		3.69	7.85	
	3600	Perforated aluminum sheets, .024" thick, corrugated, painted	.016		1.76	4.27	
	3700	Plain	.016		3.07	6.50	
	3720	Mineral fiber, 24" x 24" or 48", reveal edge, painted, 5/8" thick	.013		.98	2.71	
	3740	3/4" thick	.014	↓	1.60	3.81	
	9000	Minimum labor/equipment charge	2	Job		155	
760	0010	**SUSPENDED CEILINGS, COMPLETE** Including standard					**760**
	0100	suspension system but not incl. 1-1/2" carrier channels					
	0600	Fiberglass ceiling board, 2' x 4' x 5/8", plain faced,	.016	S.F.	.92	2.82	
	0700	Offices, 2' x 4' x 3/4"	.021		1.55	4.27	
	1800	Tile, Z bar suspension, 5/8" mineral fiber tile	.053		1.55	6.80	
	1900	3/4" mineral fiber tile	.053	↓	1.65	6.95	
	9000	Minimum labor/equipment charge	4	Job		310	

09600 | Flooring

09620 | Specialty Flooring

			LABOR-HOURS	UNIT	BARE COSTS MAT.	TOTAL INCL. O&P	
100	0010	**ATHLETIC FLOORING**					**100**
	3700	Polyethylene, in rolls, no base incl., landscape surfaces	.029	S.F.	2.45	6.25	
	3800	Nylon action surface, 1/8" thick	.029		2.63	6.55	
	3900	1/4" thick	.029		3.79	8.50	
	4000	3/8" thick	.029		4.76	10.20	
	5500	Polyvinyl chloride, sheet goods for gyms, 1/4" thick	.100		3.85	13.65	
	5600	3/8" thick	.133	↓	4.34	16.75	

09651 | Resilient Base & Access.

100	0010	**STAIR TREADS AND RISERS** See index for materials other					**100**
	0100	than rubber and vinyl					
	0300	Rubber, molded tread, 12" wide, 5/16" thick, black	.070	L.F.	8.90	20	
	0400	Colors	.070		8.25	18.95	
	0600	1/4" thick, black	.070		7.95	18.45	
	0700	Colors	.070		8.25	18.95	
	0900	Grip strip safety tread, colors, 5/16" thick	.070		10.95	23.50	
	1000	3/16" thick	.067	↓	7.75	17.90	

		**09651	Resilient Base & Access.**	LABOR-HOURS	UNIT	BARE COSTS MAT.	TOTAL INCL O&P	
100	1200	Landings, smooth sheet rubber, 1/8″ thick	.067	S.F.	3.71	11	**100**
	1300	3/16″ thick	.067	"	4.77	12.85	
	1500	Nosings, 3″ wide, 3/16″ thick, black	.057	L.F.	2.67	8.55	
	1600	Colors	.057		2.60	8.45	
	1800	Risers, 7″ high, 1/8″ thick, flat	.032		3.14	7.65	
	1900	Coved	.032		2.83	7.05	
	2100	Vinyl, molded tread, 12″ wide, colors, 1/8″ thick	.070		3.13	10.25	
	2200	1/4″ thick	.070	↓	5.05	13.50	
	2300	Landing material, 1/8″ thick	.040	S.F.	3.45	8.75	
	2400	Riser, 7″ high, 1/8″ thick, coved	.046	L.F.	2.02	6.65	
	2500	Tread and riser combined, 1/8″ thick	.100	"	5.85	16.95	
	9000	Minimum labor/equipment charge	2.667	Job		188	
200	0010	**RESILIENT BASE**					**200**
	0800	Base, cove, rubber or vinyl, .080″ thick					
	1100	Standard colors, 2-1/2″ high	.025	L.F.	.44	2.52	
	1150	4″ high	.025		.49	2.61	
	1200	6″ high	.025		.80	3.14	
	1450	1/8″ thick, standard colors, 2-1/2″ high	.025		.51	2.65	
	1500	4″ high	.025		.67	2.92	
	1550	6″ high	.025	↓	.88	3.28	
	1600	Corners, 2-1/2″ high	.025	Ea.	1.11	3.67	
	1630	4″ high	.025		1.16	3.76	
	1660	6″ high	.025	↓	1.51	4.35	

		**09653	Resilient Sheet Flooring**					
100	0010	**RESILIENT SHEET FLOORING**					**100**
	5900	Rubber, sheet goods, 36″ wide, 1/8″ thick	.067	S.F.	3.37	10.45	
	5950	3/16″ thick	.080		4.79	13.75	
	6000	1/4″ thick	.089		5.55	15.70	
	8000	Vinyl sheet goods, backed, .065″ thick, minimum	.032		1.97	5.60	
	8050	Maximum	.040		2.52	7.10	
	8100	.080″ thick, minimum	.035		2.12	6.05	
	8150	Maximum	.040		3.03	7.95	
	8200	.125″ thick, minimum	.035		2.39	6.50	
	8250	Maximum	.040	↓	3.79	9.30	
	8700	Adhesive cement, 1 gallon does 200 to 300 S.F.		Gal.	15.55	26.50	
	8800	Asphalt primer, 1 gallon per 300 S.F.			9.20	15.65	
	8900	Emulsion, 1 gallon per 140 S.F.			11.65	19.90	
	8950	Latex underlayment, liquid, fortified		↓	31.50	53.50	

		**09658	Resilient Tile Flooring**					
100	0010	**RESILIENT TILE FLOORING**					**100**
	0050	Color group B	.020	S.F.	.98	3.08	
	0100	Color group C & D	.020		1.08	3.25	
	0300	For wood subfloor, add to above for felt underlayment			.18	.31	
	0500	For less than 500 S.F., add	.016			1.12	
	0600	For over 5000 S.F., deduct	.005			.36	
	2200	Cork tile, standard finish, 1/8″ thick	.025		3.57	7.90	
	2250	3/16″ thick	.025		3.56	7.85	
	2300	5/16″ thick	.025		4.78	9.90	
	2350	1/2″ thick	.025		5.50	11.15	
	2500	Urethane finish, 1/8″ thick	.025		4.50	9.45	
	2550	3/16″ thick	.025		4.88	10.10	
	2600	5/16″ thick	.025		6.05	12.10	
	2650	1/2″ thick	.025	↓	8.50	16.30	

09658 | Resilient Tile Flooring

		LABOR-HOURS	UNIT	BARE COSTS MAT.	TOTAL INCL O&P	
100	6050	Tile, marbleized colors, 12" x 12", 1/8" thick	.020	S.F.	3.85	8	**100**
	6100	3/16" thick	.020		5.20	10.25	
	6300	Special tile, plain colors, 1/8" thick	.020		4.18	8.55	
	6350	3/16" thick	.020		5.65	11.05	
	7000	Vinyl composition tile, 12" x 12", 1/16" thick	.016		.75	2.41	
	7050	Embossed	.016		.95	2.75	
	7100	Marbleized	.016		.95	2.75	
	7150	Solid	.016		1.06	2.94	
	7200	3/32" thick, embossed	.016		.96	2.77	
	7250	Marbleized	.016		1.07	2.95	
	7300	Solid	.016		1.56	3.79	
	7350	1/8" thick, marbleized	.016		1.02	2.86	
	7400	Solid	.016		1.91	4.38	
	7500	Vinyl tile, 12" x 12", .050" thick, minimum	.016		1.68	3.99	
	7550	Maximum	.016		3.29	6.75	
	7600	1/8" thick, minimum	.016		2.12	4.74	
	7650	Solid colors	.016		4.57	8.95	
	7700	Marbleized or Travertine pattern	.016		3.39	6.90	
	7750	Florentine pattern	.016		3.90	7.80	
	7800	Maximum	.016		8	14.80	
	9500	Minimum labor/equipment charge	2	Job		141	

09662 | Static Control Flooring

		LABOR-HOURS	UNIT	BARE COSTS MAT.	TOTAL INCL O&P	
100	0010	**CONDUCTIVE RESILIENT FLOORING**					**100**
	1700	Conductive flooring, rubber tile, 1/8" thick	.025	S.F.	2.74	6.45	
	1800	Homogeneous vinyl tile, 1/8" thick	.025	"	3.73	8.15	

09680 | Carpet

		LABOR-HOURS	UNIT	BARE COSTS MAT.	TOTAL INCL O&P	
600	0010	**CARPET PAD**, commercial grade					**600**
	9000	Sponge rubber pad, minimum	.053	S.Y.	3.18	9.20	
	9100	Maximum	.053		8.25	17.90	
	9200	Felt pad, minimum	.053		3.47	9.65	
	9300	Maximum	.053		6.40	14.70	
	9400	Bonded urethane pad, minimum	.053		3.76	10.15	
	9500	Maximum	.053		6.45	14.80	
	9600	Prime urethane pad, minimum	.053		2.16	7.40	
	9700	Maximum	.053		3.99	10.50	
800	0010	**CARPET** Commercial grades, direct cement					**800**
	0700	Nylon, level loop, 26 oz., light to medium traffic	.107	S.Y.	14.95	32.50	
	0720	28 oz., light to medium traffic	.107		16.15	35	
	0900	32 oz., medium traffic	.107		21	44	
	1100	40 oz., medium to heavy traffic	.107		31.50	61.50	
	2100	Nylon, plush, 20 oz., light traffic	.140		9.95	27	
	2800	24 oz., light to medium traffic	.140		10.50	28	
	2900	30 oz., medium traffic	.107		15.65	34	
	3000	36 oz., medium traffic	.107		19.70	41	
	3100	42 oz., medium to heavy traffic	.114		20.50	42.50	
	3200	46 oz., medium to heavy traffic	.114		27	54.50	
	3300	54 oz., heavy traffic	.114		30.50	60	
	3500	Olefin, 15 oz., light traffic	.140		5.30	18.95	
	3650	22 oz., light traffic	.140		6.35	20.50	
	4500	Wool, 50 oz., medium to heavy traffic	.107		69.50	126	
	4700	Patterned, 32 oz., medium to heavy traffic	.114		68.50	125	

09680 | Carpet

		LABOR-HOURS	UNIT	BARE COSTS MAT.	TOTAL INCL O&P	
800	4900	48 oz., heavy traffic	.114	S.Y.	69.50	126	**800**
	5600	For bound carpet baseboard, add	.027	L.F.	1.09	3.74	
	5610	For stairs, not incl. price of carpet, add	.267	Riser		18.70	
	5620	For borders and patterns, add to labor					
	8950	For tackless, stretched installation, add padding to above					
	9900	Carpet cleaning machine, rent		Day		49.70	
	9910	Minimum labor/equipment charge	2.667	Job		188	

09910 | Paints

			LABOR-HOURS	UNIT	BARE COSTS MAT.	TOTAL INCL. O&P	
300	0010	**DOORS AND WINDOWS, EXTERIOR**					**300**
	0100	Door frames & trim, only					
	0110	Brushwork, primer	.016	L.F.	.05	1.14	
	0120	Finish coat, exterior latex	.016		.06	1.15	
	0130	Primer & 1 coat, exterior latex	.027	↓	.11	1.98	
	0135	2 coats, exterior latex, both sides	.533	Ea.	5.10	45	
	0140	Primer & 2 coats, exterior latex	.030	L.F.	.16	2.31	
	0150	Doors, flush, both sides, incl. frame & trim					
	0160	Roll & brush, primer	.800	Ea.	3.82	60.50	
	0170	Finish coat, exterior latex	.800		4.35	62	
	0180	Primer & 1 coat, exterior latex	1.143		8.15	90.50	
	0190	Primer & 2 coats, exterior latex	1.600		12.50	130	
	0200	Brushwork, stain, sealer & 2 coats polyurethane	2	↓	15.55	161	
	0210	Doors, French, both sides, 10-15 lite, incl. frame & trim					
	0220	Brushwork, primer	1.333	Ea.	1.91	92.50	
	0230	Finish coat, exterior latex	1.333		2.17	93.50	
	0240	Primer & 1 coat, exterior latex	2.667		4.09	185	
	0250	Primer & 2 coats, exterior latex	4		6.15	281	
	0260	Brushwork, stain, sealer & 2 coats polyurethane	3.200	↓	5.60	225	
	0270	Doors, louvered, both sides, incl. frame & trim					
	0280	Brushwork, primer	1.143	Ea.	3.82	83	
	0290	Finish coat, exterior latex	1.143		4.35	84.50	
	0300	Primer & 1 coat, exterior latex	2		8.15	149	
	0310	Primer & 2 coats, exterior latex	2.667		12.25	201	
	0320	Brushwork, stain, sealer & 2 coats polyurethane	1.778	↓	15.55	147	
	0330	Doors, panel, both sides, incl. frame & trim					
	0340	Roll & brush, primer	1.333	Ea.	3.82	96	
	0350	Finish coat, exterior latex	1.333		4.35	97.50	
	0360	Primer & 1 coat, exterior latex	2.667		8.15	193	
	0370	Primer & 2 coats, exterior latex	3.200		12.25	238	
	0380	Brushwork, stain, sealer & 2 coats polyurethane	2.667	↓	15.55	206	
	0400	Windows, per ext. side, based on 15 SF					
	0410	1 to 6 lite					
	0420	Brushwork, primer	.615	Ea.	.75	43	
	0430	Finish coat, exterior latex	.615		.86	43	
	0440	Primer & 1 coat, exterior latex	1		1.61	70.50	
	0450	Primer & 2 coats, exterior latex	1.333		2.42	93.50	
	0460	Stain, sealer & 1 coat varnish	1.143	↓	2.20	80.50	

09910	Paints	LABOR-HOURS	UNIT	BARE COSTS MAT.	TOTAL INCL O&P	
300	0470	7 to 10 lite					**300**
	0480	Brushwork, primer	.727	Ea.	.75	50.50	
	0490	Finish coat, exterior latex	.727		.86	50.50	
	0500	Primer & 1 coat, exterior latex	1.143		1.61	80	
	0510	Primer & 2 coats, exterior latex	1.600		2.42	112	
	0520	Stain, sealer & 1 coat varnish	1.333		2.20	93.50	
	0530	12 lite					
	0540	Brushwork, primer	.800	Ea.	.75	55.50	
	0550	Finish coat, exterior latex	.800		.86	55.50	
	0560	Primer & 1 coat, exterior latex	1.333		1.61	92.50	
	0570	Primer & 2 coats, exterior latex	1.600		2.42	112	
	0580	Stain, sealer & 1 coat varnish	1.333		2.20	93.50	
	0590	For oil base paint, add			10%		
310	0010	**DOORS & WINDOWS, INTERIOR LATEX**					**310**
	0100	Doors flush, both sides, incl. frame & trim					
	0110	Roll & brush, primer	.800	Ea.	3.33	59.50	
	0120	Finish coat, latex	.800		3.60	60	
	0130	Primer & 1 coat latex	1.143		6.90	88.50	
	0140	Primer & 2 coats latex	1.600		10.30	126	
	0160	Spray, both sides, primer	.400		3.50	33	
	0170	Finish coat, latex	.400		3.77	33	
	0180	Primer & 1 coat latex	.727		7.30	61.50	
	0190	Primer & 2 coats latex	1		10.90	86	
	0200	Doors, French, both sides, 10-15 lite, incl. frame & trim					
	0210	Roll & brush, primer	1.333	Ea.	1.66	92.50	
	0220	Finish coat, latex	1.333		1.80	92.50	
	0230	Primer & 1 coat latex	2.667		3.46	185	
	0240	Primer & 2 coats latex	4		5.15	279	
	0260	Doors, louvered, both sides, incl. frame & trim					
	0270	Roll & brush, primer	1.143	Ea.	3.33	82	
	0280	Finish coat, latex	1.143		3.60	83	
	0290	Primer & 1 coat, latex	2		6.75	146	
	0300	Primer & 2 coats, latex	2.667		10.50	198	
	0320	Spray, both sides, primer	.400		3.50	33	
	0330	Finish coat, latex	.400		3.77	33	
	0340	Primer & 1 coat, latex	.727		7.30	61.50	
	0350	Primer & 2 coats, latex	1		11.15	87	
	0360	Doors, panel, both sides, incl. frame & trim					
	0370	Roll & brush, primer	1.333	Ea.	3.50	96	
	0380	Finish coat, latex	1.333		3.60	96	
	0390	Primer & 1 coat, latex	2.667		6.90	191	
	0400	Primer & 2 coats, latex	3.200		10.50	235	
	0420	Spray, both sides, primer	.800		3.50	60	
	0430	Finish coat, latex	.800		3.77	60.50	
	0440	Primer & 1 coat, latex	1.600		7.30	120	
	0450	Primer & 2 coats, latex	2		11.15	154	
	0460	Windows, per interior side, based on 15 SF					
	0470	1 to 6 lite					
	0480	Brushwork, primer	.615	Ea.	.66	42	
	0490	Finish coat, enamel	.615		.71	43	
	0500	Primer & 1 coat enamel	1		1.37	70	
	0510	Primer & 2 coats enamel	1.333		2.08	93.50	
	0530	7 to 10 lite					
	0540	Brushwork, primer	.727	Ea.	.66	50	
	0550	Finish coat, enamel	.727		.71	50.50	

315

09910	Paints	LABOR-HOURS	UNIT	BARE COSTS MAT.	TOTAL INCL O&P	
310	0560	Primer & 1 coat enamel	1.143	Ea.	1.37	79	**310**
	0570	Primer & 2 coats enamel	1.600	↓	2.08	112	
	0590	12 lite					
	0600	Brushwork, primer	.800	Ea.	.66	55	
	0610	Finish coat, enamel	.800		.71	55.50	
	0620	Primer & 1 coat enamel	1.333		1.37	92	
	0630	Primer & 2 coats enamel	1.600	↓	2.08	112	
	0650	For oil base paint, add		↓	10%		
700	0010	**SIDING EXTERIOR,** Alkyd (oil base)					**700**
	0450	Steel siding, oil base, paint 1 coat, brushwork	.008	S.F.	.05	.63	
	0500	Spray	.004		.08	.38	
	0800	Paint 2 coats, brushwork	.012		.11	1.01	
	1000	Spray	.004		.14	.47	
	1200	Stucco, rough, oil base, paint 2 coats, brushwork	.012		.11	1.01	
	1400	Roller	.010		.11	.87	
	1600	Spray	.005		.12	.57	
	1800	Texture 1-11 or clapboard, oil base, primer coat, brushwork	.012		.09	.98	
	2000	Spray	.004		.09	.40	
	2400	Paint 2 coats, brushwork	.020		.16	1.60	
	2600	Spray	.006		.18	.71	
	3400	Stain 2 coats, brushwork	.017		.10	1.29	
	4000	Spray	.005		.11	.54	
	4200	Wood shingles, oil base primer coat, brushwork	.012		.08	.96	
	4400	Spray	.004		.08	.41	
	5000	Paint 2 coats, brushwork	.020		.13	1.55	
	5200	Spray	.007		.12	.69	
	6500	Stain 2 coats, brushwork	.017		.10	1.29	
	7000	Spray	.006		.13	.63	
	8000	For latex paint, deduct		↓	10%		
800	0010	**TRIM, EXTERIOR**					**800**
	0100	Door frames & trim (see Doors, interior or exterior)					
	0110	Fascia, latex paint, one coat coverage					
	0120	1" x 4", brushwork	.013	L.F.	.02	.87	
	0130	Roll	.006		.02	.46	
	0140	Spray	.004		.01	.29	
	0150	1" x 6" to 1" x 10", brushwork	.013		.06	.95	
	0160	Roll	.007		.06	.55	
	0170	Spray	.004		.05	.33	
	0180	1" x 12", brushwork	.013		.06	.95	
	0190	Roll	.008		.06	.62	
	0200	Spray	.004	↓	.05	.32	
	0210	Gutters & downspouts, metal, zinc chromate paint					
	0220	Brushwork, gutters, 5", first coat	.013	L.F.	.05	.93	
	0230	Second coat	.008		.05	.65	
	0240	Third coat	.006		.04	.50	
	0250	Downspouts, 4", first coat	.013		.05	.93	
	0260	Second coat	.008		.05	.65	
	0270	Third coat	.006	↓	.04	.50	
	0280	Gutters & downspouts, wood					
	0290	Brushwork, gutters, 5", primer	.013	L.F.	.05	.93	
	0300	Finish coat, exterior latex	.013		.05	.93	
	0310	Primer & 1 coat exterior latex	.020		.11	1.54	
	0320	Primer & 2 coats exterior latex	.025		.16	1.94	
	0330	Downspouts, 4", primer	.013		.05	.93	
	0340	Finish coat, exterior latex	.013	↓	.05	.93	

		**09910	Paints**	LABOR-HOURS	UNIT	BARE COSTS MAT.	TOTAL INCL O&P	
800	0350	Primer & 1 coat exterior latex	.020	L.F.	.11	1.54	800
	0360	Primer & 2 coats exterior latex	.025	↓	.08	1.80	
	0370	Molding, exterior, up to 14" wide					
	0380	Brushwork, primer	.013	L.F.	.06	.95	
	0390	Finish coat, exterior latex	.013		.06	.95	
	0400	Primer & 1 coat exterior latex	.020		.13	1.57	
	0410	Primer & 2 coats exterior latex	.025		.13	1.92	
	0420	Stain & fill	.008		.06	.61	
	0430	Shellac	.004		.07	.39	
	0440	Varnish	.006	↓	.07	.54	
920	0010	**WALLS AND CEILINGS**, Interior					920
	0100	Concrete, dry wall or plaster, oil base, primer or sealer coat					
	0200	Smooth finish, brushwork	.007	S.F.	.05	.55	
	0240	Roller	.004		.04	.35	
	0300	Sand finish, brushwork	.008		.04	.63	
	0340	Roller	.007		.05	.55	
	0380	Spray	.004		.04	.30	
	0800	Paint 2 coats, smooth finish, brushwork	.012		.09	.95	
	0840	Roller	.010		.09	.82	
	0880	Spray	.005		.08	.47	
	0900	Sand finish, brushwork	.013		.09	1.05	
	0940	Roller	.008		.09	.68	
	0980	Spray	.005		.08	.45	
	1600	Glaze coating, 5 coats, spray, clear	.009		.60	1.62	
	1640	Multicolor	.009	↓	.83	2.01	
	2000	Masonry or concrete block, oil base, primer or sealer coat					
	2100	Smooth finish, brushwork	.007	S.F.	.05	.52	
	2180	Spray	.003		.07	.34	
	2200	Sand finish, brushwork	.007		.07	.62	
	2280	Spray	.003		.07	.34	
	2800	Paint 2 coats, smooth finish, brushwork	.011		.15	.96	
	2880	Spray	.006		.14	.63	
	2900	Sand finish, brushwork	.012		.15	1.04	
	2980	Spray	.006		.14	.63	
	3600	Glaze coating, 5 coats, spray, clear	.009		.60	1.62	
	3620	Multicolor	.009		.83	2.01	
	4000	Block filler, 1 coat, brushwork	.019		.12	1.47	
	4100	Silicone, water repellent, 2 coats, spray	.004		.25	.69	
	4120	For latex paint, deduct		↓	10%		
940	0010	**DRY FALL PAINTING**					940
	0100	Walls					
	0200	Wallboard and smooth plaster, one coat, brush	.009	S.F.	.04	.67	
	0210	Roll	.005		.04	.42	
	0220	Spray	.003		.04	.29	
	0230	Two coats, brush	.015		.08	1.18	
	0240	Roll	.009		.08	.75	
	0250	Spray	.005		.08	.48	
	0260	Concrete or textured plaster, one coat, brush	.011		.04	.80	
	0270	Roll	.006		.04	.49	
	0280	Spray	.005		.04	.42	
	0290	Two coats, brush	.019		.08	1.42	
	0300	Roll	.011		.08	.86	
	0310	Spray	.006		.08	.55	
	0320	Concrete block, one coat, brush	.011		.04	.80	
	0330	Roll	.006	↓	.04	.49	

09910	Paints	LABOR-HOURS	UNIT	BARE COSTS MAT.	TOTAL INCL O&P	
940	0340	Spray	.005	S.F.	.04	.42	**940**
	0350	Two coats, brush	.019		.08	1.42	
	0360	Roll	.011		.08	.86	
	0370	Spray	.006		.08	.55	
	0380	Wood, one coat, brush	.011		.04	.80	
	0390	Roll	.006		.04	.49	
	0400	Spray	.009		.04	.69	
	0410	Two coats, brush	.016		.08	1.25	
	0420	Roll	.011		.08	.86	
	0430	Spray	.012	↓	.08	.96	
	0440	Ceilings					
	0450	Wallboard and smooth plaster, one coat, brush	.013	S.F.	.04	.97	
	0460	Roll	.008		.04	.59	
	0470	Spray	.005		.04	.42	
	0480	Two coats, brush	.023		.08	1.72	
	0490	Roll	.012		.08	.96	
	0500	Spray	.006		.08	.55	
	0510	Concrete or textured plaster, one coat, brush	.016		.04	1.19	
	0520	Roll	.009		.04	.69	
	0530	Spray	.005		.04	.42	
	0540	Two coats, brush	.029		.08	2.09	
	0550	Roll	.015		.08	1.18	
	0560	Spray	.006		.08	.55	
	0570	Structural steel, bar joists or metal deck, one coat, spray	.005		.04	.42	
	0580	Two coats, spray	.008	↓	.08	.65	

10160	Metal Toilet Compartments	LABOR-HOURS	UNIT	BARE COSTS MAT.	TOTAL INCL. O&P	
100	0010	**TOILET PARTITIONS, METAL**					**100**
	0110	Cubicles, ceiling hung					
	0200	Painted metal	4	Ea.	430	1,050	
	0500	Stainless steel	4		1,075	2,150	
	0600	For handicap units, incl. 52" grab bars, add		↓	320	550	
	0900	Floor and ceiling anchored					
	1000	Painted metal	3.200	Ea.	390	915	
	1300	Stainless steel	3.200		1,225	2,350	
	1400	For handicap units, incl. 52" grab bars, add		↓	273	465	
	1610	Floor mounted					
	1700	Painted metal	2.286	Ea.	390	845	
	2000	Stainless steel	2.286		1,300	2,425	
	2100	For handicap units, incl. 52" grab bars, add			273	465	
	2200	For juvenile units, deduct		↓	37.50	63.50	
	2450	Floor mounted, headrail braced					
	2500	Painted metal	2.667	Ea.	390	875	
	2800	Stainless steel	2.667		1,225	2,300	
	2900	For handicap units, incl. 52" grab bars, add			273	465	
	3000	Wall hung partitions, painted metal	2.286		495	1,025	
	3300	Stainless steel	2.286	↓	1,200	2,225	

10160 | Metal Toilet Compartments

			LABOR-HOURS	UNIT	BARE COSTS MAT.	TOTAL INCL O&P	
100	3400	For handicap units, incl. 52" grab bars, add		Ea.	273	465	100
	4000	Screens, entrance, floor mounted, 58" high, 48" wide					
	4200	Painted metal	1.067	Ea.	177	385	
	4500	Stainless steel	1.067	"	650	1,200	
	4650	Urinal screen, 18" wide					
	4700	Painted metal	2	Ea.	147	405	
	5000	Stainless steel	2	"	465	955	
	5100	Floor mounted, head rail braced					
	5300	Painted metal	2	Ea.	162	430	
	5600	Stainless steel	2	"	545	1,075	
	5750	Pilaster, flush					
	5800	Painted metal	1.600	Ea.	238	530	
	6100	Stainless steel	1.600		410	825	
	6300	Urinal screen, post braced, painted metal	1.600		245	545	
	6600	Stainless steel	1.600	▼	410	825	
	6700	Wall hung, bracket supported					
	6800	Painted metal	1.600	Ea.	247	545	
	7100	Stainless steel	1.600		375	765	
	7400	Flange supported, painted metal	1.600		182	435	
	7700	Stainless steel	1.600		435	865	
	7800	Wedge type, painted metal	1.600		215	495	
	8100	Stainless steel	1.600	▼	445	890	
	9000	Minimum labor/equipment charge	3.200	Job		249	

10165 | Plastic Laminate Toilet Compartment

			LABOR-HOURS	UNIT	BARE COSTS MAT.	TOTAL INCL O&P	
100	0010	**TOILET PARTITIONS, PLASTIC LAMINATE**					100
	0110	Cubicles, ceiling hung					
	0300	Plastic laminate on particle board	4	Ea.	575	1,300	
	0600	For handicap units, incl. 52" grab bars, add		"	320	550	
	0900	Floor and ceiling anchored					
	1100	Plastic laminate on particle board	3.200	Ea.	630	1,325	
	1400	For handicap units, incl. 52" grab bars, add		"	273	465	
	1610	Floor mounted					
	1800	Plastic laminate on particle board	2.286	Ea.	535	1,100	
	2450	Floor mounted, headrail braced					
	2600	Plastic laminate on particle board	2.667	Ea.	735	1,450	
	3400	For handicap units, incl. 52" grab bars, add			273	465	
	4300	Entrance screen, floor mtd., plas. lam., 58" high, 48" wide	1.067		415	790	
	4800	Urinal screen, 18" wide, ceiling braced, plastic laminate	2		279	630	
	5400	Floor mounted, headrail braced	2		255	590	
	5900	Pilaster, flush, plastic laminate	1.600		345	715	
	6400	Post braced, plastic laminate	1.600	▼	345	715	
	6700	Wall hung, bracket supported					
	6900	Plastic laminate on particle board	1.600	Ea.	131	350	
	7450	Flange supported					
	7500	Plastic laminate on particle board	1.600	Ea.	375	770	

10170 | Plastic Toilet Compartments

			LABOR-HOURS	UNIT	BARE COSTS MAT.	TOTAL INCL O&P	
100	0010	**TOILET PARTITIONS, PLASTIC**					100
	0110	Cubicles, ceiling hung					
	0250	Phenolic	4	Ea.	820	1,725	
	0600	For handicap units, incl. 52" grab bars, add		"	320	550	
	0900	Floor and ceiling anchored					
	1050	Phenolic	3.200	Ea.	865	1,725	

319

10170	Plastic Toilet Compartments	LABOR-HOURS	UNIT	BARE COSTS MAT.	TOTAL INCL O&P	
100	1400	For handicap units, incl. 52" grab bars, add		Ea.	273	465	100
	1610	Floor mounted					
	1750	Phenolic	2.286	Ea.	875	1,650	
	2100	For handicap units, incl. 52" grab bars, add			273	465	
	2200	For juvenile units, deduct		↓	37.50	63.50	
	2450	Floor mounted, headrail braced					
	2550	Phenolic	2.667	Ea.	840	1,650	

10180	Stone Toilet Compartments	LABOR-HOURS	UNIT	BARE COSTS MAT.	TOTAL INCL O&P	
100	0010	TOILET PARTITIONS, STONE					100
	0100	Cubicles, ceiling hung, marble	8	Ea.	1,425	3,050	
	0600	For handicap units, incl. 52" grab bars, add			320	550	
	0800	Floor & ceiling anchored, marble	6.400		1,550	3,125	
	1400	For handicap units, incl. 52" grab bars, add			273	465	
	1600	Floor mounted, marble	5.333		920	1,950	
	2400	Floor mounted, headrail braced, marble	5.333		875	1,900	
	2900	For handicap units, incl. 52" grab bars, add			273	465	
	4100	Entrance screen, floor mounted marble, 58" high, 48" wide	1.778		575	1,100	
	4600	Urinal screen, 18" wide, ceiling braced, marble	2.667	↓	575	1,175	
	5100	Floor mounted, head rail braced					
	5200	Marble	2.667	Ea.	505	1,050	
	5700	Pilaster, flush, marble	1.778		645	1,225	
	6200	Post braced, marble	1.778	↓	635	1,200	
	9000	Minimum labor/equipment charge	3.200	Job		249	

10410	Directories	LABOR-HOURS	UNIT	BARE COSTS MAT.	TOTAL INCL. O&P	
100	0010	DIRECTORY BOARDS					100
	0050	Plastic, glass covered, 30" x 20"	5.333	Ea.	234	815	
	0100	36" x 48"	8		720	1,850	
	0900	Outdoor, weatherproof, black plastic, 36" x 24"	8		670	1,775	
	1000	36" x 36"	10.667	↓	775	2,125	
	9000	Minimum labor/equipment charge	8	Job		625	

10430	Exterior Signage	LABOR-HOURS	UNIT	BARE COSTS MAT.	TOTAL INCL. O&P	
200	0012	PLAQUES					200
	3900	Plaques, custom, 20" x 30", for up to 450 letters, cast aluminum	4	Ea.	815	1,725	
	4000	Cast bronze	4		1,075	2,150	
	4200	30" x 36", up to 900 letters cast aluminum	5.333		1,725	3,375	
	4300	Cast bronze	5.333		2,225	4,225	
	5100	Exit signs, 24 ga. alum., 14" x 12" surface mounted	.267		26.50	65.50	
	5200	10" x 7"	.400	↓	10.40	49	
	6400	Replacement sign faces, 6" or 8"	.160		22	47	
	9000	Minimum labor/equipment charge	2	Job		155	

10530 | Protective Covers

10535 | Awnings & Canopies

200			LABOR-HOURS	UNIT	BARE COSTS MAT.	TOTAL INCL. O&P	200
	0010	**CANOPIES** Wall hung, .032", aluminum, prefinished, 8' x 10'	18.462	Ea.	1,525	4,375	
	0300	8' x 20'	21.818		3,075	7,325	
	1000	12' x 20'	24	↓	3,925	9,050	
	1050						
	2300	Aluminum entrance canopies, flat soffit, .032"					
	2500	3'-6" x 4'-0", clear anodized	4	Ea.	645	1,425	
	4700	Canvas awnings, including canvas, frame & lettering					
	5000	Minimum	.160	S.F.	44	88	
	5300	Average	.178		53.50	104	
	5500	Maximum	.200	↓	108	200	
	9000	Minimum labor/equipment charge	8	Job		625	

10750 | Telephone Specialties

10755 | Telephone Enclosures

400			LABOR-HOURS	UNIT	BARE COSTS MAT.	TOTAL INCL. O&P	400
	0010	**TELEPHONE ENCLOSURE**					
	0300	Shelf type, wall hung, minimum	3.200	Ea.	940	1,850	
	0400	Maximum	3.200		2,425	4,400	
	0600	Booth type, painted steel, indoor or outdoor, minimum	10.667		2,975	5,900	
	0700	Maximum (stainless steel)	10.667		9,900	17,800	
	1900	Outdoor, drive-up type, wall mounted	4		795	1,675	
	2000	Post mounted, stainless steel posts	5.333	↓	1,225	2,525	
	9000	Minimum labor/equipment charge	4	Job		310	

10800 | Toilet/Bath/Laundry Accessories

10810 | Toilet Accessories

100			LABOR-HOURS	UNIT	BARE COSTS MAT.	TOTAL INCL. O&P	100
	0010	**COMMERCIAL TOILET ACCESSORIES**					
	0200	Curtain rod, stainless steel, 5' long, 1" diameter	.615	Ea.	30.50	99.50	
	0300	1-1/4" diameter	.615		29	96.50	
	0400	Diaper changing station, horizontal, wall mounted, plastic	.800	↓	183	375	
	0500	Dispenser units, combined soap & towel dispensers,					
	0510	mirror and shelf, flush mounted	.800	Ea.	330	630	
	0600	Towel dispenser and waste receptacle,					
	0610	18 gallon capacity	.800	Ea.	256	500	
	0800	Grab bar, straight, 1-1/4" diameter, stainless steel, 18" long	.333		21	62	
	1100	36" long	.400		26.50	76.50	
	3000	Mirror, with stainless steel 3/4" square frame, 18" x 24"	.400		58.50	131	
	3300	72" x 24"	1.333		196	440	
	3500	With 5" stainless steel shelf, 18" x 24"	.400		90.50	186	
	3800	72" x 24"	1.333		255	540	
	4200	Napkin/tampon dispenser, recessed	.533		310	575	
	4300	Robe hook, single, regular	.222	↓	4.76	25.50	

10800 | Toilet/Bath/Laundry Accessories

10810 | Toilet Accessories

			LABOR-HOURS	UNIT	BARE COSTS MAT.	TOTAL INCL. O&P	
100	4400	Heavy duty, concealed mounting	.222	Ea.	10.80	35.50	**100**
	4600	Soap dispenser, chrome, surface mounted, liquid	.400		41.50	102	
	5600	Shelf, stainless steel, 5" wide, 18 ga., 24" long	.333		38.50	91.50	
	6100	Toilet tissue dispenser, surface mounted, SS, single roll	.267		11.15	40	
	6200	Double roll	.333		15.55	52.50	
	6290	Toilet seat	.200		19.35	51	
	6400	Towel bar, stainless steel, 18" long	.348		30	78	
	6500	30" long	.381		51	117	
	6700	Towel dispenser, stainless steel, surface mounted	.500		36.50	101	
	6800	Flush mounted, recessed	.800		170	355	
	7400	Tumbler holder, tumbler only	.267		25	63.50	
	7500	Soap, tumbler & toothbrush	.267		22.50	59.50	
	7700	Wall urn ash receiver, surface mount, 11" long	.667		99.50	222	
	8000	Waste receptacles, stainless steel, with top, 13 gallon	.800	↓	161	335	
	9000	Minimum labor/equipment charge	1.600	Job		124	

13700 | Security Access and Surveillance

13710 | Security Access

			LABOR-HOURS	UNIT	BARE COSTS MAT.	TOTAL INCL. O&P	
300	0010	**ACCESS CONTROL**					**300**
	0020	Card type, 1 time zone, minimum		Ea.	310	525	
	0040	Maximum			1,050	1,775	
	0060	3 time zones, minimum			765	1,300	
	0080	Maximum		↓	1,750	2,975	
	0100	System with printer, and control console, 3 zones		Total	8,550	14,600	
	0120	6 zones		"	11,300	19,200	
	0140	For each door, minimum, add		Ea.	1,250	2,125	
	0160	Maximum, add		"	1,875	3,175	

13720 | Detection & Alarm

			LABOR-HOURS	UNIT	BARE COSTS MAT.	TOTAL INCL. O&P	
065	0010	**DETECTION SYSTEMS**, not including wires & conduits					**065**
	0100	Burglar alarm, battery operated, mechanical trigger	2	Ea.	249	595	
	0200	Electrical trigger	2		297	675	
	0400	For outside key control, add	1		70.50	206	
	0600	For remote signaling circuitry, add	1		112	278	
	0800	Card reader, flush type, standard	2.963		835	1,675	
	1000	Multi-code	2.963		1,075	2,075	
	1200	Door switches, hinge switch	1.509		52.50	221	
	1400	Magnetic switch	1.509		62	236	
	1600	Exit control locks, horn alarm	2		310	700	
	1800	Flashing light alarm	2		350	770	
	2000	Indicating panels, 1 channel	2.963		330	820	
	2200	10 channel	5		1,125	2,375	
	2400	20 channel	8		2,200	4,425	
	2600	40 channel	14.035		4,000	8,025	
	2800	Ultrasonic motion detector, 12 volt	3.478		206	655	
	3000	Infrared photoelectric detector	3.478		170	595	
	3200	Passive infrared detector	3.478	↓	254	735	

13700 | Security Access and Surveillance

13720 | Detection & Alarm

			LABOR-HOURS	UNIT	BARE COSTS MAT.	TOTAL INCL O&P	
065	3400	Glass break alarm switch	1	Ea.	42.50	160	065
	3420	Switchmats, 30" x 5'	1.509		76	260	
	3440	30" x 25'	2		182	480	
	3460	Police connect panel	2		219	545	
	3480	Telephone dialer	1.509		345	715	
	3500	Alarm bell	2		69.50	291	
	3520	Siren	2		131	395	
	3540	Microwave detector, 10' to 200'	4		600	1,375	
	3560	10' to 350'	4	↓	1,750	3,350	
	3594	Fire, alarm control panel					
	3600	4 zone	8	Ea.	920	2,225	
	3800	8 zone	16		1,400	3,800	
	4000	12 zone	23.988		1,825	5,175	
	4200	Battery and rack	2		690	1,350	
	4400	Automatic charger	1		445	845	
	4600	Signal bell	1		49.50	171	
	4800	Trouble buzzer or manual station	1		37	149	
	5000	Detector, rate of rise	1		33.50	143	
	5100	Fixed temperature	1		28	134	
	5200	Smoke detector, ceiling type	1.290		75	240	
	5240	Smoke detector (addressable type)	1.333		150	370	
	5400	Duct type	2.500		250	640	
	5600	Strobe and horn	1.509		95	294	
	5800	Fire alarm horn	1.194		36.50	166	
	6000	Door holder, electro-magnetic	2		77.50	305	
	6200	Combination holder and closer	2.500		430	950	
	6400	Code transmitter	2		690	1,350	
	6600	Drill switch	1		86.50	234	
	6800	Master box	2.963		3,100	5,525	
	7000	Break glass station	1		50	172	
	7800	Remote annunciator, 8 zone lamp	4.444		175	680	
	8000	12 zone lamp	6.154		300	1,050	
	8200	16 zone lamp	7.273		300	1,125	
	8400	Standpipe or sprinkler alarm, alarm device	1		125	300	
	8600	Actuating device	1	↓	290	580	
	9410	Minimum labor/equipment charge	2	Job		172	

14200 | Elevators

14210 | Electric Traction Elevators

			LABOR-HOURS	UNIT	BARE COSTS MAT.	TOTAL INCL. O&P	
100	0012	**ELEVATOR SYSTEMS**					100
	7000	Residential, cab type, 1 floor, 2 stop, minimum	80	Ea.	8,075	21,200	
	7100	Maximum	160		13,700	37,900	
	7200	2 floor, 3 stop, minimum	133		12,000	32,700	
	7300	Maximum	266	↓	19,600	57,500	

14400 | Lifts

14420	Wheelchair Lifts	LABOR-HOURS	UNIT	BARE COSTS MAT.	TOTAL INCL O&P	
100	0010 **CHAIR / WHEELCHAIR LIFT**					100
	7700 Stair climber (chair lift), single seat, minimum	16	Ea.	3,925	8,125	
	7800 Maximum	80		5,375	16,500	
	8000 Wheelchair lift, minimum	16		5,375	10,600	
	8500 Maximum	32		12,700	24,700	
	8700 Stair lift, minimum	16		10,600	19,600	
	8900 Maximum	80	↓	16,800	36,100	

15050 | Basic Materials & Methods

15055	Selective Mech Demolition	LABOR-HOURS	UNIT	BARE COSTS MAT.	TOTAL INCL. O&P	
600	0010 **PLUMBING DEMOLITION**					600
	1020 Fixtures, including 10' piping					
	1100 Bath tubs, cast iron	2	Ea.		175	
	1120 Fiberglass	1.333			117	
	1140 Steel	1.600			139	
	1200 Lavatory, wall hung	.800			69.50	
	1220 Counter top	1			87.50	
	1300 Sink, steel or cast iron, single	1			87.50	
	1320 Double	1.143			99.50	
	1400 Water closet, floor mounted	1			87.50	
	1420 Wall mounted	1.143			99.50	
	1500 Urinal, floor mounted	2			175	
	1520 Wall mounted	1.143			99.50	
	1600 Water fountains, free standing	1			87.50	
	1620 Recessed	1.333	↓		117	
	2000 Piping, metal, up thru 1-1/2" diameter	.040	L.F.		3.50	
	2050 2" thru 3-1/2" diameter	.053			4.65	
	2100 4" thru 6" diameter	.160			13.95	
	2150 8" thru 14" diameter	.267			23	
	2153 16" thru 20" diameter	.291	↓		25.50	
	2250 Water heater, 40 gal.	1.333	Ea.		117	
	6000 Remove and reset fixtures, minimum	1.333			117	
	6100 Maximum	2	↓		175	
	9000 Minimum labor/equipment charge	4	Job		350	

15100 | Building Services Piping

15107	Metal Pipe & Fittings	LABOR-HOURS	UNIT	BARE COSTS MAT.	TOTAL INCL. O&P	
420	0010 **PIPE, COPPER** Solder joints					420
	1000 Type K tubing, couplings & clevis hangers 10' O.C.					
	1100 1/4" diameter	.095	L.F.	.99	10	
	1200 1" diameter	.121	↓	2.27	14.45	

15107 | Metal Pipe & Fittings

		LABOR-HOURS	UNIT	BARE COSTS MAT.	TOTAL INCL O&P	
420	1260	2" diameter	.200	L.F.	5.30	26.50	**420**
	2000	Type L tubing, couplings & hangers 10' O.C.					
	2100	1/4" diameter	.091	L.F.	.73	9.20	
	2120	3/8" diameter	.095		.90	9.85	
	2140	1/2" diameter	.099		1.01	10.30	
	2160	5/8" diameter	.101		1.35	11.15	
	2180	3/4" diameter	.105		1.42	11.60	
	2200	1" diameter	.118		1.87	13.40	
	2220	1-1/4" diameter	.138		2.44	16.20	
	2240	1-1/2" diameter	.154		3.02	18.55	
	2260	2" diameter	.190		4.55	24.50	
	2280	2-1/2" diameter	.258		6.90	32.50	
	2300	3" diameter	.286		9.25	38.50	
	2320	3-1/2" diameter	.372		12.65	51	
	2340	4" diameter	.410		15.50	58.50	
	2360	5" diameter	.471		40.50	106	
	2380	6" diameter	.600	↓	50.50	135	
	3000	Type M tubing, couplings & hangers 10' O.C.					
	3140	1/2" diameter	.095	L.F.	.86	9.80	
	3180	3/4" diameter	.103		1.22	11.05	
	3200	1" diameter	.114	↓	1.96	13.30	
	4000	Type DWV tubing, couplings & hangers 10' O.C.					
	4100	1-1/4" diameter	.133	L.F.	2.08	15.20	
	4120	1-1/2" diameter	.148		2.53	17.30	
	4140	2" diameter	.182		3.29	21.50	
	4160	3" diameter	.276		5.90	32	
	4180	4" diameter	.400		10.10	48.50	
	4200	5" diameter	.444		28.50	83	
	4220	6" diameter	.571	↓	40.50	116	
	9000	Minimum labor/equipment charge	2	Job		175	
460	0010	**PIPE, COPPER, FITTINGS** Wrought unless otherwise noted					**460**
	0040	Solder joints, copper x copper					
	0070	90° elbow, 1/4"	.364	Ea.	.94	33.50	
	0100	1/2"	.400		.30	35.50	
	0120	3/4"	.421		.67	38	
	0130	1"	.500		1.65	46.50	
	0250	45° elbow, 1/4"	.364		1.86	35	
	0270	3/8"	.364		1.52	34.50	
	0280	1/2"	.400		.55	35.50	
	0290	5/8"	.421		2.89	41.50	
	0300	3/4"	.421		.96	38.50	
	0310	1"	.500		3.45	49.50	
	0320	1-1/4"	.533		3.45	53	
	0330	1-1/2"	.615		4.14	61	
	0340	2"	.727		6.95	75	
	0350	2-1/2"	1.231		14.75	122	
	0360	3"	1.231		22	134	
	0370	3-1/2"	1.600		39	193	
	0380	4"	1.778		46.50	220	
	0390	5"	2.667		181	515	
	0400	6"	2.667		271	680	
	0450	Tee, 1/4"	.571		2.10	53.50	
	0470	3/8"	.571		1.61	53	
	0480	1/2"	.615		.51	55	
	0490	5/8"	.667		3.48	64.50	
	0500	3/4"	.667	↓	1.24	60.50	

15107	Metal Pipe & Fittings	LABOR-HOURS	UNIT	BARE COSTS MAT.	TOTAL INCL O&P	
460	0510	1"	.800	Ea.	3.81	76	**460**
	0520	1-1/4"	.889		6.10	87.50	
	0530	1-1/2"	1		8.45	102	
	0540	2"	1.143		13.20	122	
	0550	2-1/2"	2		26.50	204	
	0560	3"	2.286		40.50	249	
	0570	3-1/2"	2.667		118	410	
	0580	4"	3.200		98.50	420	
	0590	5"	4		305	830	
	0600	6"	4		420	1,050	
	0612	Tee, reducing on the outlet, 1/4"	.533		3.37	52.50	
	0613	3/8"	.533		3.17	52.50	
	0614	1/2"	.571		2.78	54.50	
	0615	5/8"	.615		5.60	63.50	
	0616	3/4"	.667		1.20	60.50	
	0617	1"	.727		4.03	70	
	0618	1-1/4"	.800		5.95	80	
	0619	1-1/2"	.889		6.30	88.50	
	0620	2"	1		10.30	105	
	0621	2-1/2"	1.778		24.50	183	
	0622	3"	2		34	217	
	0623	4"	2.667		66.50	320	
	0624	5"	3.200		305	770	
	0625	6"	3.429		420	990	
	0626	8"	4		1,700	3,225	
	0630	Tee, reducing on the run, 1/4"	.533		4.22	54	
	0631	3/8"	.533		5.65	56.50	
	0632	1/2"	.571		5.05	58.50	
	0633	5/8"	.615		5.65	63.50	
	0634	3/4"	.667		2.88	63	
	0635	1"	.727		4.80	72	
	0636	1-1/4"	.800		7.60	83	
	0637	1-1/2"	.889		13.85	101	
	0638	2"	1		18.45	119	
	0639	2-1/2"	1.778		44.50	215	
	0640	3"	2		62.50	265	
	0641	4"	2.667		132	435	
	0642	5"	3.200		305	770	
	0643	6"	3.429		420	990	
	0644	8"	4		1,700	3,225	
	0650	Coupling, 1/4"	.333		.23	30	
	0670	3/8"	.333		.31	30	
	0680	1/2"	.364		.23	32.50	
	0690	5/8"	.381		.71	35	
	0700	3/4"	.381		.45	34	
	0710	1"	.444		.91	40.50	
	0715	1-1/4"	.471		1.71	44.50	
	0716	1-1/2"	.533		2.25	51	
	0718	2"	.615		3.76	60.50	
	0721	2-1/2"	1.067		8	98	
	0722	3"	1.231		12	117	
	0724	3-1/2"	2		23	198	
	0726	4"	2.286		24	221	
	0728	5"	2.667		56	305	
	0731	6"	3	▼	92.50	405	
	2000	DWV, solder joints, copper x copper					

		LABOR-HOURS	UNIT	BARE COSTS MAT.	TOTAL INCL O&P	
15107 | Metal Pipe & Fittings						
460						**460**
2030	90° Elbow, 1-1/4"	.615	Ea.	3.06	59.50	
2050	1-1/2"	.667		4.12	65.50	
2070	2"	.800		5.95	80	
2090	3"	1.600		14.75	151	
2100	4"	1.778		71	262	
2250	Tee, Sanitary, 1-1/4"	.889		6.05	87.50	
2270	1-1/2"	1		7.50	101	
2290	2"	1.143		8.75	114	
2310	3"	2.286		32	233	
2330	4"	2.667		81	345	
2400	Coupling, 1-1/4"	.571		1.43	52	
2420	1-1/2"	.615		1.78	57	
2440	2"	.727		2.46	67	
2460	3"	1.455		4.77	123	
2480	4"	1.600	↓	15.20	151	
9000	Minimum labor/equipment charge	2	Job		175	
15108 | Plastic Pipe & Fittings						
520						**520**
0010	**PIPE, PLASTIC**					
0020	Fiberglass reinforced, couplings 10' O.C., hangers 3 per 10'					
0240	2" diameter	.276	L.F.	9.75	38.50	
0280	4" diameter	.340		16.70	55	
0300	6" diameter	.421	↓	25.50	77	
1800	PVC, couplings 10' O.C., hangers 3 per 10'					
1820	Schedule 40					
1860	1/2" diameter	.148	L.F.	1.41	15.35	
1870	3/4" diameter	.157		1.49	16.30	
1880	1" diameter	.174		1.66	18.10	
1890	1-1/4" diameter	.190		1.86	19.80	
1900	1-1/2" diameter	.222		1.97	22.50	
1910	2" diameter	.271		2.30	25.50	
1920	2-1/2" diameter	.286		3	27.50	
1930	3" diameter	.302		3.77	30.50	
1940	4" diameter	.333		4.76	34	
1950	5" diameter	.372		6.15	40	
1960	6" diameter	.410	↓	8.10	46	
4100	DWV type, schedule 40, couplings 10' O.C., hangers 3 per 10'					
4120	ABS					
4140	1-1/4" diameter	.190	L.F.	1.57	19.35	
4150	1-1/2" diameter	.222		1.60	22	
4160	2" diameter	.271	↓	1.71	24.50	
4400	PVC					
4410	1-1/4" diameter	.190	L.F.	1.70	19.50	
4460	2" diameter	.271		1.90	24.50	
4470	3" diameter	.302		3.19	29	
4480	4" diameter	.333		4.04	33.50	
4490	6" diameter	.410	↓	7.10	44.50	
5360	CPVC, couplings 10' O.C., hangers 3 per 10'					
5380	Schedule 40					
5460	1/2" diameter	.148	L.F.	2.60	17.35	
5470	3/4" diameter	.157		3.29	19.30	
5480	1" diameter	.174		3.97	22	
5490	1-1/4" diameter	.190		4.56	24.50	
5500	1-1/2" diameter	.222		5.05	28	
5510	2" diameter	.271		6.10	32	
5520	2-1/2" diameter	.286	↓	9.65	39	

15108	Plastic Pipe & Fittings	LABOR-HOURS	UNIT	BARE COSTS MAT.	TOTAL INCL O&P	
520	5530	3" diameter	.302	L.F.	11.75	43.50	**520**
	9900	Minimum labor/equipment charge	2	Job		175	
560	0010	**PIPE, PLASTIC, FITTINGS**					**560**
	2700	PVC (white), schedule 40, socket joints					
	2760	90° elbow, 1/2"	.240	Ea.	.29	21.50	
	2810	2"	.440	"	1.68	38	
	4500	DWV, ABS, non pressure, socket joints					
	4540	1/4 Bend, 1-1/4"	.396	Ea.	2.94	39.50	
	4560	1-1/2"	.440		1.03	40	
	4570	2"	.483	↓	1.55	40	
	4800	Tee, sanitary					
	4820	1-1/4"	.593	Ea.	1.85	55	
	4830	1-1/2"	.661		1.56	60	
	4840	2"	.800	↓	2.49	66.50	
	5000	DWV, PVC, schedule 40, socket joints					
	5040	1/4 bend, 1-1/4"	.396	Ea.	1.46	37	
	5060	1-1/2"	.440		1.03	40	
	5070	2"	.483		1.78	41	
	5080	3"	.769		4.37	68	
	5090	4"	.970		6.95	87.50	
	5100	6"	1.584		33.50	181	
	5105	8"	2.581		52	299	
	5110	1/4 bend, long sweep, 1-1/2"	.440		1.94	41.50	
	5112	2"	.483		1.76	41	
	5114	3"	.769		4.35	68	
	5116	4"	.970		8.10	90	
	5215	8"	2.581		56.50	305	
	5250	Tee, sanitary 1-1/4"	.593		2.13	56	
	5254	1-1/2"	.661		1.17	60	
	5255	2"	.800		1.90	65.50	
	5256	3"	1.151		4	97	
	5257	4"	1.455		7.35	127	
	5259	6"	2.388		38	253	
	5261	8"	3.871		114	505	
	5264	2" x 1-1/2"	.727		4.21	64	
	5266	3" x 1-1/2"	1.032		2.67	85.50	
	5268	4" x 3"	1.322		12.05	125	
	5271	6" x 4"	2.319		58	281	
	5314	Combination Y & 1/8 bend, 1-1/2"	.661		2.91	62.50	
	5315	2"	.800		3.76	69	
	5317	3"	1.151		6.65	102	
	5318	4"	1.455	↓	12.90	136	
	5324	Combination Y & 1/8 bend, reducing					
	5325	2" x 2" x 1-1/2"	.727	Ea.	4.17	64	
	5327	3" x 3" x 1-1/2"	1.032		7.25	93.50	
	5328	3" x 3" x 2"	1.046		5.05	91	
	5329	4" x 4" x 2"	1.311		10.10	120	
	5331	Wye, 1-1/4"	.593		3.21	57.50	
	5332	1-1/2"	.661		1.96	60.50	
	5333	2"	.800		2.30	66.50	
	5334	3"	1.151		5.90	100	
	5335	4"	1.455		9.05	130	
	5336	6"	2.388		35.50	248	
	5337	8"	3.871		64	425	
	5341	2" x 1-1/2"	.727		3.21	62.50	
	5342	3" x 1-1/2"	1.032	↓	4.25	88	

		15108	Plastic Pipe & Fittings	LABOR-HOURS	UNIT	BARE COSTS MAT.	TOTAL INCL O&P		
560	5343		4" x 3"	1.322	Ea.	7.20	116	560
	5344		6" x 4"	2.319		28.50	230	
	5345		8" x 6"	3.750		48.50	390	
	5347		Double wye, 1-1/2"	.879		3.87	83.50	
	5348		2"	.964		4.97	84.50	
	5349		3"	1.538		12.85	143	
	5350		4"	1.939		26	197	
	5354		2" x 1-1/2"	.952		4.54	82.50	
	5355		3" x 2"	1.509		9.55	135	
	5356		4" x 3"	1.893		20.50	185	
	5357		6" x 4"	2.207		43	246	
	5410		Reducer bushing, 2" x 1-1/4"	.438		.69	35.50	
	5412		3" x 1-1/2"	.586		3.02	51.50	
	5414		4" x 2"	.879		6.35	80	
	5416		6" x 4"	1.441		18.20	144	
	5418		8" x 6"	2.353	↓	36.50	254	
	5500		CPVC, Schedule 80, threaded joints					
	5540		90° Elbow, 1/4"	.250	Ea.	7.05	33.50	
	5560		1/2"	.264		4.10	29.50	
	5570		3/4"	.308		6.15	37	
	5580		1"	.352		8.60	45.50	
	5590		1-1/4"	.396		16.55	63.50	
	5600		1-1/2"	.440		17.85	68.50	
	5610		2"	.483		24	79	
	5620		2-1/2"	.661		74	178	
	5630		3"	.769		79.50	196	
	6000		Coupling, 1/4"	.250		9	37.50	
	6020		1/2"	.264		7.40	36	
	6030		3/4"	.308		11.95	47	
	6040		1"	.352		13.55	54	
	6050		1-1/4"	.396		14.40	59.50	
	6060		1-1/2"	.440		15.45	65	
	6070		2"	.483		18.25	68.50	
	6080		2-1/2"	.661		32.50	108	
	6090		3"	.769	↓	38	126	
	9900		Minimum labor/equipment charge	2	Job		175	

300	0010	**FLOOR AND AREA DRAINS**						300
	0400	Deck, auto park, C.I., 13" top						
	0440	3", 4", 5", and 6" pipe size	2	Ea.	550	1,100	
	0480	For galvanized body, add		"	262	445	
	2000	Floor, medium duty, C.I., deep flange, 7" dia top					
	2040	2" and 3" pipe size	1.333	Ea.	77.50	237	
	2080	For galvanized body, add			32.50	55	
	2120	For polished bronze top, add		↓	39	66.50	
	2500	Heavy duty, cleanout & trap w/bucket, C.I., 15" top					
	2540	2", 3", and 4" pipe size	2.667	Ea.	2,025	3,675	
	2560	For galvanized body, add			570	970	
	2580	For polished bronze top, add		↓	630	1,075	

500	0010	**STORM AREA DRAINS**						500
	0140	Cornice, C.I., 45° or 90° outlet					
	0200	3" and 4" pipe size	1.333	Ea.	127	320	
	0260	For galvanized body, add		↓	25	42.50	

15160	Storm Drainage Piping	LABOR-HOURS	UNIT	BARE COSTS MAT.	TOTAL INCL O&P	
500	**0280** For polished bronze dome, add		Ea.	21.50	36.50	**500**
	3860 Roof, flat metal deck, C.I. body, 12" C.I. dome					
	3890 3" pipe size	1.143	Ea.	151	345	
	3900 4" pipe size	1.231	"	151	355	
	4620 Main, all aluminum, 12" low profile dome					
	4640 2", 3 and 4" pipe size	1.143	Ea.	168	375	

15410	Plumbing Fixtures	LABOR-HOURS	UNIT	BARE COSTS MAT.	TOTAL INCL. O&P	
300	**0010 FAUCETS/FITTINGS**					**300**
	0150 Bath, faucets, diverter spout combination, sweat	1	Ea.	69.50	206	
	0200 For integral stops, IPS unions, add			73	125	
	0420 Bath, press-bal mix valve w/diverter, spout, shower hd, arm/flange	1		109	274	
	0500 Drain, central lift, 1-1/2" IPS male	.400		38	100	
	0600 Trip lever, 1-1/2" IPS male	.400		38.50	101	
	0810 Bidet					
	0812 Fitting, over the rim, swivel spray/pop-up drain	1	Ea.	136	320	
	0840 Flush valves, with vacuum breaker					
	0850 Water closet					
	0860 Exposed, rear spud	1	Ea.	109	274	
	0870 Top spud	1		100	259	
	0880 Concealed, rear spud	1		143	330	
	0890 Top spud	1		164	365	
	0900 Wall hung	1		126	300	
	0920 Urinal					
	0930 Exposed, stall	1	Ea.	100	259	
	0940 Wall, (washout)	1		100	259	
	0950 Pedestal, top spud	1		102	262	
	0960 Concealed, stall	1		115	284	
	0970 Wall (washout)	1		118	290	
	1000 Kitchen sink faucets, top mount, cast spout	.800		51.50	158	
	1100 For spray, add	.333		9.50	45.50	
	2000 Laundry faucets, shelf type, IPS or copper unions	.667		42.50	130	
	2100 Lavatory faucet, centerset, without drain	.800		37.50	133	
	2200 With pop-up drain	.500		52	133	
	2800 Self-closing, center set	.800		112	260	
	3000 Service sink faucet, cast spout, pail hook, hose end	.571		72	172	
	4000 Shower by-pass valve with union	.444		50.50	125	
	4200 Shower thermostatic mixing valve, concealed	1		233	490	
	4300 For inlet strainer, check, and stops, add			31	52.50	
	5000 Sillcock, compact, brass, IPS or copper to hose	.333		4.74	37	
	9000 Minimum labor/equipment charge	2	Job		175	

15411	Commercial/Indust Fixtures					
700	**0010 URINALS**					**700**
	3000 Wall hung, vitreous china, with hanger & self-closing valve					
	3100 Siphon jet type	5.333	Ea.	291	915	
	3120 Blowout type	5.333		310	955	

15411 | Commercial/Indust Fixtures

		LABOR-HOURS	UNIT	BARE COSTS MAT.	TOTAL INCL O&P	
700	3300	Rough-in, supply, waste & vent	5.654	Ea.	92	605	700
	5000	Stall type, vitreous china, includes valve	6.400		460	1,300	
	6980	Rough-in, supply, waste and vent	8.040	↓	119	835	
	9000	Minimum labor/equipment charge	4	Job		315	
840	0010	**WASH FOUNTAINS**					840
	1900	Group, foot control					
	2000	Precast terrazzo, circular, 36" diam., 5 or 6 persons	8	Ea.	2,650	5,150	
	2100	54" diameter for 8 or 10 persons	9.600		3,200	6,250	
	2400	Semi-circular, 36" diam. for 3 persons	8		2,350	4,675	
	2500	54" diam. for 4 or 5 persons	9.600	↓	2,925	5,750	
	5610	Group, infrared control, barrier free					
	5614	Precast terrazzo					
	5620	Semi-circular 36" diam. for 3 persons	8	Ea.	3,475	6,575	
	5630	46" diam. for 4 persons	8.571		3,900	7,350	
	5640	Circular, 54" diam. for 8 persons, button control	9.600		5,675	10,500	
	5700	Rough-in, supply, waste and vent for above wash fountains	8.791	↓	90.50	845	
	9000	Minimum labor/equipment charge	8	Job		655	

15412 | Drinking Fountains

		LABOR-HOURS	UNIT	BARE COSTS MAT.	TOTAL INCL O&P	
200	0010	**DRINKING FOUNTAIN** For connection to cold water supply					200
	1000	Wall mounted, non-recessed					
	2700	Stainless steel, single bubbler, no back	2	Ea.	825	1,575	
	2740	With back	2		980	1,850	
	2780	Dual handle & wheelchair projection type	2		505	1,050	
	2820	Dual level for handicapped type	2.500		1,025	1,950	
	3980	For rough-in, supply and waste, add	3.620	↓	42	390	
	4000	Wall mounted, semi-recessed					
	4200	Poly-marble, single bubbler	2	Ea.	680	1,325	
	4600	Stainless steel, satin finish, single bubbler	2	"	785	1,500	
	6000	Wall mounted, fully recessed					
	6400	Poly-marble, single bubbler	2	Ea.	765	1,475	
	6800	Stainless steel, single bubbler	2		660	1,300	
	7580	For rough-in, supply and waste, add	4.372	↓	42	455	
	7600	Floor mounted, pedestal type					
	8600	Enameled iron, heavy duty service, 2 bubblers	4	Ea.	1,100	2,225	
	8880	For freeze-proof valve system, add	4		335	920	
	8900	For rough-in, supply and waste, add	4.372	↓	42	455	
	9000	Minimum labor/equipment charge	4	Job		350	

15413 | Electric Water Coolers

		LABOR-HOURS	UNIT	BARE COSTS MAT.	TOTAL INCL O&P	
900	0010	**WATER COOLER**					900
	0100	Wall mounted, non-recessed					
	0140	4 GPH	4	Ea.	500	1,175	
	1000	Dual height, 8.2 GPH	4.211		805	1,725	
	1040	14.3 GPH	4.211		855	1,800	
	3300	Semi-recessed, 8.1 GPH	4	↓	840	1,750	
	4600	Floor mounted, flush-to-wall					
	4640	4 GPH	2.667	Ea.	510	1,100	
	4980	For stainless steel cabinet, add			96	164	
	5000	Dual height, 8.2 GPH	4	↓	870	1,825	
	9000	Minimum labor/equipment charge	4	Job		350	

		15418	Resi/Comm/Industrial Fixtures	LABOR-HOURS	UNIT	BARE COSTS MAT.	TOTAL INCL O&P	
100	0010	**BATHS**						100
	0100	Tubs, recessed porcelain enamel on cast iron, with trim					
	0180	48" x 42"	4	Ea.	1,450	2,800	
	0220	72" x 36"	5.333		1,350	2,725	
	0300	Mat bottom, 4' long	2.909		990	1,925	
	0380	5' long	3.636		375	920	
	0560	Corner 48" x 44"	3.636		1,400	2,650	
	2000	Enameled formed steel, 4'-6" long	2.759		295	725	
	2200	5' long	2.909	↓	286	715	
	4600	Module tub & showerwall surround, molded fiberglass					
	4610	5' long x 34" wide x 76" high	4	Ea.	490	1,150	
	4750	Handicap with 1-1/2" OD grab bar, antiskid bottom					
	4760	60" x 32-3/4" x 72" high	4	Ea.	665	1,450	
	4770	60" x 30" x 71" high with molded seat	4.571	"	825	1,800	
	6000	Whirlpool, bath with vented overflow, molded fiberglass					
	6100	66" x 48" x 24"	16	Ea.	2,225	5,050	
	6400	72" x 36" x 24"	16		2,175	4,975	
	6500	60" x 30" x 21"	16		1,900	4,475	
	6600	72" x 42" x 22"	16		3,075	6,525	
	6700	83" x 65"	53.333		4,150	11,300	
	6710	For color add			10%		
	6711	For designer colors and trim add		↓	25%		
	7000	Redwood hot tub system					
	7050	4' diameter x 4' deep	16	Ea.	1,300	3,450	
	7150	6' diameter x 4' deep	20		2,000	4,975	
	7200	8' diameter x 4' deep	20		3,025	6,725	
	9600	Rough-in, supply, waste and vent, for all above tubs, add	7.729	↓	115	800	
	9900	Minimum labor/equipment charge	5.333	Job		420	
450	0010	**LAVATORIES** With trim, white unless noted otherwise						450
	0500	Vanity top, porcelain enamel on cast iron					
	0600	20" x 18"	2.500	Ea.	206	550	
	0640	33" x 19" oval	2.500	"	405	890	
	0860	For color, add			25%		
	1000	Cultured marble, 19" x 17", single bowl	2.500	Ea.	160	470	
	1120	25" x 22", single bowl	2.500		160	470	
	1900	Stainless steel, self-rimming, 25" x 22", single bowl, ledge	2.500		224	580	
	1960	17" x 22", single bowl	2.500		222	580	
	2600	Steel, enameled, 20" x 17", single bowl	2.759		129	440	
	2900	Vitreous china, 20" x 16", single bowl	2.963		222	615	
	2960	20" x 17", single bowl	2.963		146	485	
	3580	Rough-in, supply, waste and vent for all above lavatories	6.957	↓	63	655	
	4000	Wall hung					
	4040	Porcelain enamel on cast iron, 16" x 14", single bowl	2	Ea.	315	690	
	4180	20" x 18", single bowl	2	"	237	560	
	4580	For color, add			30%		
	6000	Vitreous china, 18" x 15", single bowl with backsplash	2.286	Ea.	184	495	
	6960	Rough-in, supply, waste and vent for above lavatories	9.639	"	220	1,125	
	9000	Minimum labor/equipment charge	2.667	Job		233	
500	0011	**SHOWERS**, Stall, with drain only						500
	1520	32" square	8	Ea.	325	1,175	
	1530	36" square	8		410	1,325	
	1540	Terrazzo receptor, 32" square	8		690	1,800	
	1580	36" corner angle	8.889		755	2,000	
	3000	Fiberglass, one piece, with 3 walls, 32" x 32" square	6.667		350	1,125	
	3100	36" x 36" square	6.667		395	1,200	
	4200	Rough-in, supply, waste and vent for above showers	7.805	↓	62.50	720	

15400 | Plumbing Fixtures & Equipment

		15418	Resi/Comm/Industrial Fixtures	LABOR-HOURS	UNIT	BARE COSTS MAT.	TOTAL INCL O&P	
500	5500		Head, water economizer, 3.0 GPM	.333	Ea.	65	140	500
	9000		Minimum labor/equipment charge	4	Job		315	
900	0010		**WATER CLOSETS**					900
	0150		Tank type, vitreous china, incl. seat, supply pipe w/stop					
	0200		Wall hung, one piece	3.019	Ea.	390	900	
	0400		Two piece, close coupled	3.019		470	1,025	
	0960		For rough-in, supply, waste, vent and carrier	5.861		264	910	
	1000		Floor mounted, one piece	3.019		485	1,075	
	1100		Two piece, close coupled, water saver	3.019		160	515	
	1960		For color, add			30%		
	1980		For rough-in, supply, waste and vent	5.246	Ea.	130	635	
	3000		Bowl only, with flush valve, seat					
	3100		Wall hung	2.759	Ea.	340	800	
	3200		For rough-in, supply, waste and vent, single WC	6.250	"	274	955	
	9000		Minimum labor/equipment charge	4	Job		315	

16050 | Basic Electrical Materials & Methods

		16055	Selective Demolition	LABOR-HOURS	UNIT	BARE COSTS MAT.	TOTAL INCL. O&P	
300	0010		**ELECTRICAL DEMOLITION**					300
	0020		Conduit to 15' high, including fittings & hangers					
	0100		Rigid galvanized steel, 1/2" to 1" diameter	.033	L.F.		2.84	
	0120		1-1/4" to 2"	.040			3.44	
	0140		2-1/2" to 3-1/2"	.053			4.56	
	0200		Electric metallic tubing (EMT) 1/2" to 1"	.020			1.75	
	0220		1-1/4" to 1-1/2"	.025			2.12	
	0240		2" to 3"	.034			2.91	
	0400		Wiremold raceway, including fittings & hangers					
	0420		No. 3000	.032	L.F.		2.76	
	0440		No. 4000	.037	"		3.17	
	0500		Channels, steel, including fittings & hangers					
	0520		3/4" x 1-1/2"	.026	L.F.		2.23	
	0540		1-1/2" x 1-1/2"	.030	"		2.56	
	0600		Copper bus duct, indoor, 3 phase					
	0610		Including hangers & supports					
	0620		225 amp	.119	L.F.		10.25	
	0640		400 amp	.151			13	
	0660		600 amp	.186			16	
	0680		1000 amp	.267			23	
	0720		3000 amp	1.600			138	
	0800		Plug-in switches, 600V 3 ph, incl. disconnecting					
	0820		wire, conduit terminations, 30 amp	.516	Ea.		45	
	0840		60 amp	.576			50	
	0850		100 amp	.769			66	
	0860		200 amp	1.290			111	
	0940		1200 amp	8			690	
	0960		1600 amp	9.412			810	
	1010		Safety switches, 250 or 600V, incl. disconnection					
	1050		of wire & conduit terminations					

16055	Selective Demolition	LABOR-HOURS	UNIT	BARE COSTS MAT.	TOTAL INCL O&P	
300						**300**
1100	30 amp	.650	Ea.		56	
1120	60 amp	.909			78	
1140	100 amp	1.096			94.50	
1160	200 amp	1.600	↓		138	
1210	Panel boards, incl. removal of all breakers,					
1220	conduit terminations & wire connections					
1230	3 wire, 120/240V, 100A, to 20 circuits	3.077	Ea.		264	
1240	200 amps, to 42 circuits	6.154			535	
1260	4 wire, 120/208V, 125A, to 20 circuits	3.333			287	
1270	200 amps, to 42 circuits	6.667	↓		575	
1300	Transformer, dry type, 1 ph, incl. removal of					
1320	supports, wire & conduit terminations					
1340	1 kVA	1.039	Ea.		89.50	
1360	5 kVA	1.702			146	
1420	75 kVA	6.400	↓		555	
1440	3 Phase to 600V, primary					
1460	3 kVA	2.078	Ea.		179	
1480	15 kVA	3.810			330	
1500	30 kVA	4.571			395	
1530	112.5 kVA	6.897			670	
1560	500 kVA	14.286			1,400	
1570	750 kVA	18.182	↓		1,750	
1600	Pull boxes & cabinets, sheet metal, incl. removal					
1620	of supports and conduit terminations					
1640	6" x 6" x 4"	.257	Ea.		22	
1660	12" x 12" x 4"	.343			30	
1720	Junction boxes, 4" sq. & oct.	.100			8.60	
1740	Handy box	.075			6.45	
1760	Switch box	.075			6.45	
1780	Receptacle & switch plates	.031	↓		2.67	
1800	Wire, THW-THWN-THHN, removed from					
1810	in place conduit, to 15' high					
1830	#14	.123	C.L.F.		10.60	
1840	#12	.145			12.50	
1850	#10	.176			15.15	
1880	#4	.302			26	
1890	#3	.320			27.50	
1910	1/0	.482			41	
1920	2/0	.548			47	
1930	3/0	.640			55.50	
1980	400 kcmil	.941			81	
1990	500 kcmil	.988	↓		85.50	
2000	Interior fluorescent fixtures, incl. supports					
2010	& whips, to 15' high					
2100	Recessed drop-in 2' x 2', 2 lamp	.457	Ea.		39	
2120	2' x 4', 2 lamp	.485			42	
2140	2' x 4', 4 lamp	.533			46	
2160	4' x 4', 4 lamp	.800	↓		69	
2180	Surface mount, acrylic lens & hinged frame					
2200	1' x 4', 2 lamp	.364	Ea.		31.50	
2220	2' x 2', 2 lamp	.364			31.50	
2260	2' x 4', 4 lamp	.485			42	
2280	4' x 4', 4 lamp	.400	↓		34	
2300	Strip fixtures, surface mount					
2320	4' long, 1 lamp	.302	Ea.		26	
2340	4' long, 2 lamp	.320	↓		27.50	

16050 | Basic Electrical Materials & Methods

		16055	Selective Demolition	LABOR-HOURS	UNIT	BARE COSTS MAT.	TOTAL INCL O&P	
300	2360		8' long, 1 lamp	.381	Ea.		32.50	300
	2380		8' long, 2 lamp	.400	↓		34	
	2400		Pendant mount, industrial, incl. removal					
	2410		of chain or rod hangers, to 15' high					
	2420		4' long, 2 lamp	.457	Ea.		39	
	2440		8' long, 2 lamp	.593	"		51	
	2460		Interior incandescent, surface, ceiling					
	2470		or wall mount, to 12' high					
	2480		Metal cylinder type, 75 Watt	.258	Ea.		22.50	
	2500		150 Watt	.258	"		22.50	
	2520		Metal halide, high bay					
	2540		400 Watt	1.067	Ea.		91.50	
	2560		1000 Watt	1.333	↓		114	
	2580		150 Watt, low bay	.800	↓		69	
	2600		Exterior fixtures, incandescent, wall mount					
	2620		100 Watt	.320	Ea.		27.50	
	2640		Quartz, 500 Watt	.485			42	
	2660		1500 Watt	.593	↓		51	
	2680		Wall pack, mercury vapor					
	2700		175 Watt	.640	Ea.		55.50	
	2720		250 Watt	.640	"		55.50	
	9000		Minimum labor/equipment charge	2	Job		172	

16100 | Wiring Methods

		16120	Conductors & Cables	LABOR-HOURS	UNIT	BARE COSTS MAT.	TOTAL INCL. O&P	
550	0010		**NON-METALLIC SHEATHED CABLE** 600 volt					550
	0100		Copper with ground wire, (Romex)					
	0152		#14, 2 wire	3.200	C.L.F.	12.25	296	
	0202		3 wire	3.478		21.50	335	
	0252		#12, 2 wire	3.636		17.25	340	
	0302		3 wire	4		31	395	
	0352		#10, 2 wire	4		34	400	
	0402		3 wire	5.714		47.50	570	
	0452		#8, 3 conductor	6.154		91	690	
	0502		#6, 3 wire	6.667	↓	145	825	
	0550		SE type SER aluminum cable, 3 RHW and					
	0602		1 bare neutral, 3 #8 & 1 #8	5.333	C.L.F.	92.50	615	
	0652		3 #6 & 1 #6	6.154		105	710	
	0702		3 #4 & 1 #6	7.273		117	825	
	0752		3 #2 & 1 #4	8		173	985	
	0802		3 #1/0 & 1 #2	8.889		261	1,200	
	0852		3 #2/0 & 1 #1	10		310	1,400	
	0902		3 #4/0 & 1 #2/0	11.429	↓	440	1,725	
	6500		Service entrance cap for copper SEU					
	6700		150 amp	.800	Ea.	12.10	90	
	6800		200 amp	1	"	18.30	117	
	9000		Minimum labor/equipment charge	2	Job		172	
900	0010		**WIRE**					900
	0020		600 volt type THW, copper, solid, #14	.615	C.L.F.	2.95	58.50	

		16120	Conductors & Cables	LABOR-HOURS	UNIT	BARE COSTS MAT.	TOTAL INCL O&P	
900	0030		#12	.727	C.L.F.	4.15	69.50	900
	0040		#10	.800	↓	6.50	80	
	0051		Wire, 600 volt, stranded					
	0160		#6	1.231	C.L.F.	20	140	
	0180		#4	1.509		30.50	182	
	0200		#3	1.600		37.50	203	
	0220		#2	1.778		47	234	
	0240		#1	2		59.50	274	
	0260		1/0	2.424		72	330	
	0280		2/0	2.759		88.50	390	
	0300		3/0	3.200		109	465	
	0350		4/0	3.636		138	545	
	0400		250 kcmil	4		168	630	
	0420		300 kcmil	4.211		199	705	
	0450		350 kcmil	4.444		230	770	
	0480		400 kcmil	4.706		273	870	
	0490		500 kcmil	5		320	975	
	0510		750 kcmil	7.273		520	1,500	
	0530		Aluminum, stranded, #8	.889		12.45	97.50	
	0540		#6	1		17.05	115	
	0560		#4	1.231		21	142	
	0580		#2	1.509		29	179	
	0600		#1	1.778		42	225	
	0620		1/0	2		50.50	259	
	0640		2/0	2.222		59.50	294	
	0680		3/0	2.424		74	335	
	0700		4/0	2.581		82.50	365	
	0720		250 kcmil	2.759		100	410	
	0740		300 kcmil	2.963		139	490	
	0760		350 kcmil	3.200		141	520	
	0780		400 kcmil	3.478		165	585	
	0800		500 kcmil	4		182	650	
	0850		600 kcmil	4.211		230	760	
	0880		700 kcmil	4.706	↓	262	855	
	9000		Minimum labor/equipment charge	2	Job		172	

		16132	Conduit & Tubing					
205	0010		**CONDUIT** To 15' high, includes 2 terminations, 2 elbows and					205
	0020		11 beam clamps per 100 L.F.					
	2500		Steel, intermediate conduit (IMC), 1/2" diameter	.080	L.F.	1.37	9.20	
	2530		3/4" diameter	.089		1.62	10.45	
	2550		1" diameter	.114		2.26	13.70	
	2570		1-1/4" diameter	.123		2.98	15.70	
	2600		1-1/2" diameter	.133		3.44	17.35	
	2630		2" diameter	.160		4.44	21.50	
	2650		2-1/2" diameter	.200		8.95	32.50	
	2670		3" diameter	.267		11.50	43	
	2700		3-1/2" diameter	.296		13.15	48	
	2730		4" diameter	.320		16.35	55.50	
	5000		Electric metallic tubing (EMT), 1/2" diameter	.047		.40	4.72	
	5020		3/4" diameter	.062		.60	6.35	
	5040		1" diameter	.070		1.02	7.75	
	5060		1-1/4" diameter	.080		1.53	9.55	
	5080		1-1/2" diameter	.089		1.95	11	
	5100		2" diameter	.100	↓	2.50	12.85	

16132	Conduit & Tubing	LABOR-HOURS	UNIT	BARE COSTS MAT.	TOTAL INCL O&P	
205	5120	2-1/2" diameter	.133	L.F.	5.80	21.50	**205**
	5140	3" diameter	.160		6.35	24.50	
	5160	3-1/2" diameter	.178		8.10	29	
	5180	4" diameter	.200	↓	9	32.50	
	5200	Field bends, 45° to 90°, 1/2" diameter	.090	Ea.		7.75	
	5220	3/4" diameter	.100			8.60	
	5240	1" diameter	.110			9.45	
	5260	1-1/4" diameter	.211			18.15	
	5280	1-1/2" diameter	.222			19.10	
	5300	2" diameter	.308			26.50	
	5320	Offsets, 1/2" diameter	.123			10.60	
	5340	3/4" diameter	.129			11.10	
	5360	1" diameter	.151			13	
	5380	1-1/4" diameter	.267			23	
	5400	1-1/2" diameter	.286			24.50	
	7600	EMT, "T" fittings with covers, 1/2" diameter, set screw	.500	↓	12.40	64.50	
	9990	Minimum labor/equipment charge	2	Job		172	
260	0010	**CUTTING AND DRILLING**					**260**
	0100	Hole drilling to 10' high, concrete wall					
	0110	8" thick, 1/2" pipe size	.667	Ea.	3.26	69.50	
	0120	3/4" pipe size	.667		3.26	69.50	
	0130	1" pipe size	.842		6.70	92	
	0140	1-1/4" pipe size	.842		6.70	92	
	0150	1-1/2" pipe size	.842		6.70	92	
	0160	2" pipe size	1.818		7.25	186	
	0170	2-1/2" pipe size	1.818		7.25	186	
	0180	3" pipe size	1.818		7.25	186	
	0190	3-1/2" pipe size	2.424		7.75	244	
	0200	4" pipe size	2.424		7.75	244	
	0500	12" thick, 1/2" pipe size	.851		4.98	89.50	
	0520	3/4" pipe size	.851		4.98	89.50	
	0540	1" pipe size	1.096		9.45	121	
	0560	1-1/4" pipe size	1.096		9.45	121	
	0570	1-1/2" pipe size	1.096		9.45	121	
	0580	2" pipe size	2.222		11.10	232	
	0590	2-1/2" pipe size	2.222		11.10	232	
	0600	3" pipe size	2.222		11.10	232	
	0610	3-1/2" pipe size	2.857		12.60	295	
	0630	4" pipe size	3.200		12.60	325	
	0650	16" thick, 1/2" pipe size	1.053		6.70	111	
	0670	3/4" pipe size	1.143		6.70	120	
	0690	1" pipe size	1.333		12.25	148	
	0710	1-1/4" pipe size	1.455		12.25	160	
	0730	1-1/2" pipe size	1.455		12.25	160	
	0750	2" pipe size	2.667		14.95	279	
	0770	2-1/2" pipe size	2.963		14.95	310	
	0790	3" pipe size	3.200		14.95	330	
	0810	3-1/2" pipe size	3.478		17.45	360	
	0830	4" pipe size	4		17.45	410	
	0850	20" thick, 1/2" pipe size	1.250		8.40	133	
	0870	3/4" pipe size	1.333		8.40	141	
	0890	1" pipe size	1.600		15	178	
	0910	1-1/4" pipe size	1.667		15	184	
	0930	1-1/2" pipe size	1.739		15	191	
	0950	2" pipe size	2.963	↓	18.75	315	

16132	Conduit & Tubing	LABOR-HOURS	UNIT	BARE COSTS MAT.	TOTAL INCL O&P	
260						**260**
0970	2-1/2" pipe size	3.333	Ea.	18.75	350	
0990	3" pipe size	3.636		18.75	380	
1010	3-1/2" pipe size	4		22.50	420	
1030	4" pipe size	4.706		22.50	485	
1050	24" thick, 1/2" pipe size	1.455		10.15	156	
1070	3/4" pipe size	1.569		10.15	166	
1090	1" pipe size	1.860		17.75	209	
1110	1-1/4" pipe size	2		17.75	221	
1130	1-1/2" pipe size	2		17.75	221	
1150	2" pipe size	3.333		22.50	355	
1170	2-1/2" pipe size	3.636		22.50	385	
1190	3" pipe size	4		22.50	420	
1210	3-1/2" pipe size	4.444		27	475	
1230	4" pipe size	5.333		27	560	
1500	Brick wall, 8" thick, 1/2" pipe size	.444		3.26	48	
1520	3/4" pipe size	.444		3.26	48	
1540	1" pipe size	.601		6.70	69	
1560	1-1/4" pipe size	.601		6.70	69	
1580	1-1/2" pipe size	.601		6.70	69	
1600	2" pipe size	1.404		7.25	146	
1620	2-1/2" pipe size	1.404		7.25	146	
1640	3" pipe size	1.404		7.25	146	
1660	3-1/2" pipe size	1.818		7.75	186	
1680	4" pipe size	2		7.75	204	
1700	12" thick, 1/2" pipe size	.552		4.98	61	
1720	3/4" pipe size	.552		4.98	61	
1740	1" pipe size	.727		9.45	85	
1760	1-1/4" pipe size	.727		9.45	85	
1780	1-1/2" pipe size	.727		9.45	85	
1800	2" pipe size	1.600		11.10	172	
1820	2-1/2" pipe size	1.600		11.10	172	
1840	3" pipe size	1.600		11.10	172	
1860	3-1/2" pipe size	2.105		12.60	222	
1880	4" pipe size	2.424		12.60	253	
1900	16" thick, 1/2" pipe size	.650		6.70	73.50	
1920	3/4" pipe size	.650		6.70	73.50	
1940	1" pipe size	.860		12.25	103	
1960	1-1/4" pipe size	.860		12.25	103	
1980	1-1/2" pipe size	.860		12.25	103	
2000	2" pipe size	1.818		14.95	198	
2010	2-1/2" pipe size	1.818		14.95	198	
2030	3" pipe size	1.818		14.95	198	
2050	3-1/2" pipe size	2.424		17.45	261	
2070	4" pipe size	2.667		17.45	284	
2090	20" thick, 1/2" pipe size	.748		8.40	86	
2110	3/4" pipe size	.748		8.40	86	
2130	1" pipe size	1		15	121	
2150	1-1/4" pipe size	1		15	121	
2170	1-1/2" pipe size	1		15	121	
2190	2" pipe size	2		18.75	223	
2210	2-1/2" pipe size	2		18.75	223	
2230	3" pipe size	2		18.75	223	
2250	3-1/2" pipe size	2.667		22.50	292	
2270	4" pipe size	2.963		22.50	320	
2290	24" thick, 1/2" pipe size	.851		10.15	98	
2310	3/4" pipe size	.851		10.15	98	

16132	Conduit & Tubing	LABOR-HOURS	UNIT	BARE COSTS MAT.	TOTAL INCL O&P	
260	2330	1" pipe size	1.127	Ea.	17.75	138	260
	2350	1-1/4" pipe size	1.127		17.75	138	
	2370	1-1/2" pipe size	1.127		17.75	138	
	2390	2" pipe size	2.222		22.50	252	
	2410	2-1/2" pipe size	2.222		22.50	252	
	2430	3" pipe size	2.222		22.50	252	
	2450	3-1/2" pipe size	2.857		27	320	
	2470	4" pipe size	3.200	↓	27	350	
	2480						
	3000	Knockouts to 8' high, metal boxes & enclosures					
	3020	With hole saw, 1/2" pipe size	.151	Ea.		13	
	3040	3/4" pipe size	.170			14.65	
	3050	1" pipe size	.200			17.20	
	3060	1-1/4" pipe size	.222			19.10	
	3070	1-1/2" pipe size	.250			21.50	
	3080	2" pipe size	.296			25.50	
	3090	2-1/2" pipe size	.400			34	
	4010	3" pipe size	.500			43.50	
	4030	3-1/2" pipe size	.615			53.50	
	4050	4" pipe size	.727			62.50	
	4070	With hand punch set, 1/2" pipe size	.200			17.20	
	4090	3/4" pipe size	.250			21.50	
	4110	1" pipe size	.267			23	
	4130	1-1/4" pipe size	.286			24.50	
	4150	1-1/2" pipe size	.308			26.50	
	4170	2" pipe size	.400			34	
	4190	2-1/2" pipe size	.471			40.50	
	4200	3" pipe size	.533			46	
	4220	3-1/2" pipe size	.667			57.50	
	4240	4" pipe size	.800			69	
	4260	With hydraulic punch, 1/2" pipe size	.182			15.65	
	4280	3/4" pipe size	.211			18.15	
	4300	1" pipe size	.211			18.15	
	4320	1-1/4" pipe size	.211			18.15	
	4340	1-1/2" pipe size	.211			18.15	
	4360	2" pipe size	.250			21.50	
	4380	2-1/2" pipe size	.296			25.50	
	4400	3" pipe size	.348			30	
	4420	3-1/2" pipe size	.400	↓		34	

16133	Multi-outlet Assemblies					
800	0010	**SURFACE RACEWAY**					800
	0090	Metal, straight section					
	0100	No. 500	.080	L.F.	.73	8.15	
	0110	No. 700	.080		.82	8.30	
	0400	No. 1500, small pancake	.089		1.50	10.25	
	0600	No. 2000, base & cover, blank	.089		1.47	10.15	
	0800	No. 3000, base & cover, blank	.107		2.95	14.20	
	1000	No. 4000, base & cover, blank	.123		4.80	18.80	
	1200	No. 6000, base & cover, blank	.160	↓	8.05	27.50	
	2400	Fittings, elbows, No. 500	.200	Ea.	1.33	19.45	
	2800	Elbow cover, No. 2000	.200		2.55	21.50	
	3000	Switch box, No. 500	.500		8.40	58	
	3400	Telephone outlet, No. 1500	.500		9.65	59.50	
	3600	Junction box, No. 1500	.500	↓	6.70	55	

16100 | Wiring Methods

16133 | Multi-outlet Assemblies

			LABOR-HOURS	UNIT	BARE COSTS MAT.	TOTAL INCL O&P	
800	3800	Plugmold wired sections, No. 2000					800
	4000	1 circuit, 6 outlets, 3 ft. long	1	Ea.	24.50	128	
	4100	2 circuits, 8 outlets, 6 ft. long	1.509	"	41	201	
	9990	Minimum labor/equipment charge	1.600	Job		138	

16140 | Wiring Devices

			LABOR-HOURS	UNIT	BARE COSTS MAT.	TOTAL INCL O&P	
910	0010	**WIRING DEVICES**					910
	0200	Toggle switch, quiet type, single pole, 15 amp	.200	Ea.	4.69	25	
	0500	20 amp	.296		6.80	37	
	0550	Rocker, 15 amp	.200		5.05	26	
	0560	20 amp	.296		11.70	45.50	
	0600	3 way, 15 amp	.348		6.70	41.50	
	0850	Rocker, 15 amp	.348		7.10	42	
	0860	20 amp	.444		17	67	
	0900	4 way, 15 amp	.533		20	80.50	
	1030	Rocker, 15 amp	.533		27.50	93	
	1040	20 amp	.727		42	134	
	1650	Dimmer switch, 120 volt, incandescent, 600 watt, 1 pole	.500		10.80	62	
	2460	Receptacle, duplex, 120 volt, grounded, 15 amp	.200		1.14	19.15	
	2470	20 amp	.296		7.05	37.50	
	2490	Dryer, 30 amp	.533		10.35	64	
	2500	Range, 50 amp	.727		10.75	81	
	2600	Wall plates, stainless steel, 1 gang	.100		1.80	11.70	
	2800	2 gang	.151		4.10	20	
	3200	Lampholder, keyless	.308		9.20	42.50	
	3400	Pullchain with receptacle	.364		8.90	47	
	9000	Minimum labor/equipment charge	2	Job		172	

16500 | Lighting

16520 | Exterior Luminaires

			LABOR-HOURS	UNIT	BARE COSTS MAT.	TOTAL INCL. O&P	
300	0010	**EXTERIOR FIXTURES** With lamps					300
	0200	Wall mounted, incandescent, 100 watt	1	Ea.	28	134	
	0400	Quartz, 500 watt	1.509		53.50	222	
	1100	Wall pack, low pressure sodium, 35 watt	2		214	535	
	1150	55 watt	2		255	605	
	1160	High pressure sodium, 70 watt	2		185	490	
	1170	150 watt	2		228	560	
	1180	Metal Halide, 175 watt	2		255	605	
	1190	250 watt	2		260	615	
	1200	Floodlights with ballast and lamp,					
	1400	pole mounted, pole not included					
	2250	Low pressure sodium, 55 watt	2.963	Ea.	485	1,075	
	2270	90 watt	4	"	535	1,250	
	9000	Minimum labor/equipment charge	2.133	Job		183	
320	0010	**EXIT AND EMERGENCY LIGHTING**					320
	0080	Exit light ceiling or wall mount, incandescent, single face	1	Ea.	38.50	152	
	0100	Double face	1.194	"	44	178	
	0300	Emergency light units, battery operated					

		16520	Exterior Luminaires	LABOR-HOURS	UNIT	BARE COSTS MAT.	TOTAL INCL O&P	
320	0350		Twin sealed beam light, 25 watt, 6 volt each					320
	0500		Lead battery operated	2	Ea.	110	360	
	0700		Nickel cadmium battery operated	2	"	505	1,025	

Part Five
Location Adjustment Factors

Costs shown in *Means ADA Compliance Pricing Guide* are based on national averages for materials and installation. To adjust these costs to a specific location, simply multiply the total project cost by the factor listed for the particular city.

The data is arranged alphabetically by state and postal zip code numbers. For a city not listed, use the factor for a nearby city with similar economic characteristics.

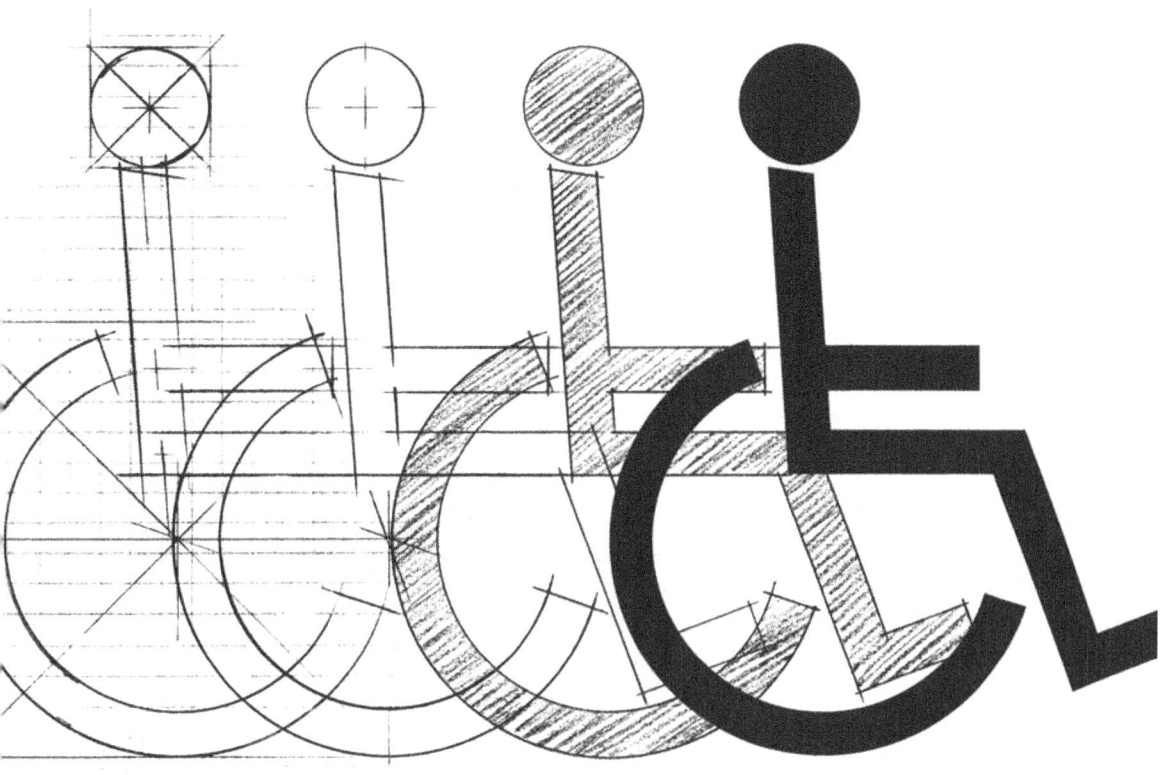

Location Adjustment Factors

Costs shown are based on National Averages for materials and installation. To adjust these costs to a specific location, simply multiply the base cost by the factor for that city. The data is arranged alphabetically by state and postal zip code numbers. For a city not listed, use the factor for a nearby city with similar economic characteristics.

STATE/ZIP	CITY	Residential	Commercial
ALABAMA			
350-352	Birmingham	.86	.87
354	Tuscaloosa	.73	.78
355	Jasper	.70	.77
356	Decatur	.76	.79
357-358	Huntsville	.84	.85
359	Gadsden	.73	.80
360-361	Montgomery	.75	.79
362	Anniston	.68	.74
363	Dothan	.74	.75
364	Evergreen	.71	.76
365-366	Mobile	.79	.81
367	Selma	.72	.76
368	Phenix City	.73	.78
369	Butler	.71	.75
ALASKA			
995-996	Anchorage	1.28	1.26
997	Fairbanks	1.30	1.25
998	Juneau	1.28	1.25
999	Ketchikan	1.30	1.31
ARIZONA			
850,853	Phoenix	.87	.88
852	Mesa/Tempe	.85	.85
855	Globe	.81	.83
856-857	Tucson	.85	.86
859	Show Low	.84	.84
860	Flagstaff	.85	.86
863	Prescott	.82	.84
864	Kingman	.82	.84
865	Chambers	.82	.82
ARKANSAS			
716	Pine Bluff	.77	.80
717	Camden	.66	.69
718	Texarkana	.71	.72
719	Hot Springs	.64	.68
720-722	Little Rock	.82	.82
723	West Memphis	.75	.77
724	Jonesboro	.73	.77
725	Batesville	.72	.74
726	Harrison	.73	.74
727	Fayetteville	.62	.66
728	Russellville	.71	.73
729	Fort Smith	.77	.78
CALIFORNIA			
900-902	Los Angeles	1.07	1.08
903-905	Inglewood	1.04	1.04
906-908	Long Beach	1.03	1.04
910-912	Pasadena	1.04	1.05
913-916	Van Nuys	1.07	1.07
917-918	Alhambra	1.08	1.06
919-921	San Diego	1.03	1.05
922	Palm Springs	1.00	1.03
923-924	San Bernardino	1.01	1.02
925	Riverside	1.07	1.08
926-927	Santa Ana	1.05	1.04
928	Anaheim	1.08	1.08
930	Oxnard	1.08	1.08
931	Santa Barbara	1.07	1.08
932-933	Bakersfield	1.05	1.06
934	San Luis Obispo	1.07	1.06
935	Mojave	1.04	1.03
936-938	Fresno	1.11	1.09
939	Salinas	1.12	1.11
940-941	San Francisco	1.22	1.24
942,956-958	Sacramento	1.12	1.11
943	Palo Alto	1.17	1.16
944	San Mateo	1.21	1.18
945	Vallejo	1.14	1.12
946	Oakland	1.19	1.18
947	Berkeley	1.22	1.17
948	Richmond	1.23	1.17
949	San Rafael	1.21	1.18
950	Santa Cruz	1.15	1.14
951	San Jose	1.20	1.19
952	Stockton	1.10	1.09
953	Modesto	1.09	1.09

STATE/ZIP	CITY	Residential	Commercial
CALIFORNIA (CONT'D)			
954	Santa Rosa	1.15	1.15
955	Eureka	1.11	1.07
959	Marysville	1.11	1.10
960	Redding	1.12	1.10
961	Susanville	1.11	1.10
COLORADO			
800-802	Denver	.95	.96
803	Boulder	.94	.93
804	Golden	.91	.94
805	Fort Collins	.90	.93
806	Greeley	.79	.86
807	Fort Morgan	.94	.93
808-809	Colorado Springs	.92	.95
810	Pueblo	.93	.93
811	Alamosa	.90	.93
812	Salida	.92	.93
813	Durango	.94	.93
814	Montrose	.89	.92
815	Grand Junction	.93	.93
816	Glenwood Springs	.92	.95
CONNECTICUT			
060	New Britain	1.08	1.08
061	Hartford	1.07	1.08
062	Willimantic	1.07	1.08
063	New London	1.07	1.06
064	Meriden	1.07	1.08
065	New Haven	1.08	1.09
066	Bridgeport	1.08	1.08
067	Waterbury	1.08	1.08
068	Norwalk	1.08	1.09
069	Stamford	1.09	1.11
D.C.			
200-205	Washington	.92	.95
DELAWARE			
197	Newark	1.01	1.03
198	Wilmington	1.01	1.03
199	Dover	1.01	1.03
FLORIDA			
320,322	Jacksonville	.78	.81
321	Daytona Beach	.85	.87
323	Tallahassee	.72	.76
324	Panama City	.66	.71
325	Pensacola	.76	.80
326,344	Gainesville	.76	.80
327-328,347	Orlando	.79	.83
329	Melbourne	.86	.90
330-332,340	Miami	.83	.87
333	Fort Lauderdale	.84	.86
334,349	West Palm Beach	.83	.84
335-336,346	Tampa	.87	.88
337	St. Petersburg	.76	.81
338	Lakeland	.84	.87
339,341	Fort Myers	.81	.83
342	Sarasota	.85	.85
GEORGIA			
300-303,399	Atlanta	.89	.90
304	Statesboro	.67	.73
305	Gainesville	.73	.78
306	Athens	.74	.80
307	Dalton	.70	.76
308-309	Augusta	.76	.79
310-312	Macon	.77	.79
313-314	Savannah	.79	.80
315	Waycross	.71	.77
316	Valdosta	.71	.75
317	Albany	.73	.77
318-319	Columbus	.76	.77
HAWAII			
967	Hilo	1.23	1.21
968	Honolulu	1.24	1.22

STATE/ZIP	CITY	Residential	Commercial
STATES & POSS.			
969	Guam	1.60	1.29
IDAHO			
832	Pocatello	.89	.91
833	Twin Falls	.73	.76
834	Idaho Falls	.72	.80
835	Lewiston	.99	.99
836-837	Boise	.90	.92
838	Coeur d'Alene	.85	.86
ILLINOIS			
600-603	North Suburban	1.11	1.09
604	Joliet	1.13	1.09
605	South Suburban	1.10	1.08
606	Chicago	1.15	1.13
609	Kankakee	1.03	1.02
610-611	Rockford	1.04	1.04
612	Rock Island	.97	.97
613	La Salle	1.02	.99
614	Galesburg	.99	.99
615-616	Peoria	1.02	1.01
617	Bloomington	.98	.99
618-619	Champaign	.99	1.01
620-622	East St. Louis	.99	.99
623	Quincy	.98	.96
624	Effingham	.99	.97
625	Decatur	.99	.98
626-627	Springfield	.99	.98
628	Centralia	.97	.97
629	Carbondale	.96	.95
INDIANA			
460	Anderson	.92	.91
461-462	Indianapolis	.96	.94
463-464	Gary	1.02	1.00
465-466	South Bend	.92	.91
467-468	Fort Wayne	.92	.91
469	Kokomo	.94	.91
470	Lawrenceburg	.88	.88
471	New Albany	.86	.86
472	Columbus	.93	.90
473	Muncie	.92	.91
474	Bloomington	.96	.92
475	Washington	.91	.92
476-477	Evansville	.91	.93
478	Terre Haute	.91	.93
479	Lafayette	.92	.90
IOWA			
500-503,509	Des Moines	.93	.92
504	Mason City	.78	.82
505	Fort Dodge	.77	.80
506-507	Waterloo	.81	.82
508	Creston	.83	.84
510-511	Sioux City	.87	.88
512	Sibley	.74	.77
513	Spencer	.76	.77
514	Carroll	.76	.78
515	Council Bluffs	.82	.88
516	Shenandoah	.74	.76
520	Dubuque	.86	.89
521	Decorah	.77	.78
522-524	Cedar Rapids	.94	.93
525	Ottumwa	.84	.86
526	Burlington	.88	.87
527-528	Davenport	.98	.97
KANSAS			
660-662	Kansas City	.95	.96
664-666	Topeka	.78	.84
667	Fort Scott	.85	.83
668	Emporia	.73	.80
669	Belleville	.74	.79
670-672	Wichita	.81	.85
673	Independence	.74	.77
674	Salina	.73	.80
675	Hutchinson	.68	.74
676	Hays	.74	.78
677	Colby	.76	.79
678	Dodge City	.74	.79
679	Liberal	.68	.74
KENTUCKY			
400-402	Louisville	.92	.91
403-405	Lexington	.84	.83

STATE/ZIP	CITY	Residential	Commercial
KENTUCKY (CONT'D)			
406	Frankfort	.82	.84
407-409	Corbin	.67	.71
410	Covington	.95	.95
411-412	Ashland	.93	.96
413-414	Campton	.68	.72
415-416	Pikeville	.78	.81
417-418	Hazard	.67	.72
420	Paducah	.89	.91
421-422	Bowling Green	.89	.89
423	Owensboro	.83	.86
424	Henderson	.92	.93
425-426	Somerset	.67	.71
427	Elizabethtown	.88	.89
LOUISIANA			
700-701	New Orleans	.86	.86
703	Thibodaux	.82	.85
704	Hammond	.80	.83
705	Lafayette	.78	.80
706	Lake Charles	.80	.82
707-708	Baton Rouge	.80	.80
710-711	Shreveport	.79	.80
712	Monroe	.75	.79
713-714	Alexandria	.74	.77
MAINE			
039	Kittery	.81	.87
040-041	Portland	.89	.90
042	Lewiston	.89	.90
043	Augusta	.84	.88
044	Bangor	.87	.90
045	Bath	.82	.87
046	Machias	.82	.87
047	Houlton	.87	.89
048	Rockland	.80	.86
049	Waterville	.76	.85
MARYLAND			
206	Waldorf	.84	.87
207-208	College Park	.85	.90
209	Silver Spring	.85	.89
210-212	Baltimore	.90	.91
214	Annapolis	.86	.89
215	Cumberland	.85	.87
216	Easton	.69	.72
217	Hagerstown	.86	.88
218	Salisbury	.75	.77
219	Elkton	.83	.82
MASSACHUSETTS			
010-011	Springfield	1.05	1.03
012	Pittsfield	1.00	1.00
013	Greenfield	1.00	1.00
014	Fitchburg	1.09	1.06
015-016	Worcester	1.10	1.08
017	Framingham	1.11	1.09
018	Lowell	1.12	1.10
019	Lawrence	1.12	1.10
020-022, 024	Boston	1.17	1.15
023	Brockton	1.11	1.09
025	Buzzards Bay	1.09	1.06
026	Hyannis	1.08	1.07
027	New Bedford	1.11	1.08
MICHIGAN			
480,483	Royal Oak	1.05	1.02
481	Ann Arbor	1.05	1.03
482	Detroit	1.10	1.07
484-485	Flint	.99	.98
486	Saginaw	.96	.96
487	Bay City	.96	.96
488-489	Lansing	.98	.98
490	Battle Creek	.94	.94
491	Kalamazoo	.93	.92
492	Jackson	.94	.95
493,495	Grand Rapids	.84	.85
494	Muskegon	.90	.91
496	Traverse City	.83	.86
497	Gaylord	.85	.89
498-499	Iron Mountain	.92	.94
MINNESOTA			
550-551	Saint Paul	1.14	1.12
553-555	Minneapolis	1.18	1.13

STATE/ZIP	CITY	Residential	Commercial
556-558	Duluth	1.10	1.05
559	Rochester	1.05	1.03
560	Mankato	1.02	1.01
561	Windom	.85	.90
562	Willmar	.86	.93
563	St. Cloud	1.10	1.09
564	Brainerd	.98	1.00
565	Detroit Lakes	.99	1.00
566	Bemidji	.96	.99
567	Thief River Falls	.94	.97
MISSISSIPPI			
386	Clarksdale	.60	.66
387	Greenville	.69	.75
388	Tupelo	.63	.70
389	Greenwood	.63	.67
390-392	Jackson	.72	.75
393	Meridian	.66	.74
394	Laurel	.62	.68
395	Biloxi	.75	.78
396	McComb	.73	.76
397	Columbus	.64	.70
MISSOURI			
630-631	St. Louis	1.01	1.02
633	Bowling Green	.91	.94
634	Hannibal	.88	.91
635	Kirksville	.81	.89
636	Flat River	.94	.97
637	Cape Girardeau	.87	.94
638	Sikeston	.84	.91
639	Poplar Bluff	.85	.91
640-641	Kansas City	1.01	1.02
644-645	St. Joseph	.95	.96
646	Chillicothe	.86	.85
647	Harrisonville	.96	.98
648	Joplin	.83	.85
650-651	Jefferson City	.89	.92
652	Columbia	.89	.94
653	Sedalia	.87	.92
654-655	Rolla	.89	.88
656-658	Springfield	.85	.88
MONTANA			
590-591	Billings	.88	.89
592	Wolf Point	.85	.88
593	Miles City	.87	.87
594	Great Falls	.89	.89
595	Havre	.82	.87
596	Helena	.89	.88
597	Butte	.84	.88
598	Missoula	.84	.86
599	Kalispell	.83	.86
NEBRASKA			
680-681	Omaha	.90	.90
683-685	Lincoln	.79	.85
686	Columbus	.69	.76
687	Norfolk	.78	.83
688	Grand Island	.78	.84
689	Hastings	.76	.80
690	Mccook	.69	.75
691	North Platte	.75	.80
692	Valentine	.66	.72
693	Alliance	.65	.71
NEVADA			
889-891	Las Vegas	1.01	1.04
893	Ely	.92	.93
894-895	Reno	.97	.99
897	Carson City	.97	.98
898	Elko	.93	.92
NEW HAMPSHIRE			
030	Nashua	.91	.94
031	Manchester	.91	.94
032-033	Concord	.88	.93
034	Keene	.73	.77
035	Littleton	.81	.82
036	Charleston	.71	.75
037	Claremont	.72	.75
038	Portsmouth	.85	.90

STATE/ZIP	CITY	Residential	Commercial
NEW JERSEY			
070-071	Newark	1.13	1.11
072	Elizabeth	1.15	1.10
073	Jersey City	1.12	1.10
074-075	Paterson	1.13	1.11
076	Hackensack	1.12	1.10
077	Long Branch	1.12	1.09
078	Dover	1.12	1.09
079	Summit	1.12	1.10
080,083	Vineland	1.10	1.08
081	Camden	1.11	1.08
082,084	Atlantic City	1.14	1.08
085-086	Trenton	1.12	1.10
087	Point Pleasant	1.11	1.09
088-089	New Brunswick	1.12	1.09
NEW MEXICO			
870-872	Albuquerque	.86	.89
873	Gallup	.86	.90
874	Farmington	.86	.90
875	Santa Fe	.86	.89
877	Las Vegas	.86	.89
878	Socorro	.86	.89
879	Truth/Consequences	.84	.86
880	Las Cruces	.83	.84
881	Clovis	.85	.89
882	Roswell	.86	.89
883	Carrizozo	.86	.90
884	Tucumcari	.86	.89
NEW YORK			
100-102	New York	1.37	1.34
103	Staten Island	1.30	1.30
104	Bronx	1.32	1.29
105	Mount Vernon	1.18	1.19
106	White Plains	1.21	1.19
107	Yonkers	1.22	1.21
108	New Rochelle	1.23	1.19
109	Suffern	1.15	1.14
110	Queens	1.30	1.30
111	Long Island City	1.33	1.31
112	Brooklyn	1.34	1.31
113	Flushing	1.32	1.31
114	Jamaica	1.32	1.30
115,117,118	Hicksville	1.22	1.24
116	Far Rockaway	1.31	1.31
119	Riverhead	1.23	1.25
120-122	Albany	.96	.97
123	Schenectady	.96	.97
124	Kingston	1.04	1.09
125-126	Poughkeepsie	1.08	1.11
127	Monticello	1.05	1.08
128	Glens Falls	.88	.92
129	Plattsburgh	.93	.92
130-132	Syracuse	.96	.96
133-135	Utica	.93	.94
136	Watertown	.92	.95
137-139	Binghamton	.92	.93
140-142	Buffalo	1.06	1.02
143	Niagara Falls	1.04	1.02
144-146	Rochester	.99	.99
147	Jamestown	.91	.93
148-149	Elmira	.89	.92
NORTH CAROLINA			
270,272-274	Greensboro	.74	.75
271	Winston-Salem	.74	.75
275-276	Raleigh	.75	.76
277	Durham	.74	.75
278	Rocky Mount	.64	.68
279	Elizabeth City	.62	.69
280	Gastonia	.74	.74
281-282	Charlotte	.75	.75
283	Fayetteville	.72	.75
284	Wilmington	.72	.74
285	Kinston	.62	.67
286	Hickory	.62	.67
287-288	Asheville	.72	.74
289	Murphy	.66	.67
NORTH DAKOTA			
580-581	Fargo	.81	.85
582	Grand Forks	.76	.82
583	Devils Lake	.81	.83
584	Jamestown	.75	.80
585	Bismarck	.81	.85

STATE/ZIP	CITY	Residential	Commercial
NORTH DAKOTA (CONT'D)			
586	Dickinson	.78	.84
587	Minot	.81	.86
588	Williston	.78	.83
OHIO			
430-432	Columbus	.96	.95
433	Marion	.94	.93
434-436	Toledo	1.02	1.00
437-438	Zanesville	.91	.91
439	Steubenville	.96	.96
440	Lorain	1.03	1.00
441	Cleveland	1.03	1.02
442-443	Akron	1.00	.99
444-445	Youngstown	.97	.97
446-447	Canton	.95	.95
448-449	Mansfield	.97	.95
450	Hamilton	.96	.93
451-452	Cincinnati	.96	.94
453-454	Dayton	.93	.91
455	Springfield	.94	.92
456	Chillicothe	.97	.95
457	Athens	.88	.89
458	Lima	.91	.94
OKLAHOMA			
730-731	Oklahoma City	.81	.83
734	Ardmore	.79	.81
735	Lawton	.83	.83
736	Clinton	.78	.81
737	Enid	.78	.81
738	Woodward	.77	.80
739	Guymon	.67	.67
740-741	Tulsa	.80	.81
743	Miami	.83	.82
744	Muskogee	.72	.72
745	Mcalester	.75	.76
746	Ponca City	.78	.80
747	Durant	.77	.80
748	Shawnee	.77	.80
749	Poteau	.78	.80
OREGON			
970-972	Portland	1.02	1.04
973	Salem	1.02	1.04
974	Eugene	1.01	1.03
975	Medford	1.00	1.03
976	Klamath Falls	1.00	1.04
977	Bend	1.02	1.04
978	Pendleton	.99	1.00
979	Vale	.98	.95
PENNSYLVANIA			
150-152	Pittsburgh	.98	1.01
153	Washington	.94	.99
154	Uniontown	.92	.97
155	Bedford	.88	.94
156	Greensburg	.95	.98
157	Indiana	.91	.96
158	Dubois	.90	.96
159	Johnstown	.91	.96
160	Butler	.94	.97
161	New Castle	.94	.97
162	Kittanning	.95	.98
163	Oil City	.91	.95
164-165	Erie	.97	.96
166	Altoona	.90	.94
167	Bradford	.90	.96
168	State College	.93	.95
169	Wellsboro	.89	.94
170-171	Harrisburg	.93	.95
172	Chambersburg	.89	.93
173-174	York	.89	.93
175-176	Lancaster	.91	.92
177	Williamsport	.86	.89
178	Sunbury	.90	.93
179	Pottsville	.90	.94
180	Lehigh Valley	.99	1.01
181	Allentown	1.02	1.00
182	Hazleton	.90	.95
183	Stroudsburg	.93	.97
184-185	Scranton	.96	.97
186-187	Wilkes-Barre	.92	.94
188	Montrose	.90	.95
189	Doylestown	1.04	1.05

STATE/ZIP	CITY	Residential	Commercial
PENNSYLVANIA (CONT'D)			
190-191	Philadelphia	1.13	1.12
193	Westchester	1.07	1.06
194	Norristown	1.06	1.07
195-196	Reading	.95	.97
PUERTO RICO			
009	San Juan	.84	.86
RHODE ISLAND			
028	Newport	1.07	1.04
029	Providence	1.07	1.05
SOUTH CAROLINA			
290-292	Columbia	.73	.74
293	Spartanburg	.72	.72
294	Charleston	.72	.75
295	Florence	.67	.72
296	Greenville	.71	.72
297	Rock Hill	.65	.66
298	Aiken	.84	.85
299	Beaufort	.67	.69
SOUTH DAKOTA			
570-571	Sioux Falls	.77	.81
572	Watertown	.73	.78
573	Mitchell	.75	.77
574	Aberdeen	.76	.80
575	Pierre	.76	.79
576	Mobridge	.74	.77
577	Rapid City	.76	.78
TENNESSEE			
370-372	Nashville	.84	.87
373-374	Chattanooga	.77	.79
375,380-381	Memphis	.83	.87
376	Johnson City	.72	.80
377-379	Knoxville	.74	.79
382	Mckenzie	.70	.75
383	Jackson	.71	.78
384	Columbia	.73	.78
385	Cookeville	.68	.76
TEXAS			
750	Mckinney	.77	.80
751	Waxahackie	.78	.80
752-753	Dallas	.84	.84
754	Greenville	.70	.73
755	Texarkana	.75	.78
756	Longview	.69	.73
757	Tyler	.76	.80
758	Palestine	.69	.72
759	Lufkin	.74	.74
760-761	Fort Worth	.84	.82
762	Denton	.77	.77
763	Wichita Falls	.80	.79
764	Eastland	.73	.72
765	Temple	.76	.75
766-767	Waco	.78	.79
768	Brownwood	.69	.71
769	San Angelo	.73	.75
770-772	Houston	.86	.87
773	Huntsville	.69	.72
774	Wharton	.71	.75
775	Galveston	.84	.85
776-777	Beaumont	.83	.82
778	Bryan	.74	.81
779	Victoria	.75	.76
780	Laredo	.73	.76
781-782	San Antonio	.79	.81
783-784	Corpus Christi	.78	.78
785	Mc Allen	.76	.75
786-787	Austin	.80	.80
788	Del Rio	.66	.68
789	Giddings	.69	.71
790-791	Amarillo	.79	.80
792	Childress	.76	.77
793-794	Lubbock	.77	.79
795-796	Abilene	.76	.78
797	Midland	.77	.78
798-799,885	El Paso	.76	.77
UTAH			
840-841	Salt Lake City	.84	.89
842,844	Ogden	.82	.87

STATE/ZIP	CITY	Residential	Commercial
843	Logan	.83	.88
845	Price	.73	.78
846-847	Provo	.84	.88
VERMONT			
050	White River Jct.	.74	.76
051	Bellows Falls	.75	.76
052	Bennington	.74	.76
053	Brattleboro	.75	.77
054	Burlington	.80	.85
056	Montpelier	.81	.84
057	Rutland	.82	.85
058	St. Johnsbury	.75	.77
059	Guildhall	.74	.76
VIRGINIA			
220-221	Fairfax	.86	.90
222	Arlington	.87	.90
223	Alexandria	.90	.91
224-225	Fredericksburg	.76	.83
226	Winchester	.72	.78
227	Culpeper	.78	.79
228	Harrisonburg	.68	.75
229	Charlottesville	.73	.81
230-232	Richmond	.82	.84
233-235	Norfolk	.79	.82
236	Newport News	.77	.81
237	Portsmouth	.75	.81
238	Petersburg	.80	.83
239	Farmville	.69	.72
240-241	Roanoke	.73	.75
242	Bristol	.68	.75
243	Pulaski	.66	.72
244	Staunton	.69	.74
245	Lynchburg	.70	.76
246	Grundy	.68	.73
WASHINGTON			
980-981,987	Seattle	1.01	1.04
982	Everett	1.03	1.02
983-984	Tacoma	.99	1.02
985	Olympia	.99	1.01
986	Vancouver	.98	1.02
988	Wenatchee	.92	.95
989	Yakima	.95	.97
990-992	Spokane	1.00	.96
993	Richland	.99	.97
994	Clarkston	.98	.96
WEST VIRGINIA			
247-248	Bluefield	.89	.88
249	Lewisburg	.89	.91
250-253	Charleston	.97	.94
254	Martinsburg	.85	.88
255-257	Huntington	.96	.95
258-259	Beckley	.90	.92
260	Wheeling	.92	.95
261	Parkersburg	.92	.94
262	Buckhannon	.91	.94
263-264	Clarksburg	.91	.94
265	Morgantown	.92	.94
266	Gassaway	.92	.94
267	Romney	.87	.91
268	Petersburg	.89	.93
WISCONSIN			
530,532	Milwaukee	1.06	1.01
531	Kenosha	1.06	1.00
534	Racine	1.04	1.00
535	Beloit	1.02	.99
537	Madison	1.01	.98
538	Lancaster	1.00	.96
539	Portage	.98	.95
540	New Richmond	1.00	.97
541-543	Green Bay	1.03	.97
544	Wausau	.96	.94
545	Rhinelander	.97	.96
546	La Crosse	.96	.96
547	Eau Claire	1.00	.97
548	Superior	.99	.97
549	Oshkosh	.96	.94
WYOMING			
820	Cheyenne	.76	.79
821	Yellowstone Nat. Pk.	.71	.76
822	Wheatland	.72	.76
823	Rawlins	.70	.75

STATE/ZIP	CITY	Residential	Commercial
824	Worland	.69	.74
825	Riverton	.70	.76
826	Casper	.77	.82
827	Newcastle	.69	.74
828	Sheridan	.74	.79
829-831	Rock Springs	.74	.77
CANADIAN FACTORS (reflect Canadian currency)			
ALBERTA			
	Calgary	1.05	1.05
	Edmonton	1.05	1.05
	Fort McMurray	1.03	1.01
	Lethbridge	1.04	1.01
	Lloydminster	1.03	1.01
	Medicine Hat	1.03	1.01
	Red Deer	1.03	1.01
BRITISH COLUMBIA			
	Kamloops	1.01	1.03
	Prince George	1.02	1.04
	Vancouver	1.07	1.07
	Victoria	1.02	1.04
NS	Bridgewater	.93	.95
Quebec	Granby	1.09	1.01
Quebec	Hull	1.09	1.01
Quebec	Joliette	1.10	1.01
Quebec	Rimouski	1.10	1.01
Quebec	Rouyn-Noranda	1.09	1.00
Quebec	Saint Hyacinthe	1.09	1.00
ON	Sault Ste Marie	1.03	1.01
Quebec	Sorel	1.10	1.01
Quebec	St Jerome	1.09	1.00
Ontario	Timmins	1.08	1.04
Nova Scotia	Truro	.93	.95
MANITOBA			
	Brandon	.99	.96
	Portage la Prairie	.99	.96
	Winnipeg	.99	.98
NEW BRUNSWICK			
	Bathurst	.91	.91
	Dalhousie	.91	.91
	Fredericton	.97	.93
	Moncton	.91	.91
	Newcastle	.91	.91
	Saint John	.97	.94
NEWFOUNDLAND			
	Corner Brook	.92	.94
	St. John's	.93	.94
NORTHWEST TERRITORIES			
	Yellowknife	.99	.99
NOVA SCOTIA			
	Dartmouth	.93	.95
	Halifax	.94	.95
	New Glasgow	.93	.95
	Sydney	.92	.93
	Yarmouth	.93	.95
ONTARIO			
	Barrie	1.10	1.05
	Brantford	1.11	1.07
	Cornwall	1.10	1.05
	Hamilton	1.11	1.08
	Kingston	1.11	1.06
	Kitchener	1.05	1.02
	London	1.09	1.06
	North Bay	1.08	1.04
	Oshawa	1.10	1.06
	Ottawa	1.11	1.06
	Owen Sound	1.08	1.05
	Peterborough	1.09	1.05
	Sarnia	1.12	1.07
	St. Catharines	1.05	1.01
	Sudbury	1.03	1.01
	Thunder Bay	1.07	1.02
	Toronto	1.14	1.10
	Windsor	1.08	1.03

STATE/ZIP	CITY	Residential	Commercial
PRINCE EDWARD ISLAND			
	Charlottetown	.88	.91
	Summerside	.88	.91
QUEBEC			
	Cap-de-la-Madeleine	1.10	1.01
	Charlesbourg	1.10	1.01
	Chicoutimi	1.10	1.00
	Gatineau	1.09	1.00
	Laval	1.09	1.00
	Montreal	1.09	1.02
	Quebec	1.11	1.02
	Sherbrooke	1.09	1.01
	Trois Rivieres	1.10	1.01
SASKATCHEWAN			
	Moose Jaw	.91	.92
	Prince Albert	.90	.91
	Regina	.91	.92
	Saskatoon	.90	.91
YUKON			
	Whitehorse	.89	.91

Part Six
Resources, Glossary, and Index

Table of Contents

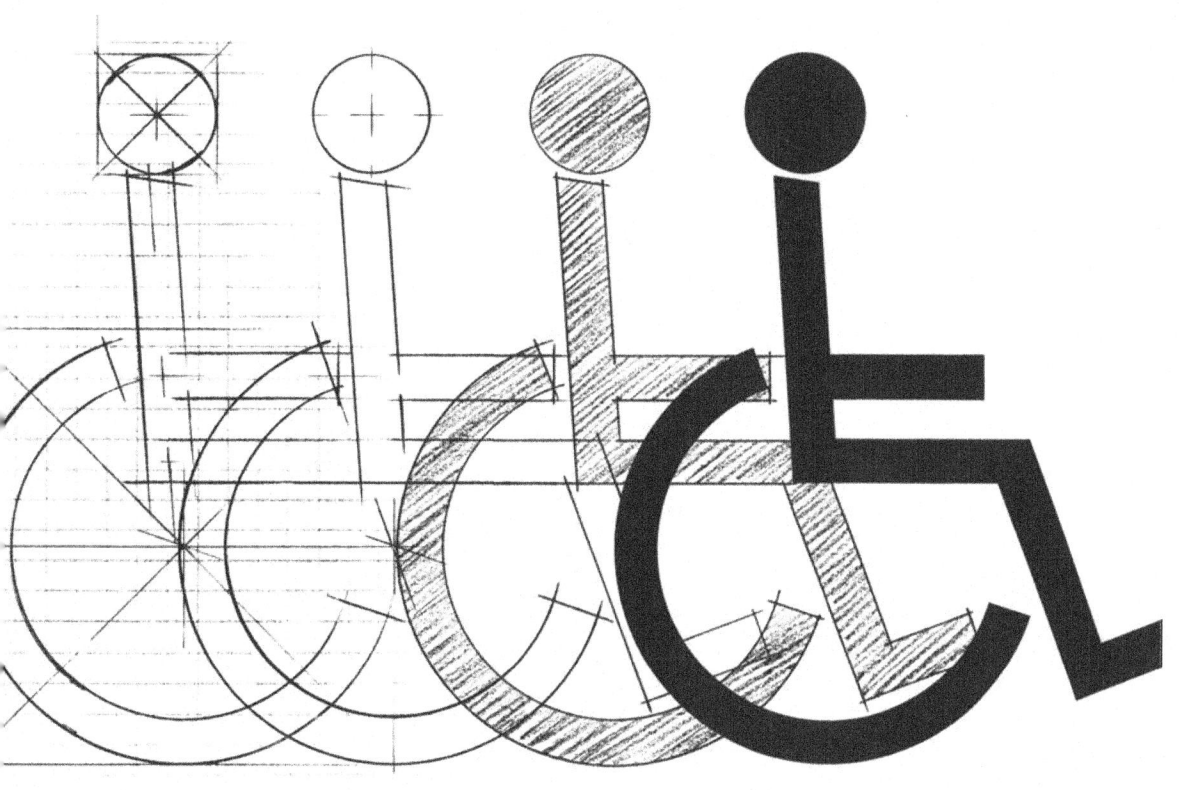

Resources

The following is an annotated list of sources of additional ADA information. Many of the organizations and agencies provide publications (with accessible formats upon request) and toll-free technical assistance by telephone.

Note: In addition to reviewing the ADA requirements, check with your state or local access board or building inspector to find out what state or local codes or guidelines might apply.

Accessibility Regulations and Standards

ADA Standards for Accessible Design

As published in the Title III regulations (28 CFR Part 36, revised July 1, 1994) issued by the Department of Justice. The *ADA Standards for Accessible Design* are in Appendix A of the Title III Regulations, and are in effect as of the publishing of this book. It is anticipated that they will be replaced by standards based on the 2004 ADAAG. *See U.S. Department of Justice*

ADA Regulation for Title III 28 C.F.R. Part 36

The Department of Justice's regulation implementing Title III of the ADA, which prohibits discrimination on the basis of disability in "places of public accommodation" (businesses and non-profit agencies that serve the public) and "commercial facilities" (other businesses). The regulation includes Appendix A to Part 36 - Standards for Accessible Design establishing minimum standards for ensuring accessibility when designing and constructing a new facility or altering an existing facility. *See U.S. Department of Justice*

ADA Regulation for Title II, 28 C.F.R. Part 35

The Department of Justice's regulation implementing Title II, subtitle A, of the ADA which prohibits discrimination on the basis of disability in all services, programs, and activities provided to the public by state and local governments, except public transportation services. *See U.S. Department of Justice*

Uniform Federal Accessibility Standards (UFAS)

These are standards for the design, construction, and alteration of federal and federally funded buildings so that people with disabilities will have ready access to and use of them in accordance with the Architectural Barriers Act. *See U.S. Access Board*

U.S. Department of Transportation Americans with Disabilities Act of 1990 Implementing Regulations

Regulations adopted by the Department of Transportation, includes standards on non-discrimination, transportation services, vehicles, facilities, detectable warnings, and bus stop accessibility. *See U.S. Department of Transportation*

Fair Housing Act Regulations

The sections of the regulations that address design and construction issues are 24 Code of Federal Regulations (C.F.R.) 100.205 and some definitions are found in Section 24 C.F.R. 100.201. Other portions of the regulations that deal with disability issues are in 24 C.F.R. 100.201-.203. *See HUD and Fair Housing FIRST*

Equal Employment Opportunity Commission (EEOC) ADA Regulations 29 C.F.R Part 1630

Regulations to implement the equal employment provisions of the Americans with Disabilities Act. *See EEOC*

Accessibility Guidelines

The Americans with Disabilities Act Accessibility Guidelines for Buildings and Facilities (ADAAG)

Under the ADA, the U.S. Access Board, an independent federal agency, is responsible for accessibility guidelines covering newly built and altered facilities. In 1991, the Board published the *ADA Accessibility Guidelines* (ADAAG), which serve as the basis for the Department of Justice's ADA Standards for Accessible Design and the

Department of Transportation's ADA Standards. On July 23, 2004, the Board issued updated accessibility guidelines more consistent with model building codes, such as the International Building Code (IBC), and with industry standards. The 2004 guidelines feature a new format and organization and have been extensively edited for greater clarity.

The Board also has guidance and reports for different kinds of facilities, for transportation vehicles, and for communication elements (e.g. *Accessible Rights-of-Way, Visual Alarm Technical Bulletin, Guides on Accessible Recreation*). *See U.S. Access Board*

Fair Housing Accessibility Guidelines 24 CFR Part 100

Guidelines adopted by the Department of Housing and Urban Development to provide builders and developers with technical guidance on how to comply with the specific accessibility requirements of the Fair Housing Amendments Act of 1988. *See HUD and Fair Housing FIRST*

Designing Sidewalks and Trails for Access: Parts I and II (July 1999)

In an effort to determine when ADAAG applies to sidewalks and trails, and to bridge the remaining gaps, the Federal Highway Administration sponsored a project to research existing conditions on sidewalks for people with disabilities. Part I presents the result of extensive research into existing guidelines and practice, and Part II presents a manual with guidelines and criteria for building accessible sidewalks and trails. *See U.S. Department of Transportation*

Access Checklists

Existing Facilities Survey Checklist 2.1

The checklist identifies accessibility problems and solutions for meeting ADA Title III barrier removal obligations. It was written by Adaptive Environments and is distributed through the regional ADA & Accessible IT Centers funded by the National Institute on Disability and Rehabilitation Research (NIDRR). *See ADA & IT Technical Assistance Centers*

ADAAG Checklist, Checklist for Sidewalks and Street Crossings, and UFAS Checklist

These checklists are published by the U.S. Access Board and are available on the web or by ordering. *See U.S. Access Board*

Technical Assistance Manuals

ADAAG Manual: A Guide to the Americans with Disabilities Act Accessibility Guidelines

Published in 1998, this manual clarifies ADAAG provisions in an effort to address frequently asked questions. Recommendations are provided that go beyond the minimum of ADAAG. Lots of helpful drawings and diagrams. *See U.S. Access Board*

After the 2004 ADAAG is adopted by the U.S. Department of Justice as enforceable standards, the U.S. Access Board may develop a manual addressing the new standards.

Title I Technical Assistance Manual, Title II Technical Assistance Manual, and Title III Technical Assistance Manual

These manuals explain the regulations, including questions, answers, and practical examples. Title I is published by the U.S. Equal Employment Opportunity Commission (EEOC). Title II and III manuals, are published by the *U.S. Department of Justice*.

Fair Housing Act Design Manual: A Manual to Assist Designers and Builders in Meeting the Accessibility Requirements of The Fair Housing Act (1998)

This manual provides clear and helpful guidance about ways to design and construct housing which complies with the Fair Housing Act. The Design Manual is one of the seven HUD-recognized safe harbors for compliance with the Fair Housing Act's design and construction requirements. *See HUD and Fair Housing FIRST*

ADA Title II Action Guide for State and Local Governments

Plain language explanation of state and local government ADA obligations. Provides clear and detailed worksheets for conducting a self-evaluation and developing a transition plan. *See Adaptive Environments*

Compliance with the Americans with Disabilities Act: A Self-Evaluation Guide for Public Elementary and Secondary Schools

Plain language explanation of schools obligations under the ADA and Section 504 of the Rehabilitation Act. Helpful examples plus forms for conducting a self-evaluation and developing a transition plan. *See Adaptive Environments*

Organizations for Technical Assistance

ADA & Accessible IT Centers

Toll-free information line provides technical assistance on all aspects of the ADA. The ten regional centers also distribute ADA publications, provide training, disseminate information through newsletters and keep people up-to-date on new regulations and policies through email lists. In 2001, their responsibilities expanded to include providing those same services in the area of accessible information technology. This includes building accessible web pages, ensuring that distance learning programs are accessible, and ensuring that technology purchases are those that are best able to work with assistive devices used by people with disabilities.

> 800-949-4232 (voice/TTY) will connect you to the ADA Center in your region.
> www.adata.org/dbtac.html to find the ADA Center in your region.

Fair Housing Accessibility FIRST

Fair Housing Accessibility FIRST is an initiative sponsored by the U.S. Department of Housing and Urban Development (HUD) designed to promote compliance with the Fair Housing Act design and construction requirements. The program consists of a comprehensive training curriculum, as well as a toll-free information line and a Web site designed to provide technical guidance to the public.

> 888-341-7781 (voice, TTY)
> contact@fairhousingfirst.org
> www.fairhousingfirst.org

Adaptive Environments

Adaptive Environments Center promotes human-centered design that works for everyone across the spectrum of ability and age and enhances human experience. They produce technical assistance materials and publications on accessibility, including award-winning design guidelines. They also provide education and consultation about strategies, precedents, and best practices that go beyond legal requirements to design places, things, communication, and policy that integrate solutions to the reality of human diversity across the lifespan.

> Adaptive Environments
> 374 Congress Street, Suite 301
> Boston, MA 02210
> 617-695-1225 (voice, TTY)
> 617-482-8099 (fax)
> info@AdaptiveEnvironments.org
> www.AdaptiveEnvironments.org

Job Accommodation Network

The Job Accommodation Network is a service of the Office of Disability Employment Policy (ODEP) of the U.S. Department of Labor. JAN's mission is to facilitate the employment and retention of workers with disabilities by providing employers, employment providers, people with disabilities, their family members, and other interested parties with information on job accommodations, self-employment and small business opportunities, and related subjects.

> Job Accommodation Network
> P.O. Box 6080
> Morgantown, WV 26506-6080
> 800-526-7234 (voice, TTY)
> jan@jan.wvu.edu
> www.jan.wvu.edu

Assistive Technology Projects (TAPs)

Fifty-six states and territories have assistive technology programs as authorized under the Assistive Technology Act of 1998. The TAPs are the most up-to-date resources for assistive technology information that workplaces , schools, and others can adopt. A Web site provides links to each program where you can obtain technical assistance.

> www.resna.org/taproject

Federal Agencies

U.S. Access Board

The Access Board has many ADA and related accessibility publications and resources. Key responsibilities of the Board include: developing and maintaining accessibility requirements for the built environment, transit vehicles, telecommunications equipment, and electronic and information technology; providing technical assistance and training on these guidelines and standards; and enforcing accessibility standards for federally funded facilities.

> U.S. Access Board
> 1331 F Street, NW, Suite 1000
> Washington, DC 20004-1111
> 800-872-2253 (voice)
> 800-993-2822 (TTY)
> 202-272-0081 (fax)
> ta@access-board.gov
> www.access-board.gov

U.S. Department of Justice (DOJ)

The DOJ develops and enforces the ADA regulations for Title II (state and local governments) and Title III (public accommodations and commercial facilities.) DOJ has produced technical assistance manuals and many other helpful documents. Most of their materials are available online. DOJ also provides technical assistance and certification of state and local building codes.

> The U.S. Department of Justice
> 950 Pennsylvania Avenue, NW
> Civil Rights Division

Disability Rights Section - NYAV
Washington, D.C. 20530
800-514-0301 (voice)
800-514-0383 (TTY)
202-307-1198 (fax)
www.usdoj.gov/crt/ada or
www.ada.gov

U.S. Department of Transportation (DOT)

The DOT's publications are available online. In addition to publications on laws and regulations, the Federal Highway Administration (FHA) provides written accessibility guidance. Their publications are available from the same Web site.

Office of the Secretary
Departmental Office of Civil Rights
S-30, Room 10215
400 Seventh St., SW
Washington, DC 20590
202-366-4648 (voice)
202-366-5273 (TTY)
www.dot.gov/citizen_services/
disability/disability.html

U.S. Department of Housing and Urban Development (HUD)

HUD maintains a Disability Rights and Resources web page designed to answer frequently asked questions on the housing rights of people with disabilities and the responsibilities of housing providers and building and design professionals under federal law.

U.S. Department of Housing and Urban Development
451 7th Street S.W.
Washington, DC 20410
202-708-1112 (voice)
202-708-1455 (TTY)
www.hud.gov/offices/fheo/
disabilities/index.cfm

The U.S. Equal Employment Opportunity Commission

The U.S. Equal Employment Opportunity Commission (EEOC) enforces all of the laws prohibiting job discrimination. EEOC provides oversight and coordination of all federal equal employment opportunity regulations, practices, and policies. They also provide publications, trainings, and outreach.

U.S. Equal Employment
Opportunity Commission
1801 L Street, N.W.
Washington, D.C. 20507
800-669-4000 (voice)
800-669-6820 (TTY)
www.eeoc.gov

Costing Resources

RSMeans

RSMeans is North America's leading provider of construction cost and reference data, available in book and CD-ROM formats and online. They provide 27 annual cost data publications, a library of reference publications, and online cost calculators and estimators.

R.S. Means
63 Smiths Lane
P.O. Box 800
Kingston, MA 02364-9988
800-334-3509 (voice)
800-632-6732 (fax)
www.rsmeans.com

Cross-Reference: Pricing Guide to ADAAG

The following list cross references projects in the *Means ADA Compliance Pricing Guide, Second Edition*, to sections in the 2004 edition of the *Americans with Disabilities Act Accessibility Guidelines for Buildings and Facilities*. The list contains the primary ADAAG section(s) only; consult the specific projects for a complete list of all related requirements.

Abbreviations

A	Area Square Feet; Ampere	Cab.	Cabinet
ABS	Acrylonitrile Butadiene Stryrene; Asbestos Bonded Steel	Cair.	Air Tool Laborer
A.C.	Alternating Current; Air-Conditioning; Asbestos Cement; Plywood Grade A & C	Calc	Calculated
		Cap.	Capacity
		Carp.	Carpenter
		C.B.	Circuit Breaker
		C.C.A.	Chromate Copper Arsenate
A.C.I.	American Concrete Institute	C.C.F.	Hundred Cubic Feet
AD	Plywood, Grade A & D	cd	Candela
Addit.	Additional	cd/sf	Candela per Square Foot
Adj.	Adjustable	CD	Grade of Plywood Face & Back
af	Audio-frequency	CDX	Plywood, Grade C & D, exterior glue
A.G.A.	American Gas Association		
Agg.	Aggregate	Cefi.	Cement Finisher
A.H.	Ampere Hours	Cem.	Cement
A hr.	Ampere-hour	CF	Hundred Feet
A.H.U.	Air Handling Unit	C.F.	Cubic Feet
A.I.A.	American Institute of Architects	CFM	Cubic Feet per Minute
AIC	Ampere Interrupting Capacity	c.g.	Center of Gravity
Allow.	Allowance	CHW	Chilled Water; Commercial Hot Water
alt.	Altitude		
Alum.	Aluminum	C.I.	Cast Iron
a.m.	Ante Meridiem	C.I.P.	Cast in Place
Amp.	Ampere	Circ.	Circuit
Anod.	Anodized	C.L.	Carload Lot
Approx.	Approximate	Clab.	Common Laborer
Apt.	Apartment	Clam	Common maintenance laborer
Asb.	Asbestos	C.L.F.	Hundred Linear Feet
A.S.B.C.	American Standard Building Code	CLF	Current Limiting Fuse
Asbe.	Asbestos Worker	CLP	Cross Linked Polyethylene
A.S.H.R.A.E.	American Society of Heating, Refrig. & AC Engineers	cm	Centimeter
		CMP	Corr. Metal Pipe
A.S.M.E.	American Society of Mechanical Engineers	C.M.U.	Concrete Masonry Unit
		CN	Change Notice
A.S.T.M.	American Society for Testing and Materials	Col.	Column
		CO₂	Carbon Dioxide
Attchmt.	Attachment	Comb.	Combination
Avg.	Average	Compr.	Compressor
A.W.G.	American Wire Gauge	Conc.	Concrete
AWWA	American Water Works Assoc.	Cont.	Continuous; Continued
Bbl.	Barrel	Corr.	Corrugated
B. & B.	Grade B and Better; Balled & Burlapped	Cos	Cosine
		Cot	Cotangent
B. & S.	Bell and Spigot	Cov.	Cover
B. & W.	Black and White	C/P	Cedar on Paneling
b.c.c.	Body-centered Cubic	CPA	Control Point Adjustment
B.C.Y.	Bank Cubic Yards	Cplg.	Coupling
BE	Bevel End	C.P.M.	Critical Path Method
B.F.	Board Feet	CPVC	Chlorinated Polyvinyl Chloride
Bg. cem.	Bag of Cement	C.Pr.	Hundred Pair
BHP	Boiler Horsepower; Brake Horsepower	CRC	Cold Rolled Channel
		Creos.	Creosote
B.I.	Black Iron	Crpt.	Carpet & Linoleum Layer
Bit.; Bitum.	Bituminous	CRT	Cathode-ray Tube
Bk.	Backed	CS	Carbon Steel, Constant Shear Bar Joist
Bkrs.	Breakers		
Bldg.	Building	Csc	Cosecant
Blk.	Block	C.S.F.	Hundred Square Feet
Bm.	Beam	CSI	Construction Specifications Institute
Boil.	Boilermaker		
B.P.M.	Blows per Minute	C.T.	Current Transformer
BR	Bedroom	CTS	Copper Tube Size
Brg.	Bearing	Cu	Copper, Cubic
Brhe.	Bricklayer Helper	Cu. Ft.	Cubic Foot
Bric.	Bricklayer	cw	Continuous Wave
Brk.	Brick	C.W.	Cool White; Cold Water
Brng.	Bearing	Cwt.	100 Pounds
Brs.	Brass	C.W.X.	Cool White Deluxe
Brz.	Bronze	C.Y.	Cubic Yard (27 cubic feet)
Bsn.	Basin	C.Y./Hr.	Cubic Yard per Hour
Btr.	Better	Cyl.	Cylinder
BTU	British Thermal Unit	d	Penny (nail size)
BTUH	BTU per Hour	D	Deep; Depth; Discharge
B.U.R.	Built-up Roofing	Dis.;Disch.	Discharge
BX	Interlocked Armored Cable	Db.	Decibel
c	Conductivity, Copper Sweat	Dbl.	Double
C	Hundred; Centigrade	DC	Direct Current
C/C	Center to Center, Cedar on Cedar	DDC	Direct Digital Control

Demob.	Demobilization
d.f.u.	Drainage Fixture Units
D.H.	Double Hung
DHW	Domestic Hot Water
Diag.	Diagonal
Diam.	Diameter
Distrib.	Distribution
Dk.	Deck
D.L.	Dead Load; Diesel
DLH	Deep Long Span Bar Joist
Do.	Ditto
Dp.	Depth
D.P.S.T.	Double Pole, Single Throw
Dr.	Driver
Drink.	Drinking
D.S.	Double Strength
D.S.A.	Double Strength A Grade
D.S.B.	Double Strength B Grade
Dty.	Duty
DWV	Drain Waste Vent
DX	Deluxe White, Direct Expansion
dyn	Dyne
e	Eccentricity
E	Equipment Only; East
Ea.	Each
E.B.	Encased Burial
Econ.	Economy
E.C.Y	Embankment Cubic Yards
EDP	Electronic Data Processing
EIFS	Exterior Insulation Finish System
E.D.R.	Equiv. Direct Radiation
Eq.	Equation
Elec.	Electrician; Electrical
Elev.	Elevator; Elevating
EMT	Electrical Metallic Conduit; Thin Wall Conduit
Eng.	Engine, Engineered
EPDM	Ethylene Propylene Diene Monomer
EPS	Expanded Polystyrene
Eqhv.	Equip. Oper., Heavy
Eqlt.	Equip. Oper., Light
Eqmd.	Equip. Oper., Medium
Eqmm.	Equip. Oper., Master Mechanic
Eqol.	Equip. Oper., Oilers
Equip.	Equipment
ERW	Electric Resistance Welded
E.S.	Energy Saver
Est.	Estimated
esu	Electrostatic Units
E.W.	Each Way
EWT	Entering Water Temperature
Excav.	Excavation
Exp.	Expansion, Exposure
Ext.	Exterior
Extru.	Extrusion
f.	Fiber stress
F	Fahrenheit; Female; Fill
Fab.	Fabricated
FBGS	Fiberglass
F.C.	Footcandles
f.c.c.	Face-centered Cubic
f'c.	Compressive Stress in Concrete; Extreme Compressive Stress
F.E.	Front End
FEP	Fluorinated Ethylene Propylene (Teflon)
F.G.	Flat Grain
F.H.A.	Federal Housing Administration
Fig.	Figure
Fin.	Finished
Fixt.	Fixture
Fl. Oz.	Fluid Ounces
Flr.	Floor
F.M.	Frequency Modulation; Factory Mutual
Fmg.	Framing
Fndtn.	Foundation

Fori.	Foreman, Inside	I.W.	Indirect Waste
Foro.	Foreman, Outside	J	Joule
Fount.	Fountain	J.I.C.	Joint Industrial Council
FPM	Feet per Minute	K	Thousand; Thousand Pounds;
FPT	Female Pipe Thread		Heavy Wall Copper Tubing, Kelvin
Fr.	Frame	K.A.H.	Thousand Amp. Hours
F.R.	Fire Rating	KCMIL	Thousand Circular Mils
FRK	Foil Reinforced Kraft	KD	Knock Down
FRP	Fiberglass Reinforced Plastic	K.D.A.T.	Kiln Dried After Treatment
FS	Forged Steel	kg	Kilogram
FSC	Cast Body; Cast Switch Box	kG	Kilogauss
Ft.	Foot; Feet	kgf	Kilogram Force
Ftng.	Fitting	kHz	Kilohertz
Ftg.	Footing	Kip.	1000 Pounds
Ft. Lb.	Foot Pound	KJ	Kiljoule
Furn.	Furniture	K.L.	Effective Length Factor
FVNR	Full Voltage Non-Reversing	K.L.F.	Kips per Linear Foot
FXM	Female by Male	Km	Kilometer
Fy.	Minimum Yield Stress of Steel	K.S.F.	Kips per Square Foot
g	Gram	K.S.I.	Kips per Square Inch
G	Gauss	kV	Kilovolt
Ga.	Gauge	kVA	Kilovolt Ampere
Gal.	Gallon	K.V.A.R.	Kilovar (Reactance)
Gal./Min.	Gallon per Minute	KW	Kilowatt
Galv.	Galvanized	KWh	Kilowatt-hour
Gen.	General	L	Labor Only; Length; Long;
G.F.I.	Ground Fault Interrupter		Medium Wall Copper Tubing
Glaz.	Glazier	Lab.	Labor
GPD	Gallons per Day	lat	Latitude
GPH	Gallons per Hour	Lath.	Lather
GPM	Gallons per Minute	Lav.	Lavatory
GR	Grade	lb.; #	Pound
Gran.	Granular	L.B.	Load Bearing; L Conduit Body
Grnd.	Ground	L. & E.	Labor & Equipment
H	High; High Strength Bar Joist;	lb./hr.	Pounds per Hour
	Henry	lb./L.F.	Pounds per Linear Foot
H.C.	High Capacity	lbf/sq.in.	Pound-force per Square Inch
H.D.	Heavy Duty; High Density	L.C.L.	Less than Carload Lot
H.D.O.	High Density Overlaid	L.C.Y.	Loose Cubic Yards
Hdr.	Header	Ld.	Load
Hdwe.	Hardware	LE	Lead Equivalent
Help.	Helpers Average	LED	Light Emitting Diode
HEPA	High Efficiency Particulate Air	L.F.	Linear Foot
	Filter	Lg.	Long; Length; Large
Hg	Mercury	L & H	Light and Heat
HIC	High Interrupting Capacity	LH	Long Span Bar Joist
HM	Hollow Metal	L.H.	Labor Hours
H.O.	High Output	L.L.	Live Load
Horiz.	Horizontal	L.L.D.	Lamp Lumen Depreciation
H.P.	Horsepower; High Pressure	L-O-L	Lateralolet
H.P.F.	High Power Factor	lm	Lumen
Hr.	Hour	lm/sf	Lumen per Square Foot
Hrs./Day	Hours per Day	lm/W	Lumen per Watt
HSC	High Short Circuit	L.O.A.	Length Over All
Ht.	Height	log	Logarithm
Htg.	Heating	L.P.	Liquefied Petroleum; Low Pressure
Htrs.	Heaters	L.P.F.	Low Power Factor
HVAC	Heating, Ventilation & Air-	LR	Long Radius
	Conditioning	L.S.	Lump Sum
Hvy.	Heavy	Lt.	Light
HW	Hot Water	Lt. Ga.	Light Gauge
Hyd.;Hydr.	Hydraulic	L.T.L.	Less than Truckload Lot
Hz.	Hertz (cycles)	Lt. Wt.	Lightweight
I.	Moment of Inertia	L.V.	Low Voltage
I.C.	Interrupting Capacity	M	Thousand; Material; Male;
ID	Inside Diameter		Light Wall Copper Tubing
I.D.	Inside Dimension; Identification	M²CA	Meters Squared Contact Area
I.F.	Inside Frosted	m/hr; M.H.	Man-hour
I.M.C.	Intermediate Metal Conduit	mA	Milliampere
In.	Inch	Mach.	Machine
Incan.	Incandescent	Mag. Str.	Magnetic Starter
Incl.	Included; Including	Maint.	Maintenance
Int.	Interior	Marb.	Marble Setter
Inst.	Installation	Mat; Mat'l.	Material
Insul.	Insulation/Insulated	Max.	Maximum
I.P.	Iron Pipe	MBF	Thousand Board Feet
I.P.S.	Iron Pipe Size	MBH	Thousand BTU's per hr.
I.P.T.	Iron Pipe Threaded	MC	Metal Clad Cable

M.C.F.	Thousand Cubic Feet
M.C.F.M.	Thousand Cubic Feet per Minute
M.C.M.	Thousand Circular Mils
M.C.P.	Motor Circuit Protector
MD	Medium Duty
M.D.O.	Medium Density Overlaid
Med.	Medium
MF	Thousand Feet
M.F.B.M.	Thousand Feet Board Measure
Mfg.	Manufacturing
Mfrs.	Manufacturers
mg	Milligram
MGD	Million Gallons per Day
MGPH	Thousand Gallons per Hour
MH, M.H.	Manhole; Metal Halide; Man-Hour
MHz	Megahertz
Mi.	Mile
MI	Malleable Iron; Mineral Insulated
mm	Millimeter
Mill.	Millwright
Min., min.	Minimum, minute
Misc.	Miscellaneous
ml	Milliliter, Mainline
M.L.F.	Thousand Linear Feet
Mo.	Month
Mobil.	Mobilization
Mog.	Mogul Base
MPH	Miles per Hour
MPT	Male Pipe Thread
MRT	Mile Round Trip
ms	Millisecond
M.S.F.	Thousand Square Feet
Mstz.	Mosaic & Terrazzo Worker
M.S.Y.	Thousand Square Yards
Mtd.	Mounted
Mthe.	Mosaic & Terrazzo Helper
Mtng.	Mounting
Mult.	Multi; Multiply
M.V.A.	Million Volt Amperes
M.V.A.R.	Million Volt Amperes Reactance
MV	Megavolt
MW	Megawatt
MXM	Male by Male
MYD	Thousand Yards
N	Natural; North
nA	Nanoampere
NA	Not Available; Not Applicable
N.B.C.	National Building Code
NC	Normally Closed
N.E.M.A.	National Electrical Manufacturers Assoc.
NEHB	Bolted Circuit Breaker to 600V.
N.L.B.	Non-Load-Bearing
NM	Non-Metallic Cable
nm	Nanometer
No.	Number
NO	Normally Open
N.O.C.	Not Otherwise Classified
Nose.	Nosing
N.P.T.	National Pipe Thread
NQOD	Combination Plug-on/Bolt on Circuit Breaker to 240V.
N.R.C.	Noise Reduction Coefficient
N.R.S.	Non Rising Stem
ns	Nanosecond
nW	Nanowatt
OB	Opposing Blade
OC	On Center
OD	Outside Diameter
O.D.	Outside Dimension
ODS	Overhead Distribution System
O.G.	Ogee
O.H.	Overhead
O & P	Overhead and Profit
Oper.	Operator
Opng.	Opening
Orna.	Ornamental
OSB	Oriented Strand Board

O. S. & Y.	Outside Screw and Yoke
Ovhd.	Overhead
OWG	Oil, Water or Gas
Oz.	Ounce
P.	Pole; Applied Load; Projection
p.	Page
Pape.	Paperhanger
P.A.P.R.	Powered Air Purifying Respirator
PAR	Parabolic Reflector
Pc., Pcs.	Piece, Pieces
P.C.	Portland Cement; Power Connector
P.C.F.	Pounds per Cubic Foot
P.C.M.	Phase Contract Microscopy
P.E.	Professional Engineer; Porcelain Enamel; Polyethylene; Plain End
Perf.	Perforated
Ph.	Phase
P.I.	Pressure Injected
Pile.	Pile Driver
Pkg.	Package
Pl.	Plate
Plah.	Plasterer Helper
Plas.	Plasterer
Pluh.	Plumbers Helper
Plum.	Plumber
Ply.	Plywood
p.m.	Post Meridiem
Pntd.	Painted
Pord.	Painter, Ordinary
pp	Pages
PP; PPL	Polypropylene
P.P.M.	Parts per Million
Pr.	Pair
P.E.S.B.	Pre-engineered Steel Building
Prefab.	Prefabricated
Prefin.	Prefinished
Prop.	Propelled
PSF; psf	Pounds per Square Foot
PSI; psi	Pounds per Square Inch
PSIG	Pounds per Square Inch Gauge
PSP	Plastic Sewer Pipe
Pspr.	Painter, Spray
Psst.	Painter, Structural Steel
P.T.	Potential Transformer
P. & T.	Pressure & Temperature
Ptd.	Painted
Ptns.	Partitions
Pu	Ultimate Load
PVC	Polyvinyl Chloride
Pvmt.	Pavement
Pwr.	Power
Q	Quantity Heat Flow
Quan.; Qty.	Quantity
Q.C.	Quick Coupling
r	Radius of Gyration
R	Resistance
R.C.P.	Reinforced Concrete Pipe
Rect.	Rectangle
Reg.	Regular
Reinf.	Reinforced
Req'd.	Required
Res.	Resistant
Resi.	Residential
Rgh.	Rough
RGS	Rigid Galvanized Steel
R.H.W.	Rubber, Heat & Water Resistant; Residential Hot Water
rms	Root Mean Square
Rnd.	Round
Rodm.	Rodman
Rofc.	Roofer, Composition
Rofp.	Roofer, Precast
Rohe.	Roofer Helpers (Composition)
Rots.	Roofer, Tile & Slate
R.O.W.	Right of Way
RPM	Revolutions per Minute
R.S.	Rapid Start
Rsr	Riser
RT	Round Trip
S.	Suction; Single Entrance; South
SCFM	Standard Cubic Feet per Minute
Scaf.	Scaffold
Sch.; Sched.	Schedule
S.C.R.	Modular Brick
S.D.	Sound Deadening
S.D.R.	Standard Dimension Ratio
S.E.	Surfaced Edge
Sel.	Select
S.E.R.;	Service Entrance Cable
S.E.U.	Service Entrance Cable
S.F.	Square Foot
S.F.C.A.	Square Foot Contact Area
S.F. Flr.	Square Foot of Floor
S.F.G.	Square Foot of Ground
S.F. Hor.	Square Foot Horizontal
S.F.R.	Square Feet of Radiation
S.F. Shlf.	Square Foot of Shelf
S4S	Surface 4 Sides
Shee.	Sheet Metal Worker
Sin.	Sine
Skwk.	Skilled Worker
SL	Saran Lined
S.L.	Slimline
Sldr.	Solder
SLH	Super Long Span Bar Joist
S.N.	Solid Neutral
S-O-L	Socketolet
sp	Standpipe
S.P.	Static Pressure; Single Pole; Self-Propelled
Spri.	Sprinkler Installer
spwg	Static Pressure Water Gauge
Sq.	Square; 100 Square Feet
S.P.D.T.	Single Pole, Double Throw
SPF	Spruce Pine Fir
S.P.S.T.	Single Pole, Single Throw
SPT	Standard Pipe Thread
Sq. Hd.	Square Head
Sq. In.	Square Inch
S.S.	Single Strength; Stainless Steel
S.S.B.	Single Strength B Grade
sst	Stainless Steel
Sswk.	Structural Steel Worker
Sswl.	Structural Steel Welder
St.; Stl.	Steel
S.T.C.	Sound Transmission Coefficient
Std.	Standard
STK	Select Tight Knot
STP	Standard Temperature & Pressure
Stpi.	Steamfitter, Pipefitter
Str.	Strength; Starter; Straight
Strd.	Stranded
Struct.	Structural
Sty.	Story
Subj.	Subject
Subs.	Subcontractors
Surf.	Surface
Sw.	Switch
Swbd.	Switchboard
S.Y.	Square Yard
Syn.	Synthetic
S.Y.P.	Southern Yellow Pine
Sys.	System
t.	Thickness
T	Temperature; Ton
Tan	Tangent
T.C.	Terra Cotta
T & C	Threaded and Coupled
T.D.	Temperature Difference
T.E.M.	Transmission Electron Microscopy
TFE	Tetrafluoroethylene (Teflon)
T. & G.	Tongue & Groove; Tar & Gravel
Th.; Thk.	Thick
Thn.	Thin
Thrded	Threaded
Tilf.	Tile Layer, Floor
Tilh.	Tile Layer, Helper
THHN	Nylon Jacketed Wire
THW.	Insulated Strand Wire
THWN;	Nylon Jacketed Wire
T.L.	Truckload
T.M.	Track Mounted
Tot.	Total
T-O-L	Threadolet
T.S.	Trigger Start
Tr.	Trade
Transf.	Transformer
Trhv.	Truck Driver, Heavy
Trlr	Trailer
Trlt.	Truck Driver, Light
TV	Television
T.W.	Thermoplastic Water Resistant Wire
UCI	Uniform Construction Index
UF	Underground Feeder
UGND	Underground Feeder
U.H.F.	Ultra High Frequency
U.L.	Underwriters Laboratory
Unfin.	Unfinished
URD	Underground Residential Distribution
US	United States
USP	United States Primed
UTP	Unshielded Twisted Pair
V	Volt
V.A.	Volt Amperes
V.C.T.	Vinyl Composition Tile
VAV	Variable Air Volume
VC	Veneer Core
Vent.	Ventilation
Vert.	Vertical
V.F.	Vinyl Faced
V.G.	Vertical Grain
V.H.F.	Very High Frequency
VHO	Very High Output
Vib.	Vibrating
V.L.F.	Vertical Linear Foot
Vol.	Volume
VRP	Vinyl Reinforced Polyester
W	Wire; Watt; Wide; West
w/	With
W.C.	Water Column; Water Closet
W.F.	Wide Flange
W.G.	Water Gauge
Wldg.	Welding
W. Mile	Wire Mile
W-O-L	Weldolet
W.R.	Water Resistant
Wrck.	Wrecker
W.S.P.	Water, Steam, Petroleum
WT., Wt.	Weight
WWF	Welded Wire Fabric
XFER	Transfer
XFMR	Transformer
XHD	Extra Heavy Duty
XHHW; XLPE	Cross-Linked Polyethylene Wire Insulation
XLP	Cross-linked Polyethylene
Y	Wye
yd	Yard
yr	Year
Δ	Delta
%	Percent
~	Approximately
Ø	Phase
@	At
#	Pound; Number
<	Less Than
>	Greater Than

Glossary

The following terms are defined as they apply to the Americans with Disabilities Act. Definitions reprinted directly from ADAAG for clarity appear in quotes.

28 CFR 35 Issued by the Department of Justice (26 July, 1991 Federal Register), the regulations governing nondiscrimination on the basis of disability in state and local government services.

28 CFR 36 Issued by the Department of Justice (26 July, 1991 Federal Register), the regulations governing nondiscrimination on the basis of disability in public accommodations and commercial facilities.

a.f.f. Above finished floor. The height of an element measured from the top of the existing floor surface.

Access aisle An accessible pedestrian space between elements, such as parking spaces, seating, and desks, that provides clearances appropriate for use of the elements.

Accessible A facility, building, space, or building element that can be used by people with disabilities.

Accessible route A continuous unobstructed path, meeting ADAAG, through a building or facility.

ADAAG *The Americans with Disabilities Act Accessibility Guidelines for Buildings and Facilities.* The guidelines published by the U.S. Access Board setting forth the *minimum* criteria that can be used by the Department of Justice for their access regulations.

Adaptable A space that can be easily altered to accommodate the needs of individuals with disabilities (such as a sink with cabinets below, where cabinets can be removed to provide knee space).

Addition An increase in the gross floor area or height of a building or facility.

Alteration Changes to a building that affect usage, such as remodeling, renovation, or historic restoration. Also refers to resurfacing exterior driveways and paths and changing structural parts or elements, and changes or rearrangement in the plan configuration of walls and full-height partitions.

Assistive listening system (ALS) "An amplification system utilizing transmitters, receivers, and coupling devices to bypass the acoustical space between a sound source and a listener by means of induction loop, radio frequency, infrared, or direct-wired equipment."

C-pull A C-shaped door or drawer handle that does not require grasping with fingers to operate (sometimes called a D-pull).

Clear The dimension between two opposite surfaces (such as between two walls or railings) or closest protrusion.

Clear opening The dimension between the face of a door open to 90° and the stop on the opposite jamb.

Commercial facility A facility intended for non-residential use by a private entity whose operations affect commerce. Under Title III, commercial facilities are businesses that are not public accommodations (do not fit into the twelve categories of businesses listed under Public Accommodations). Examples are warehouses, factories, and corporate office space.

Cross slope The slope of a walking surface perpendicular to the direction of travel measured as a ratio of drop to width, usually expressed in a percentage (e.g., 2% means that the sidewalk drops one inch vertically for every 50" of width).

Curb ramp A short ramp cutting into a curb to allow access between street level and sidewalk level.

Equivalent facilitation Alternative design or technology that provides substantially equivalent or greater access to a facility than required in the technical or scoping sections of the ADAAG.

Facility "All or any portion of buildings, structures, routes, or vehicular ways located on a site."

Level Flat, horizontal surface, pitched only for drainage (2% maximum).

Maximum extent feasible Applies to the occasional case where the nature of an existing facility makes it impossible to comply fully with applicable accessibility standards. In such circumstances, the alteration must provide the maximum physical accessibility possible. Any altered features of the facility that can be made accessible must be made accessible. If providing accessibility to individuals with certain disabilities (e.g., those who use wheelchairs) would not be feasible, the facility still must be made accessible to persons with other types of disabilities (e.g., those who use crutches, those who have impaired vision or hearing, or those who have other impairments).

Modification Physical change to an architectural space or element.

On center A measurement of the distance between the centers of two repeating members in a structure.

Operable part "A component of an element used to insert or withdraw objects, or to activate, deactivate, or adjust the element."

Play component "An element intended to generate specific opportunities for play, socialization, or learning. Play components are manufactured or natural and are stand-alone or part of a composite play structure."

Primary function A major activity for which a facility is intended. Areas that contain a primary function include, but are not limited to, customer service lobbies of banks, dining areas of cafeterias, meeting rooms of conference centers, and offices and other work areas in which the activities of the public accommodation or other private entity using the facility are carried out. Mechanical rooms, boiler rooms, supply storage rooms, employee lounges or locker rooms, janitorial closets, entrances, corridors, and restrooms are not areas containing a primary function.

Public accommodation Facilities or businesses operated by a private entity that fall within at least one of the following twelve categories: places of lodging, establishments serving food or drink, places of exhibition or entertainment, places of public gathering, sales or rental establishments, service places, stations used for public transportation, places for public display or collection, places of recreation, places of education, social service establishments, places of exercise and recreation.

Private building or facility "A place of public accommodation or a commercial building or facility subject to Title III of the ADA and 28 CFR part 36 or a transportation building or facility subject to Title III of the ADA and 49 CFR 37.45."

Public building or facility "A building or facility or portion of a building or facility designed, constructed, or altered by, on behalf of, or for the use of a public entity subject to Title II of the ADA and 28 CFR part 35 or to Title II of the ADA and 49 CFR 37.41 or 37.43."

Public entrance An entrance to a building that is used by the public, and not a service entrance.

Public use "Interior or exterior rooms, spaces, or elements that are made available to the public. Public use may be provided at a building or facility that is privately or publicly owned."

Ramp A surface with a running slope steeper than 1:20, but less than 1:12.

Reach range The widest measure of the reach of an adult seated in a wheelchair.

Reasonable accommodation Modifications or adjustments to a work environment, or to the manner or circumstances under which the position held or desired is customarily performed, that enable a qualified individual with a disability to perform the essential functions of that position.

Running slope Slope parallel to the direction of travel, usually measured in a ratio of rise to run (or height to length; e.g., 1:12 means that the slope rises one inch vertically for every 12" of horizontal length.)

Sans serif A lettering type without serifs (the small lines used to finish off the main stroke of a letter, such as on the top and bottom of this "I").

Signage Displayed verbal, symbolic, tactile, and pictorial information.

Simple serif A lettering type with small serifs (no firm definition; usually a non-elaborate typeface).

Slip-resistant A surface that provides traction even when wet.

Slope Pitch of walking surface. *See cross slope or running slope.*

Tactile Something detectable by touch (such as raised letters).

TDD Telecommunications Display Device. *See text telephone.*

Technically infeasible "[A]n alteration ... that ... has little likelihood of being accomplished because existing structural conditions would require removing or altering a load-bearing member that is an essential part of the structural frame; or because other existing physical or site constraints prohibit modification."

Text telephone A phone which transmits characters by means of a keyboard, and displays characters typed into it either on a screen or a roll of paper. Commonly referred to as either a TDD or a TTY.

Transfer device (system) Equipment that aids wheelchair occupants in transferring to and from another support element.

TTY Teletypewriter. *See text telephone.*

Typical Refers to identification of a single element that is repeated throughout the facility.

Walk(way) "An exterior prepared surface for pedestrian use, including pedestrian areas such as plazas and courts."

Index

V

Valet, 16
Van access, 13
Vertical lift, 62-65, 67
Vestibule, enlarge, 112-115
Visual alarms, 158-159, 237

W

Wall-mounted controls, 152-153
Wing walls, 122-123

Notes

Notes

Notes

Notes

Notes

Notes

Notes

Notes

Notes